Maturing the Snowflake Data Cloud

A Templated Approach to Delivering and Governing Snowflake in Large Enterprises

Andrew Carruthers
Sahir Ahmed

Apress®

Maturing the Snowflake Data Cloud: A Templated Approach to Delivering and Governing Snowflake in Large Enterprises

Andrew Carruthers
Birmingham, UK

Sahir Ahmed
Bradford, UK

ISBN-13 (pbk): 978-1-4842-9339-3
https://doi.org/10.1007/978-1-4842-9340-9

ISBN-13 (electronic): 978-1-4842-9340-9

Managing Director, Apress Media LLC: Welmoed Spahr
Acquisitions Editor: Jonathan Gennick
Development Editor: Laura Berendson
Editorial Assistant: Shaul Elson
Copy Editor: Mary Behr

Cover image designed by eStudioCalamar

Distributed to the book trade worldwide by Springer Science+Business Media New York, 1 New York Plaza, Suite 4600, New York, NY 10004-1562, USA. Phone 1-800-SPRINGER, fax (201) 348-4505, e-mail orders-ny@springer-sbm.com, or visit www.springeronline.com. Apress Media, LLC is a California LLC and the sole member (owner) is Springer Science + Business Media Finance Inc (SSBM Finance Inc). SSBM Finance Inc is a **Delaware** corporation.

For information on translations, please e-mail booktranslations@springernature.com; for reprint, paperback, or audio rights, please e-mail bookpermissions@springernature.com.

Apress titles may be purchased in bulk for academic, corporate, or promotional use. eBook versions and licenses are also available for most titles. For more information, reference our Print and eBook Bulk Sales web page at www.apress.com/bulk-sales.

Any source code or other supplementary material referenced by the author in this book is available to readers on GitHub. For more detailed information, please visit www.apress.com/source-code.

Printed on acid-free paper

For Diane, Esther, Josh, Verity, Violet, Jordan, and Beth
For Nazish, Amelia, and Ayat

Table of Contents

About the Authors

Andrew Carruthers is the Director for the Snowflake Corporate Data Cloud at the London Stock Exchange Group. Comprising two Snowflake accounts supporting an ingestion data lake and a consumption analytics hub, the Corporate Data Cloud services a growing customer base of over 7,000 end users. He leads both the Centre for Enablement in developing tooling, best practices, and training, and the Snowflake Landing Zone for provisioning Snowflake accounts conforming to both internal standards and best practices. This book is the distilled best practices for implementing a Snowflake Landing Zone.

Andrew has more than 30 years of hands-on relational database design, development, and implementation experience starting with Oracle in 1993. Until joining the London Stock Exchange Group, he operated as an independent IT consultant. His experience has been gained predominantly through working at major European financial institutions. Andrew is considered a visionary and thought leader within his domain, with a tight focus on delivery. Successfully bridging the gap between Snowflake technological capability and business usage of technology, he often develops proofs of concepts to showcase benefits leading to successful business outcomes.

Since early 2020 Andrew has immersed himself in Snowflake and is considered a subject matter expert. He is SnowPro Core certified, contributes to online forums, and speaks at Snowflake events on behalf of the London Stock Exchange Group. In recognition of his contribution to implementing Snowflake at the London Stock Exchange Group, Andrew recently received the Snowflake Data Driver award. This category recognizes a technology trailblazer who has pioneered the data cloud into their organization.

Andrew has two daughters, ages 19 and 21, both of whom are elite figure skaters. He has a passion for Jaguar cars, having designed and implemented modifications, and has published articles for *Jaguar Enthusiast* and *Jaguar Driver*. Andrew enjoys 3D printing and has a mechanical engineering workshop with a lathe, milling machine, and TIG welder, to name but a few tools, and enjoys developing his workshop skills.

Sahir Ahmed is a Snowflake CorePro certified developer working alongside Andrew on the Snowflake Corporate Data Cloud at the London Stock Exchange Group. Sahir began investing his time learning Snowflake prior to joining the London Stock Exchange Group. He drew upon his experience with other relational databases such as MySQL and MS Access. Sahir developed a passion for Snowflake by designing a car parts catalog using JavaScript stored procedures and SQL scripts. As Sahir's skills and knowledge expanded, he dug deeper into Snowflake and within a year he became SnowPro Core certified.

In his spare time, Sahir takes advantage of his self-taught skills in both electrical and manual car repairs, working on many brands of cars such as Jaguar, BMW, and Audi, to name a few. He also competes in amateur boxing competitions. Seen as a role model, Sahir supports and encourages youngsters both within boxing and his wider community. His biggest driving factor, and where his most precious time is spent, is with his wife and two young daughters. They keep his motivation alive and push him to be the best version of himself.

About the Technical Reviewer

Nadir Doctor is a database and data warehousing architect plus DBA who has worked in various industries with multiple OLTP and OLAP technologies as well as primary data platforms including Snowflake, Databricks, CockroachDB, DataStax, Cassandra, ScyllaDB, Redis, MS SQL Server, Oracle, Db2 Cloud, AWS, Azure, and GCP. His major focus is health check scripting for security, high availability, performance optimization, cost reduction, and operational excellence. He has presented at several technical conference events, is active in user group participation, and can be reached on LinkedIn.

Thank you to Andrew Carruthers, Sahir Ahmed, and all the staff at Springer Nature. I'm grateful for the immense support of my loving wife, children, and family during the technical review of this book. I hope you all find the content enjoyable, inspiring, and useful.

Acknowledgments

From Andy:

Once again, thanks are not enough to the Apress team for the opportunity to deliver this book. Specifically, to Sowmya Thodur and Shaul Elson: Thank you for your patient guidance, help, and assistance. Also, to Nadir Doctor: Thank you for delivering a comprehensive technical review. Your input provided both insight and valuable comments. For those unknown to me—the editors, reviewers, and production staff: Please take another bow; you are the unsung heroes who made things happen twice.

To my friends at Snowflake who continue to show me the "art of the possible," who first encouraged, advised, and guided, and are now friends, mentors, and partners, Jonathan Nicholson, Andy McCann, Will Riley, Cillian Bane, James Hunt, and more recently Alessandro Dallatomasina: You have my heartfelt thanks.

To all my colleagues at the London Stock Exchange Group (LSEG): Just as before, you all helped make this book possible, and I will be forever grateful and thankful we came together in the right place, at the right time.

For my inspiring and over-achieving colleagues delivering the Corporate Data Cloud, Nitin Rane, Srinivas Venkata, Matt Willis, Dhiraj Saxena, Gowthami Bandla, Arabella Peake, Kalpesh Parekh, Bally Gill, Radhakrishnan Leela, and Rajan Babu Selvanamasivayam: Keep on "doing the needful." There is always more needful to be done!

Also, take a bow, the Snowflake Landing Zone team, which provided inspiration for this book, Nareesh Komuravelly and Ashan Vidura Dantanarayana. I have rarely seen so much hard work and effort expended by such a small team. Keep going, onwards and upwards; you truly are formidable.

Now for my next opportunity at LSEG, so kindly provided by Matt Adams, truly a visionary leader within the Data & Analytics Division, thank you for your confidence in me. And for kindling the spark for a potential third book in this series…

For my friend Jonathan Gennick: A chance conversation produced a good friend. Your personal touch, help, and encouragement helped more than you know. Thank you.

And for my very dear friends Marco Costella, Martin Cole, and Steve Loosley: My life is all the richer for your friendship, kindness, and encouragement.

ACKNOWLEDGMENTS

To my family, Esther and Josh, Verity, and baby Violet; also to Jordan and Beth; and to my wonderful girlfriend, Diane, who has started out on her Snowflake journey too: Your constant support, encouragement, and steadfast presence every day made this second book possible. You retain the patience of a saint and remain more lovely than I can say.

Finally, to my co-author Sahir Ahmed, the son I never had, whose contribution should not be understated: Your constant presence, persistent approach, and overwhelming enthusiasm has propelled this project forward. You deserves much credit for putting up with me.

From Sahir:

I'd like to start by thanking the Apress team for the opportunity to write and publish this book. Without the team's constant support and guidance, this would not be possible.

To Nitin Rane, Drummond Field, Ashan Vidura Dantanarayana, Marco Costella, and all my colleagues at LSEG: You have been rockets attached to my career, accelerating me way further, and faster, than I could ever imagine. Thank you for your endless hours of support and commitment to help me understand and improve at each step of my journey.

A simple thank you is simply not sufficient to my co-author Andrew Carruthers. He has been a mentor and guiding beacon for me way before our endeavor into this current book. His belief and countless hours invested into me have helped make this book possible and me a better person. I would truly be lost out at sea without him.

Finally, I'd like to say a huge thank you to my beautiful wife, Nazish, and my two daughters, Amelia and Ayat. They have been my support structure and endless fuel and motivation whenever things got tough. If I ever needed a smile or laughter, they were only ever a call away. I couldn't be where I am without your support and love. Thank you, my girls; you are my world.

CHAPTER 1

Maturing the Snowflake Data Cloud

This book follows from where *Building the Snowflake Data Cloud* (Apress, 2022) left off and unpacks some themes of particular interest to organizations that are serious about Snowflake and for those people interested in the big picture who want to set themselves up for success at the outset. We rely upon code developed within *Building the Snowflake Data Cloud* to implement features discussed inside this publication

In this book, we tackle several concepts that might not be immediately obvious. These concepts will be exposed throughout this book and while it might be tempting to develop a simple checklist, please remember context is everything and there are often wider considerations. If you are looking for a playbook for your organization to implement best practices, you are holding the blueprint in your hands.

Here we focus on delivering Snowflake as a platform, which is distinctly different from delivering applications and products built using Snowflake as the core platform. We trust the distinction will become self-evident as you progress through this book.

We assume familiarity with the Snowflake architecture and that you possess good technical knowledge with hands-on SQL skills as a minimum. We do not assume Snowflake certifications but some experience is essential to get the most out of this publication.

While writing *Building the Snowflake Data Cloud*, Snowflake as a product has evolved. Some of these changes were captured as they were released because they happened to arrive with the chapter in flight. Other changes could not be written about as they were in private preview and subject to a non-disclosure agreement (NDA): we are happy to report some of these features have evolved and are now in public preview or are now generally available. We are now able to provide some deep dives with hands-on coded examples. Note that we focus on AWS as the primary cloud service provider (CSP), therefore all code examples reference AWS.

1

© Andrew Carruthers and Sahir Ahmed 2023
A. Carruthers and S. Ahmed, *Maturing the Snowflake Data Cloud*, https://doi.org/10.1007/978-1-4842-9340-9_1

What is certain is that Snowflake as a product continues to surprise and the pace of change is accelerating. A tongue-in-cheek request to any Snowflake product owners reading this book: Slow down, I need to catch up! Or as I say to my teams, "Scream if you want to go faster!"

Identical to the approach taken with *Building the Snowflake Data Cloud*, we adopt the same three questions: Why, How, and What? Simon Sinek unpacked "How great leaders inspire action" in a TEDx Talk found here: `www.youtube.com/watch?v=u4ZoJKF_VuA`. Please do take 18 minutes to watch the video, which I found transformational. This book adopts his pattern, with Why being the most important question of all.

Snowflake documentation can be found here: `https://docs.snowflake.com/en/`. The official documentation is actually rather good and we reference specific sections to support our deep dives later in this book.

Supporting our security focus, which is central to everything we do with Snowflake, we recommend reviewing the Cloud Data Management Capabilities (CDMC) Framework and joining the EDM Council at `https://edmcouncil.org/` for a thorough treatise on controls to manage data movement into cloud platforms. We provide a reference architecture for the CDMC Framework later within this book.

Let's start your new adventure to maturing your organization's Snowflake implementations with a brief discussion on provisioning a "sandpit," a Snowflake account spun up for Proofs of Concepts. We then dive into discussion on establishing a baseline platform for core Development and Production Snowflake accounts, which we refer to as a landing zone (LZ).

In common with *Building the Snowflake Data Cloud* and for use throughout this book, we anticipate you have both a Snowflake trial account from `https://signup.snowflake.com/` and an AWS account available from `https://aws.amazon.com/resources/create-account/`.

Introducing the Landing Zone

Why do you need a LZ? And having become convinced of the need for a LZ, how do you go about designing a LZ? Finally, what must you do to implement a LZ? Are there best practices, templates, and code samples you can leverage to reduce your implementation time?

We're glad you asked! All of the above are addressed in this chapter. Everything within this book has been proven in real-world conditions at client sites. This book delivers what it says.

Why Do You Need a LZ?

Different parts of organizations proceed at different paces. Often each operating division will engage vendor products according to need, budget, available skills, and business demand. You therefore may find yourself where two or more discrete parts of your organization are doing the same thing but in different ways, sometimes with the full engagement of your governance and risk management colleagues, and sometimes not.

If you are the first mover with Snowflake within your organization, it is not unreasonable to either assume or be given responsibility for ensuring that Snowflake conforms to all governance and risk management requirements. You may find that the security requirements evolve over time and new requirements emerge on your journey, but no matter; this is before a single line of code has been written to implement an application using Snowflake.

The integration of Snowflake into your organization should be a precursor activity before building your first application. If yours is the first team to implement Snowflake within your organization, and more so if you are also new to the organization yourself, how much value would a template design be in reducing your implementation timeline? Or as the writer of Proverb 19:18 says, "Where there is no vision, the people perish: but he that keeps the law, happy is he." In other words, you need a design template (or baseline) that fulfils all your organizational needs to both implement Snowflake as a platform and enable developers to do what they are good at: building applications on platforms.

Step forward the LZ, which enables consistent pattern-based delivery plus controlled maintenance, updates, and releases as new components are developed or existing components are enhanced.

If you are a second mover or later with Snowflake, you may expect to leverage the good work undertaken by your first mover colleagues. What you may find in practice, and through experience, is not every prior implementation satisfies your exact requirements, and some implementation details may run orthogonal to your needs. Regardless of which, it is highly unlikely a single template will even exist during the early year or two of Snowflake implementation.

From experience, your Snowflake Sales Engineer (SE) colleagues will not deliver a LZ template; instead, they will deliver knowledge transfer on core Snowflake capabilities and advice on best practices. You must remember that your Snowflake SE colleagues are spread very thin across a number of clients and simply unable to provide the design capacity to take into consideration every aspect of your organization-specific governance and risk management requirements.

Likewise, your Snowflake Professional Services (PS) colleagues are ill equipped to understand every vagary of your organization-specific governance and risk management requirements. It is a simple but costly mistake to expect PS colleagues to deliver anything more than a simple LZ framework ready for you to tailor to your needs. But you are holding a proven LZ design template, removing one deliverable and series of architectural conversations, thus reducing your implementation times.

If you need a final compelling reason to implement a LZ, it is this: The cost, time, and inconvenience of retrofitting a LZ into an organization is far greater than if a LZ is designed and agreed up front.

How Do You Design a LZ?

When designing a LZ, you must take into consideration all of your governance and risk management stakeholder concerns and address each in turn. In practice, you have a wide variety of considerations to address, which Figure 1-1 illustrates. Your design may have more or less components.

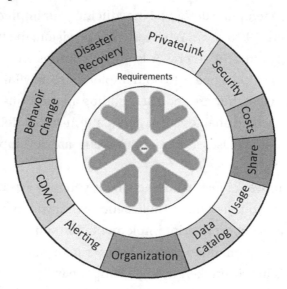

Figure 1-1. *Landing zone virtuous circle*

At the center of your LZ is Snowflake's core capabilities. They are the product features your governance and risk management stakeholders may have requirements to control as represented by the inner circle. The outer ring articulates the technical and business capabilities you are obliged to implement solutions for and we discuss them in detail later in this chapter.

Now that we have covered core deliverables, and recognizing yours may differ, you have a starting point for defining the constituent parts of each component.

What Does a LZ Need?

Each of the component identified in your LZ design has two constituent parts:

1. A core Snowflake capability with features and options

2. A suite of requirements your governance and risk management colleagues specify seeking to control the feature or option

You know Snowflake capabilities are constantly being enhanced and extended, therefore you must not assume your job is ever complete. Instead, you must be proactive in managing your LZ to reflect the latest product capabilities and constantly turn up the dial with your risk management practices.

As a technical custodian, you are best placed to advise your governance and risk management colleagues of Snowflake product changes as you are regularly notified by email and hold the relationship with Snowflake SE colleagues, so it is incumbent upon you to ensure information is distributed internally within your organization.

In accordance with the separation of roles and responsibilities, your governance and risk management colleagues are responsible for specifying controls and guard rails. This is the bedrock upon which you develop LZ components.

Having identified a Snowflake product change or enhanced capability that may affect either an existing control or mandate a new control, you are obliged to update your LZ template. In order to effectively maintain your LZ, and in accordance with industry best practices, you must deploy your updated code into a separate LZ account and test before releasing to the wider estate.

We recommend monthly review of the Snowflake Account Usage Store views along with investigation of all items mentioned within the change release bulletin as part of a robust proactive essential maintenance policy.

Snowflake Adoption

We often encounter development teams who for a variety of reasons want to try Snowflake without investing a lot of time, money, and effort before deciding Snowflake is the optimal platform of choice for their program.

For those wishing to try out Snowflake within their organizations in a controlled manner, that is, subject to some governance without the overhead of provisioning a fully-fledged LZ, we suggest a fully supported sandpit account be created.

What Is a Sandpit Account?

In simple terms, a sandpit account is a Snowflake account and corresponding CSP account intended to support a Proof of Concept (PoC). The environment should be supported by a Snowflake Subject Matter Expert (SME) who is capable of advising, assisting, and encouraging best practices.

Each PoC should be a tightly scoped, well-defined, time-limited project with specific predetermined success criteria to establish whether Snowflake and associated tooling fulfils the PoC objectives. We suggest a PoC be restricted to two weeks with the expectation of provisioned service withdrawal at the end of the time period.

The sandpit is an isolated, prebuilt Snowflake account that can be assigned on demand to users. When the PoC has ended, the Snowflake account can either be repurposed or recreated for the next PoC. Regardless of the approach, do ensure all corresponding CSP account configuration has been removed because there are components that can cause unintended billing (as I found out to my cost)!

You are delivering a platform, not a fully-fledged product or application. Your responsibilities lie with delivering capability for non-Snowflake experts to try something out in a safe and secure manner.

We do not prescribe any particular configuration but suggest as a minimum these items be considered.

Network Policy

Prevent unauthorized system access by implementing one or more network policies restricting access to known IP ranges.

Data Egress

Set Snowflake account flags to prevent data egress to Internet locations outside of your appetite.

Consumption

Implementing resource monitors and consumption limits is a sensible precaution.

Sample Configuration

To facilitate all PoCs, you should provision a suite of baseline objects and sample code. We recommend a starter database with schema and XS warehouse, along with a role with entitlement to create objects be delivered, something like this:

```
SET poc_database      = 'POC';
SET poc_owner_schema  = 'POC.poc_owner';
SET poc_warehouse     = 'poc_wh';
SET poc_owner_role    = 'poc_owner_role';

USE ROLE sysadmin;

CREATE OR REPLACE DATABASE IDENTIFIER ( $poc_database ) DATA_RETENTION_
TIME_IN_DAYS = 7;

CREATE OR REPLACE WAREHOUSE IDENTIFIER ( $poc_warehouse ) WITH
WAREHOUSE_SIZE        = 'X-SMALL'
AUTO_SUSPEND          = 60
AUTO_RESUME           = TRUE
MIN_CLUSTER_COUNT     = 1
MAX_CLUSTER_COUNT     = 4
SCALING_POLICY        = 'STANDARD'
INITIALLY_SUSPENDED = TRUE;
```

If your POC involves Snowpark, you may want to add a `WAREHOUSE_TYPE = 'SNOWPARK-OPTIMIZED'` to have a bigger memory allocation to handle large datasets.

```
CREATE OR REPLACE SCHEMA IDENTIFIER ( $poc_owner_schema   );

USE ROLE securityadmin;

CREATE OR REPLACE ROLE IDENTIFIER ( $poc_owner_role  )         COMMENT =
'POC.poc_owner Role';

GRANT ROLE IDENTIFIER ( $poc_owner_role  ) TO ROLE securityadmin;
```

```
GRANT USAGE    ON DATABASE  IDENTIFIER ( $poc_database    ) TO ROLE
IDENTIFIER ( $poc_owner_role   );
GRANT USAGE    ON WAREHOUSE IDENTIFIER ( $poc_warehouse    ) TO ROLE
IDENTIFIER ( $poc_owner_role   );
GRANT MONITOR ON WAREHOUSE IDENTIFIER ( $poc_warehouse    ) TO ROLE
IDENTIFIER ( $poc_owner_role   );
GRANT OPERATE ON WAREHOUSE IDENTIFIER ( $poc_warehouse    ) TO ROLE
IDENTIFIER ( $poc_owner_role   );
GRANT USAGE    ON SCHEMA    IDENTIFIER ( $poc_owner_schema ) TO ROLE
IDENTIFIER ( $poc_owner_role   );

GRANT USAGE                      ON SCHEMA IDENTIFIER ( $poc_owner_
schema    ) TO ROLE IDENTIFIER ( $poc_owner_role );
GRANT MONITOR                    ON SCHEMA IDENTIFIER ( $poc_owner_
schema    ) TO ROLE IDENTIFIER ( $poc_owner_role );
GRANT MODIFY                     ON SCHEMA IDENTIFIER ( $poc_owner_
schema    ) TO ROLE IDENTIFIER ( $poc_owner_role );
GRANT CREATE TABLE               ON SCHEMA IDENTIFIER ( $poc_owner_
schema    ) TO ROLE IDENTIFIER ( $poc_owner_role );
GRANT CREATE VIEW                ON SCHEMA IDENTIFIER ( $poc_owner_
schema    ) TO ROLE IDENTIFIER ( $poc_owner_role );
GRANT CREATE SEQUENCE            ON SCHEMA IDENTIFIER ( $poc_owner_
schema    ) TO ROLE IDENTIFIER ( $poc_owner_role );
GRANT CREATE FUNCTION            ON SCHEMA IDENTIFIER ( $poc_owner_
schema    ) TO ROLE IDENTIFIER ( $poc_owner_role );
GRANT CREATE PROCEDURE           ON SCHEMA IDENTIFIER ( $poc_owner_
schema    ) TO ROLE IDENTIFIER ( $poc_owner_role );
GRANT CREATE STREAM              ON SCHEMA IDENTIFIER ( $poc_owner_
schema    ) TO ROLE IDENTIFIER ( $poc_owner_role );
GRANT CREATE MATERIALIZED VIEW   ON SCHEMA IDENTIFIER ( $poc_owner_
schema    ) TO ROLE IDENTIFIER ( $poc_owner_role );
GRANT CREATE FILE FORMAT         ON SCHEMA IDENTIFIER ( $poc_owner_
schema    ) TO ROLE IDENTIFIER ( $poc_owner_role );
GRANT CREATE TAG                 ON SCHEMA IDENTIFIER ( $poc_owner_
schema    ) TO ROLE IDENTIFIER ( $poc_owner_role );
```

```
GRANT CREATE EXTERNAL TABLE      ON SCHEMA IDENTIFIER ( $poc_owner_
schema   ) TO ROLE IDENTIFIER ( $poc_owner_role );
GRANT CREATE PIPE                ON SCHEMA IDENTIFIER ( $poc_owner_
schema   ) TO ROLE IDENTIFIER ( $poc_owner_role );
GRANT CREATE STAGE               ON SCHEMA IDENTIFIER ( $poc_owner_
schema   ) TO ROLE IDENTIFIER ( $poc_owner_role );
GRANT CREATE TASK                ON SCHEMA IDENTIFIER ( $poc_owner_
schema   ) TO ROLE IDENTIFIER ( $poc_owner_role );
GRANT MONITOR ON DATABASE IDENTIFIER ( $poc_database ) TO ROLE IDENTIFIER (
$poc_owner_role );
```

We have left the creation of a user and setting their default role for your investigation.

Terms of Usage

Agree that only public or suitably anonymized data is used for all PoCs. You must not be seen to facilitate non-public data egress onto the public Internet.

Get agreement from the engaging team to reimburse credits consumed during the PoC period.

Ensure there is strict understanding that the sandpit will not be used as a tactical fix for generating Production datasets.

A precursor for delivering the sandpit account is that the engaging team conducting the PoC accepts the provisioning team's terms of usage.

Common Pitfalls

Migrating to Snowflake is not trivial. As my good friend Andy McCann says, "Good practice travels far," but the reality of developing Snowflake applications is that inexperienced people can, and do, make expensive mistakes.

You may mitigate risk and limit your exposure by provisioning Sandpit accounts using Snowflake trial accounts found here: https://signup.snowflake.com/. The challenges of using free trial accounts are self-evident; there are no proper controls, or accountability, for usage and you put your organization's reputation at risk.

Through experience, we have found the availability of seasoned, skilled, and knowledgeable Snowflake developers makes all the difference in preventing expensive mistakes.

Administrative Notes

Within this section we highlight some administrative points to facilitate you working through this chapter. We discuss how to use Snowflake via both the user interface (UI) and SnowSQL, the command line interface.

Classic User Interface (UI)

For all trial accounts, Snowflake has removed the Classic UI by default, signaling the eventual demise of the Classic UI.

However, the Classic UI is still available and is useful for loading tables via the load wizard, a tool missing from Snowsight. In the top left of your Snowsight session, click the down arrow next to your username and then click **Profile**, as shown in Figure 1-2.

Figure 1-2. *SnowSQL completion dialog*

The dialogue shown in Figure 1-3 appears. Note the default is set to Snowsight but you have the option of selecting Classic UI, as shown in Figure 1-3.

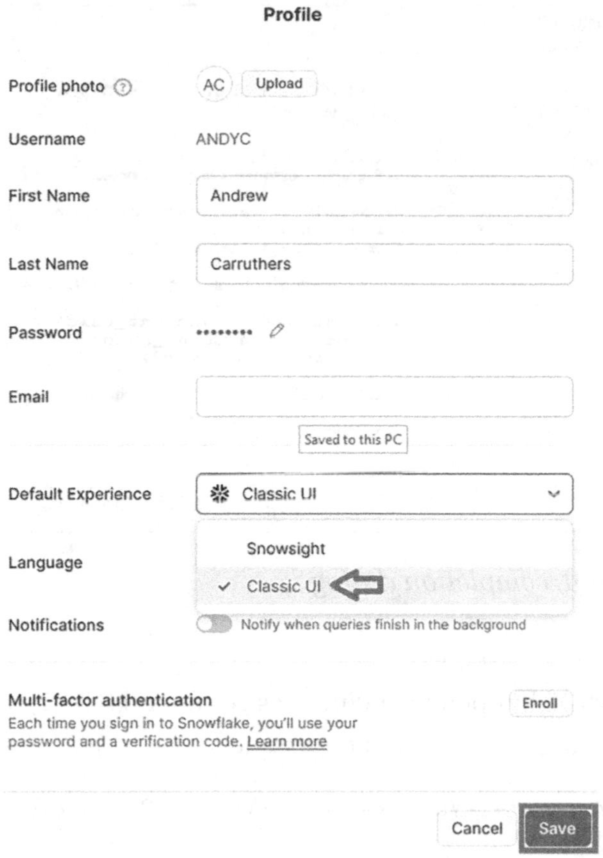

Profile

Profile photo ⑦   AC   Upload

Username   ANDYC

First Name   Andrew

Last Name   Carruthers

Password   •••••••• ✎

Email

Saved to this PC

Default Experience   ❄ Classic UI   ⌄

                      Snowsight

Language   ✓ Classic UI ⟵

Notifications   ⬤ Notify when queries finish in the background

Multi-factor authentication   Enroll
Each time you sign in to Snowflake, you'll use your
password and a verification code. Learn more

                                     Cancel   Save

Figure 1-3. *Selecting Classic UI*

Then click **Save** before logging out and logging back in.

On login, Snowflake will present the option to set Classic UI as the default user interface.

Given Snowflake's desire to remove the Classic UI, we found ourselves in a challenging position as some chapters were written using the Classic UI. We focused our attention on using SnowSQL as we authored the remainder of this book.

SnowSQL

SnowSQL is useful for automation and can be found here: `https://developers.snowflake.com/snowsql/`. We assume MS Windows for your installation. When complete, the dialog shown in Figure 1-4 appears.

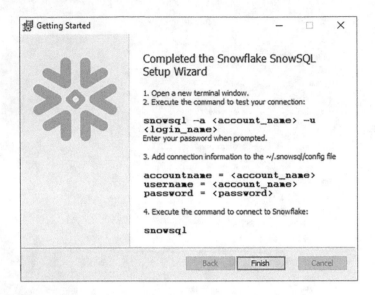

Figure 1-4. *SnowSQL completion dialog*

Click **Finish** to exit the installer.

You invoke SnowSQL by opening a Windows command (CMD) dialog, change directory to your working directory, and then issue

```
snowsql -a <Your Account>.<Your Region>.<Your CSP> -u <Your User>
```

An example login string is

```
REM snowsql -a xx11376.eu-west-1.aws -u AndyC
```

You will be prompted for your password:

    Password: <- Enter your password here

On login, you will see the SnowSQL version and prompt, noting your version may be later than v1.2.23 shown:

```
* SnowSQL * v1.2.23
Type SQL statements or !help
AndyC#(no warehouse)@(no database).(no schema)>
```

To exit, type

```
AndyC#(no warehouse)@(no database).(no schema)>!exit
Goodbye!
```

You will be returned to the Windows command prompt.

You may wish to update the template configuration file. To do so using Windows Explorer, search for **config**, which we found in `C:\Users\<Your User>\.snowsql\1.2.23` and edit using Notepad.

Within this file, lines beginning with # are commented out. To enable the lines, remove the #.

Edit those lines of interest, save the config file, and test that all changes are activated.

PrivateLink

PrivateLink is an AWS service you use to logically join AWS with your Snowflake account; it is the AWS service for creating private VPC endpoints that allow direct, secure connectivity between AWS VPCs and the Snowflake VPC without traversing the public Internet.

You can use multiple PrivateLinks to segregate products and networks for monitoring and cataloging tools onto different AWS accounts and establish PrivateLink to a single Snowflake account for each.

Chapter 2 provides a step-by-step walkthrough of how to configure your AWS account; enable, disable, and monitor your PrivateLink configuration from within Snowflake; and finally test using external tooling.

PrivateLink can only be created within the same CSP region and Snowflake region. Cross-region PrivateLink is not supported by Snowflake.

We do not consider AWS DirectConnect configuration to on-prem within this chapter. As you will see, PrivateLink has sufficient challenges of its own, but for AWS DirectConnect, please see documentation found here: `https://aws.amazon.com/directconnect/`.

Figure 1-5 illustrates several CSP accounts linked to a single Snowflake account using each CSP named equivalent to AWS PrivateLink.

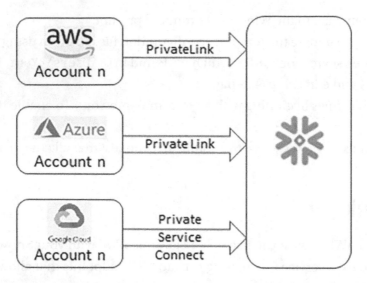

Figure 1-5. *Multiple PrivateLink connections to a single Snowflake account*

Snowflake PrivateLink self-service functions require a security federated token passed as an argument to prove and confirm ownership of the AWS account attempting to manage Snowflake's PrivateLink Service. Figure 1-6 illustrates the general interaction for each Snowflake PrivateLink self-service function.

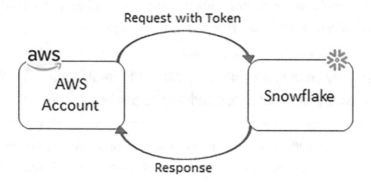

Figure 1-6. *General AWS PrivateLink connectivity pattern*

Almost all PrivateLink configurations are performed using the AWS console, and testing may be performed using your preferred ssh tool. We show how to use both Apple Mac terminal and MS Windows PuTTY, although other ssh tooling is equally usable and the supplied guides are readily adapted.

We assume you have sufficient entitlement to build, test, and deploy components using both AWS console (root) and Snowflake (accountadmin); without these entitlements, progress will be stalled. All of the supplied examples were tested using a local user's AWS account and Snowflake trial account. Your organization may implement restrictive entitlement policies so please check before attempting PrivateLink configuration.

Due to the complex nature of PrivateLink configuration, we have adopted a checklist approach to assist implementation. Accurate population of the checklist serves several purposes and is of particular value when troubleshooting. For those interested in automation using a tool such as Terraform, the checklist acts as an aide-memoire and forms the basis of system documentation for later reference.

Wherever possible, our ambition is to reduce maintenance overheads and adopt a pattern-based approach to implementation, a theme we constantly return to throughout this book.

Security Monitoring

Security monitoring utilizes components that allow you to monitor, control, and manage your Snowflake estate. They are the internal and external alerting mechanisms defined and implemented to meet your organization's security profile where generated events provide advanced warning of malicious activity and cause you to take remedial action.

You can also use third-party tooling for security monitoring because they provide another layer of protection against unauthorized system access, malicious actors, and bad intent. Despite your best intentions and despite building internal controls defined by the best and brightest people your organizations can employ, there is always someone smarter or more devious with ill intent who may be able to evade the best-in-breed defenses you have designed.

Our preferred method of protecting our systems is to implement multiple layers of defense. We discuss how to implement a layered defense, one where a single compromised component does not automatically render all of our Snowflake vulnerable but instead may isolate the compromise to a single part of the system. You will identify and develop compensating controls for security monitoring. You start by imagining your Landing Zone delivered Snowflake platform account has a Production application implemented and ask what controls you expect to have been deployed to ensure conformance to best practice.

For those organizations working with Snowflake to deploy your Landing Zone, your Snowflake Sales Engineer will be able to provide a template "Security Features Checklist". This list is generic in nature and asks many questions to prompt and direct your thinking to consider where to focus efforts and prevent security vulnerabilities from arising.

Depending upon the agreed frequency, and the agreed mechanism for delivery, we offer some pointers to aid your automation of control event information to consuming parties as outlines in Figure 1-7. The distribution channels are not exhaustive, merely indicative and all require some degree of integration with third-party tooling.

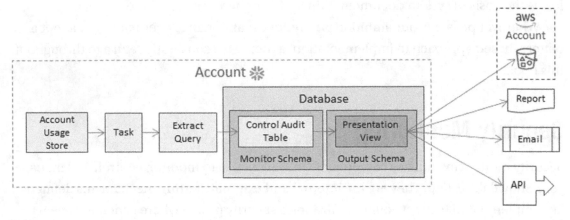

Figure 1-7. *Event detection and automated distribution*

Cost Reporting

Every organization is rightly obsessed with costs. Snowflake, with its consumption-based model, is a prime candidate for cost monitoring and reallocation of consumption costs to the consuming team, line of business, or operating division. Your Financial Operations (FinOps) colleagues will be very keen to both analyze current consumption and project future consumption using any and all metrics we expose to them.

You may also face other internal and regulatory requirements for cost monitoring, the centralized provisioning of cost monitoring lends itself to wider utilization including:

- Future Snowflake consumption trend projections.

- Identification of consumption spikes and anomalies.

- Automated inclusion of new Snowflake consumption.

Organizations that provision Snowflake accounts via a centralized Landing Zone (LZ), have an ideal opportunity to capture all consumption metrics through the creation of both a Centralized Monitoring Store (CMS), and cost monitoring data sharing from each provisioned account. Alternatively, each Snowflake account would need to deliver their metrics individually. In a decentralized, federated organization without a single reference architecture or consistent approach, considerable time would be spent reconciling disparate views of data. Far better to have a single, consistent, and centrally maintained approach.

Your objective is to deliver a "plug-and-play" cost monitoring capability, where all future LZ provisioned accounts deliver their consumption metrics into a single hub for central reporting. You might extend your cost monitoring capability beyond Snowflake into the Cloud Service Provider (CSP) tooling. We discuss extending monitoring capability later. Our approach is to deliver a step-by-step walk through of the components required to create both the CMS and capabilities to transport metrics to the CMS.

Figure 1-8 illustrates the conceptual approach to delivering a CMS for the common collation of consumption metrics:

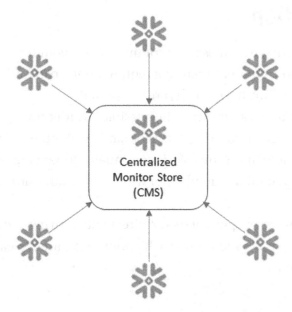

Figure 1-8. *Conceptual Centralized Monitoring Store (CMS)*

As you will read later, the concept of a CMS is readily extended to capturing other categories of information where centralized data access is required.

The CMS has two primary use-cases:

- Capturing all consumption metrics for your organization for consolidated reporting purposes.

- Capturing federated metrics. Some examples are

  - Informing Marketing and Sales teams, as explained within Chapter 5

  - Informing disaster recovery decision making, as explained within Chapter 12

Provisioning cost monitoring through your LZ enables a centralized maintenance team to build out and test enhancements. Changes will be deployed in a controlled manner to all accounts, thus ensuring all metrics are consistent and correct at the point of delivery.

Share Utilization

As Snowflake data sharing becomes more commonplace, your business colleagues will define increasingly more complex reporting requirements to identify not only the data sets consumed but also how the data sets are consumed.

In this chapter we discuss the "art of the possible" by identifying and accessing the Snowflake-supplied share monitoring data sets, along with other relevant and available information. We also note that Snowflake is continually enhancing the available data sets and we illustrate one such example later where Snowflake monitoring capability has been affected.

Let's now consider the purpose of secure direct data sharing, data exchange, and the Data Marketplace. Figure 1-9 showcases each option compared to each other, providing context for this chapter.

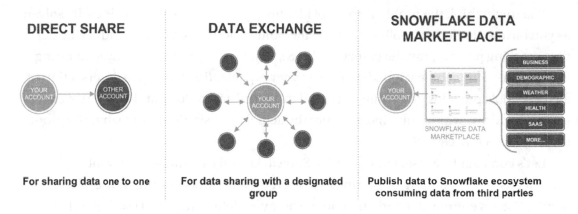

Figure 1-9. *Snowflake data interchange options*

From a data consumption perspective, the authors acknowledge Snowflake has a balancing act to perform and it must tread a fine line. Consumers do not wish to expose their bespoke algorithms while consuming providers' data, whereas for providers, having access to consuming algorithms provides business intelligence and value.

Putting aside security concerns, as they are expected and in organizations large and small therefore should be a given, a sensible compromise is in place, offering both producers and consumers some of the information they want without exposing the "secret sauce" or intellectual property of the consuming parties. And this is what we discuss here: the challenge of obtaining as much information as possible from several perspectives, along with new features presenting opportunity to monetize our data. For those of us who have been around Snowflake for a while, we realize the pace of change is accelerating and data sharing is no exception.

We also discuss data clean rooms (DCRs) but due to space constraints do not deep-dive into their implementation. Instead, we offer insights, advice, and reference materials for DCRs, leaving this for your further investigation.

Usage Reporting

In highly secure environments such as you will encounter within your organizations, at some point in the lifecycle of your data warehouse you will be asked to provision self-service of data sets. Your approach must be to provision access to only those data sets to which an end user is properly approved and entitled to view, and nothing more.

You might think the problem space of identifying "who can see what" is easily solved as you have a plethora of tooling available including role-based access control (RBAC), data masking policies, row-level security (RLS), object tagging, and tag-based masking available in your technical toolkit to protect your organizations data. The intersection of all these tools provides everything you need to control access to your data. However, the issue we address within this chapter is not the actual provisioning of data, but the proof of "how" data is accessed.

Let's consider how users interact with Snowflake and dig a little deeper into their needs while retaining a security mindset. After all, the success or failure of your applications is entirely dependent upon the how you deliver (or don't) successful business outcomes, and security is your biggest concern.

You can create the perception of having lots of data domains within your data warehouses, but from an end user perspective, and to paraphrase Samuel Taylor Coleridge, "data, data everywhere and not a drop to use." How can you deliver data in a manner acceptable to all, while proving that your data remains secure?

Your response to the question of data security, as a minimum, must address these questions:

- **Who can see what?** Roles, objects, and attributes accessible by a user.

- **Who can do what?** Roles and objects a user can manipulate.

- **Who has seen what?** The historical evidence of data access.

- **Who has done what?** The historical evidence of object manipulation.

Each question may have multiple parts, which you will progressively build up into a single unified SQL statement from which you can determine the proof your security requirements are met.

However, you must recognize there are limitations with both point-in-time and historical SQL statement capture, perhaps the "Achilles Heel" of all monitoring: SQL statement capture does not guarantee the same data will be returned when replayed later as the underlying data will change over time. In order to ensure data consistency at the point of replay, you must ensure your underlying data is not subject to updates within its temporal boundary. That is, the data cannot change without the temporal boundary changing too, and for this reason we prefer Slowly Changing Dimension 2 (SCD2) data sets as explained here: `https://en.wikipedia.org/wiki/Slowly_changing_dimension#Type_2:_add_new_row`.

SCD2 is no protection if you allow your data warehouse contents to be changed in an ad-hoc manner by users or support staff. We recommend a thorough review of all data manipulation paths for your environment before relying upon the patterns expressed within this chapter. At the initial point of creation of your data warehouse, it is far easier to establish the principle of no ad-hoc data manipulation than to attempt enforcement at some later date, by which time the integrity of your data warehouse will have been compromised. You must have fidelity back to the original received data.

Furthermore, entitlements to row-level security roles may change between the time you capture the SQL statement and time at which the SQL statement was executed. You cannot rely upon being able to exactly replicate the original scenario unless you conduct a thorough investigation and ensure no changes have been made that will affect results.

Nobody said this was going to be easy...

Your answer to self-service must be to identify and report all data access including any changes to the role used while taking into consideration the temporal nature of your data. Use the Snowflake Account Usage Store to source all required information. Whilst some answers are seemingly better answered from each database `information_schema`, you should remember the `information_schema` scope is limited to an individual database and is time-limited to 14 days. We suggest the use of `information_schema` is a poor choice for whole account data capture, unless you either cycle through each database information_schema in turn or `UNION` all required database `information_schemas` into a single accessible view.

Data Catalog

From your investigations so far, you have discovered how to identify and interrogate the Snowflake Account Usage Store to derive information useful for a variety of purposes. Next you turn your attention to investigating a different aspect of information available from the Account Usage Store: to expose the metadata relating to the data content held within Snowflake. You are mostly not interested in the attribute values themselves, but much more interested in how you define the containers that hold the data.

For every object and attribute of business significance defined within Snowflake, your data governance colleagues want to know what each object and attribute means in both business, and technical terms. To this end, you must both implement the access paths to metadata and provision the means for tooling to extract identified metadata, after which, in other tooling, both business and technical information can be added.

Fortunately, accessing Snowflake metadata is quite simple and leverages your existing knowledge from working through this book systematically, and within this chapter we show you how.

Organizations

The term "organization" is confusing as it can refer to the corporate environment many of us work in. For example, the corporate environment the authors work for is the London Stock Exchange Group (also known as LSEG). In this book, we use the word "organization" to mean any corporate environment as a place of work.

In contrast, a Snowflake organization is an optional high-level container to manage all Snowflake accounts within any corporate structure. Within this chapter, we use the term "Snowflake organization" to differentiate use. You can visualize a Snowflake organization as the outer layer to a Matryoshka Doll, commonly referred to Russian Dolls. See here for more information: `https://en.wikipedia.org/wiki/Matryoshka_doll`.

While the Matryoshka Doll analogy breaks down rather quickly, you can find the illustration in Figure 1-10 useful as a starting point in explaining Snowflake containers, with Snowflake organization being the outermost.

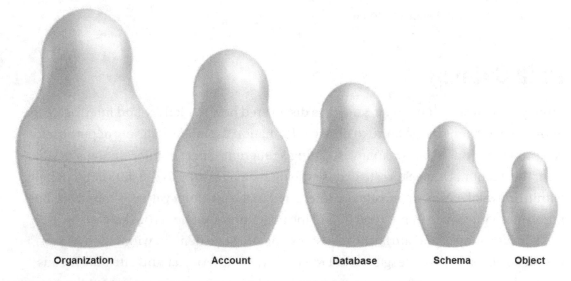

Organization   **Account**   **Database**   **Schema**   **Object**

Figure 1-10. *Matryoshka analogy for Snowflake containers*

As you work through the examples within this book, the most entitled role is `accountadmin`, which you use in the context of administering your account. So where does the Snowflake organization feature come from and how does this optional new feature work?

We provide a full explanation and detailed walk-through of the Snowflake organization features as implemented via a new role called `orgadmin`.

Alerting

Snowflake alerts is an emerging topic based upon new functionality currently available in public preview. First, you learn how to use the built-in Snowflake email capability and create wrapper stored procedures in both JavaScript and Python for later use in demonstrating alerting. You then move onto creating alerts, which provide in-built capability to create conditions, periodically monitor for when the condition is either met or exceeded, and then implement an action. You walk through each step and explore scenarios where alerts may prove useful.

For those familiar with event-based processing, Snowflake alerts are broadly comparable, noting the use of a monitoring schedule.

We suggest judicious use of Snowflake alerts. Naturally, Snowflake credit consumption will occur therefore consider the cost implementations as part of your analysis.

Within the context of centralized provisioning for a LZ, we suggest the creation of email wrapper procedures and sample alert configuration code will be sufficient. We do not advocate a centralized alerting capability except for stored procedure maintenance. We also expect each account owner or tenant will develop alerting according to their own need.

CDMC

The Cloud Data Management Capabilities (CDMC) Framework offers an extensive control-based framework to manage data migrations from on-prem to cloud. Whilst target-agnostic in nature, we suggest a Snowflake reference architecture to implement controls taking inputs from external systems, the Snowflake Account Usage Store, and unstructured documentation which may reside outside of Snowflake.

Your journey starts with proposing that your organization implement a data catalog as a baseline environment in which to store metadata for your data assets. For the purpose of illustrating how CDMC controls can be implemented, we offer an evidence repository for your later customization along with document templates to record information for later replay into a data catalog.

Your Snowflake reference architecture delivers a single design pattern, reusing components wherever possible. Figure 1-11 focusses on developing the CDMC repository for which a full explanation and deep dive walkthrough is provided within Chapter 10.

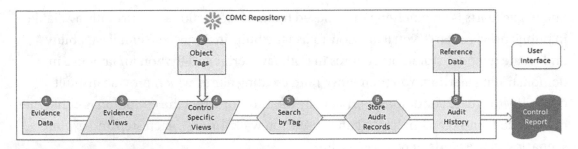

Figure 1-11. *CDMC repository overview*

Object tagging is essential for the interpretation of the reference architecture to implement the CDMC repository. You build object tagging JavaScript stored procedures which may have wider utility within your organization to facilitate self-service by end users.

We do not provide control report information as shown within Figure 1-12 and leave this for your consideration.

Behavior Change Management

Snowflake enables, by default, the easy management of changes to the data warehouse, including the ability to roll back changes if necessary, and to track and monitor the impact of changes on performance and data integrity. Additionally, Snowflake offers built-in versioning and time travel capabilities, which allow users to easily view and revert to previous versions of data and schema.

Snowflake automatically deploy releases every week in a seamless, behind-the-scenes manner. In this way, Snowflake ensures its software is maintained with the most up-to-date version with the latest features enabled. Further information can be found here: `https://docs.snowflake.com/en/user-guide/intro-releases.html`.

In addition to the automated weekly deployments, Snowflake periodically provides behavior change bundles, which are a collection of changes that are bundled together and deployed as a single unit. Snowflake Knowledge Base states,

> *The behavior change release process at Snowflake lets you test new behavior changes for two months before the changes are enabled across Snowflake accounts.*

The full article can be found here: `https://community.snowflake.com/s/article/Behavior-Change-Policy?r=398`.

Each bundle may include changes to data schema, data replication settings, and data governance policies, as well as new features or bug fixes. The goal of a behavior change bundles are to

- Minimize the impact of changes on performance and data integrity by deploying all related changes at once, rather than deploying them in separate, smaller updates

- Provide the ability to test the behavior change bundle before deployment into production environments

- Selectively enable new features and/or product capabilities for participation in private previews and provision early access for development activities

The automated weekly deployments are a way of ensuring Snowflake is updated with the latest features, security patches and bug fixes, while a behavior change bundle is a way of deploying a candidate set of related future changes for testing and rollback if needed. Once the bundle has been deemed stable and ready for deployment, it may be included in the next automated weekly deployment. With the exception of November and December every year, Snowflake typically deploy behavior changes two months after publication, usually within the third or fourth weekly release for the month. November and December are excluded due to most organizations having a change freeze period over the Christmas holiday period.

Disaster Recovery (DR)

Disaster recovery is the one capability that no one ever wants to invoke in a real-world scenario. Despite the highly unlikely event of ever needing to invoke DR, you must provision full capability along with periodic testing to demonstrate to others that your service and application capability is resilient. Many organizations conduct at least annual, and sometimes more frequent, DR testing. You must be prepared.

This chapter articulates Snowflake DR capability as available at the time of writing. The authors acknowledge Snowflake is continually improving DR capability and we expect you to periodically revise your DR plans in accordance with the latest Snowflake capabilities. We also acknowledge Snowflake may not yet implement cloud service provider-specific features from within the Snowflake environment. We discuss these features external to Snowflake and our aim is to explain the current "art of the possible" and highlight areas for further consideration. Our investigation relies upon two Snowflake accounts implemented using Snowflake Organization features explained in Chapter 8 and also reuses the CATALOG database developed in Chapter 7.

DR relies upon features are only available with Snowflake Business Critical Edition, Figure 1-13 illustrates features available within each Snowflake edition, noting Failover and Failback are only available for Business Critical Edition.

Standard
Fully functional SQL Database
Secure Data Shares
24 x 365 Support
1 day Time Travel
Encryption in transit and at rest
Virtual Warehouses
Federated Authentication
Replication Enabled

Enterprise
Standard +
Clustered Warehouses
90 Day Time Travel
Annual rekeying of data
Materialized Views

Business Critical
Enterprise +
Externally validated secure
Encrypted data everywhere
Tri-Secret Secure, BYOK
AWS PrivateLink support
Enhanced security posture
Failover and Failback support

Virtual Private Snowflake (VPS)
Business Critical +
Customer dedicated virtual servers
Customer dedicated metadata store
Some restrictions...

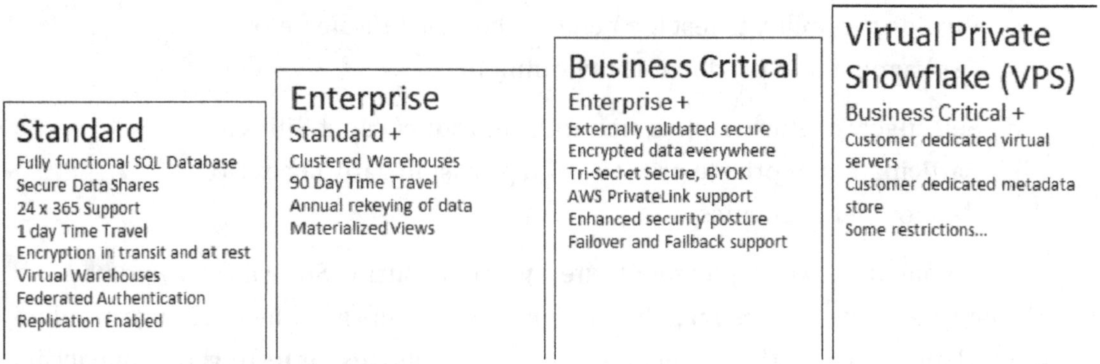

Figure 1-13. *Snowflake editions*

To upgrade Standard or Enterprise edition to Business Critical Edition, please contact Snowflake Support. Further information can be found here: https:// community.snowflake.com/s/article/How-To-Submit-a-Support-Case-in-Snowflake-Lodge.

We discuss these scenarios in turn:

- Single database replication

- Whole account replication including replication groups/failover groups

- Client redirect

It is imperative that you have a full understanding of each scenario, purpose, limitations, and options. To this end, we address each scenario in turn, providing a hands-on walk-through of each.

We start by discussing Snowflake availability and explain CSP High Availability (HA) Zones, along with providing a brief explanation of the components which constitute a CSP.

With your understanding of Snowflake availability, we discuss Business Continuity Planning (BCP) which represents your organization's appetite for both the timeliness of service recovery, potential for data loss, and a variety of other factors. Working through the features implemented by CSP HA, we distill your DR requirements down further before discussing failure scenarios and available mitigation options.

With the available information at hand, and taking into consideration your organization's preferences and other limitations, you investigate the DR options available for implementing each of the three scenarios: database replication, account replication, and replication groups/failover groups.

All testing was conducted using a free trial account available here: `https://signup.snowflake.com/`. We anticipate you may use your commercial organization's Snowflake accounts, in which case the presented steps may need amending if the event accounts already exist. Implementing DR requires the use of the `ORGADMIN` role, without which DR cannot be achieved.

During testing, we found an issue with creating accounts outside the currently logged in region where the confirmation email with initial login was not sent. The confirmation email contains an initialization link required to set the default user. Should the same issue arise for your testing, raise a support ticket with Snowflake Support.

No real-world DR scenario would be complete without considering the wider implications of single or multiple region failure. You must deliver a workable and pragmatic DR implementation catering for as many foreseeable failure scenarios as possible. We seed your thoughts by identifying a few scenarios and offering information for your further consideration.

Planning your DR test and then proving its effectiveness is essential in building confidence in both Snowflake and your capability to deal with any failure scenario.

To assist diagnosing your application failure scenario, we provide a basic troubleshooting guide to identify information useful to both yourselves and to Snowflake when raising support tickets. We note the inability to connect to a Snowflake account will obviate some checks, but partial loss of service may necessitate DR invocation even if both CSP the region, and your application, are accessible.

Lastly, we deliver a template checklist as a starting point for your DR planning.

Working through these issues will raise more questions for your organization's DR implementation than can be answered here. Nobody said designing and implementing DR was going to be easy...

Summary

In this chapter, you learned about the concept of a Snowflake landing sone and explored the range of capabilities to be developed in order to support each segment of the LZ virtuous circle. You examined some of the benefits of delivering Snowflake into your organization through common, repeatable design patterns. In order to drive Snowflake adoption, you learned about the creation of a sandpit account to provide time-limited access to a fully-functional and SME-supported Snowflake environment.

We then discussed changes to the Snowflake UI noting the move away from the Classic UI to instead using Snowsight. We also discussed installing SnowSQL, the Snowflake supplied command line interface.

Our focus moved to developing each subject for the LZ virtuous circle, noting the opportunity for reuse of the tools and techniques implemented. There is plenty of scope to cross-fertilize chapters with content from another.

Each chapter is largely self-contained. Some later chapters rely upon earlier chapters, although the content should be readily adaptable to your own environment. We have tried to keep chapters self-contained, but you will benefit from working through each chapter sequentially.

CHAPTER 2

Implementing PrivateLink

Since writing *Building the Snowflake Data Cloud* there have been enhancements to PrivateLink provisioning for both enabling and disabling self-service of PrivateLink on AWS and PrivateLink for Azure. Emails to Snowflake support to enable PrivateLink are a thing of the past but, as you will see, implementing PrivateLink remains challenging, and we make no apology for the length and breadth of material covered within this chapter, which provides an exhaustive and detailed step-by-step guide.

Self-service support for Google Cloud Private Service Connect is planned for a future release.

Configuring PrivateLink is a complex process. Within this guide, PrivateLink is created within your AWS account via an EC2 instance through to your Snowflake account. We walk you through the full process end to end, including creating and accessing the EC2 instance and then testing if you can access your Snowflake account via PrivateLink.

It is not possible to test PrivateLink without creating an EC2 instance.

We have included a troubleshooting guide to assist your implementation, recognizing not every eventuality is covered, but at least you have a starting point. Likewise, we have included a checklist which serves as a to-do list, an auditable document for recording configuration information and one that provides a starting point for later scripting of content for automated deployment.

As a reminder of what we are trying to achieve with PrivateLink, please see Figure 2-1, which articulates the full network connectivity between on-prem and CSP via Direct Connect and then CSP to Snowflake via PrivateLink.

© Andrew Carruthers and Sahir Ahmed 2023
A. Carruthers and S. Ahmed, *Maturing the Snowflake Data Cloud*, https://doi.org/10.1007/978-1-4842-9340-9_2

PrivateLink can only be created within the same AWS console region and
Snowflake region. Cross-region PrivateLink is not supported.

In this chapter we focus on the middle to right-hand side of this diagram. We do not
consider Direct Connect, which is shown for completeness only and is marked Out of Scope.

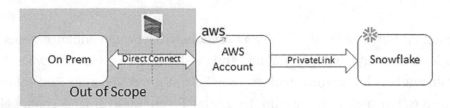

Figure 2-1. *AWS connectivity pattern*

The diagram shown in Figure 2-2 and supplied by Snowflake helps explain
PrivateLink within the context of Snowflake VPC provisioning. For further details, please
refer to the PrivateLink overview found here: `www.snowflake.com/blog/privatelink-`
`for-snowflake-no-internet-required/`.

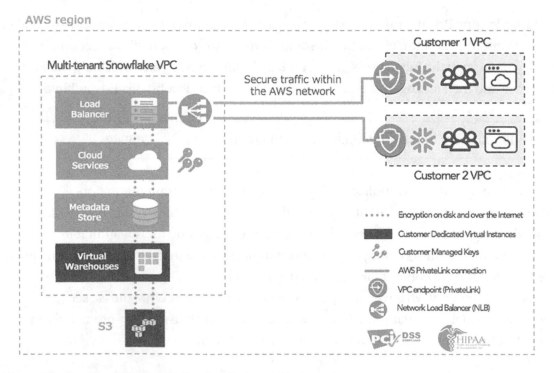

Figure 2-2. *Snowflake VPCs and PrivateLink*

You should also be aware it is possible to connect many AWS accounts to a single Snowflake account. For each AWS account, you must repeat the implementation pattern outlined within this chapter. We do not discuss automating AWS profile and user creation, nor do we discuss storage and retrieval of keys; we leave this for your engineering team to deliver.

Once PrivateLink has been provisioned within Snowflake, it is not the end of the story. To use PrivateLink, you must configure your virtual private cloud (VPC) environment (VPCE) discussed later.

Retrieving the AWS Account ID

You begin by logging into your AWS account as root and performing administration tasks to create a new user and then update an existing policy to add entitlement. You move on to configuring your Snowflake account for private connectivity.

After logging into your AWS account, please make a note of your AWS account ID. When logged in, it can be found in the top right of your screen, as shown in Figure 2-3. You will use this later. Not forgetting to remove field separators, our example account ID becomes 168745977517.

Figure 2-3. *AWS Account ID*

Snowflake PrivateLink Functions

Moving over to Snowflake, let's walk through a worked example for self service using the supplied PrivateLink SQL functions listed in Table 2-1, which are derived from Snowflake documentation.

Table 2-1. *PrivateLink Self-Service Functions*

Function	Description
system$authorize_ privatelink	Enables private connectivity to the Snowflake service for the current account
system$revoke_ privatelink	Disables private connectivity to the Snowflake service for the current account
system$get_privatelink	Verifies whether your current account is authorized for private connectivity to the Snowflake service
system$get_privatelink_ authorized_endpoints	Returns a list of the authorized endpoints for your current account to use with private connectivity to the Snowflake service

Connectivity to Snowflake Service

Snowflake PrivateLink self-service functions require a security federated token passed as an argument to prove and confirm ownership of the AWS account attempting to manage Snowflake's PrivateLink service. Assuming the token is valid, the appropriate response is returned. If the token is invalid, an error occurs. Figure 2-4 illustrates the general interaction for each Snowflake PrivateLink self-service function.

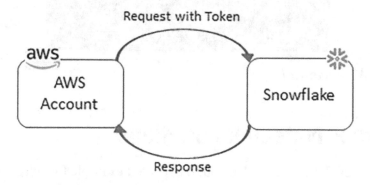

Figure 2-4. *General AWS PrivateLink connectivity pattern*

Tokens last for a finite duration, so expect to recreate tokens when making changes to Snowflake PrivateLink configuration. We show you how to do this within this chapter.

Some organizations restrict their provisioned AWS accounts to prevent command line interface (CLI) interaction, in which case representation should be made to create and entitle an IAM user within your AWS account to use AWS CloudShell and

AWS Security Token Service (STS). We use both CloudShell and STS to return a set of temporary security credentials for a federated user. See the AWS documentation on STS for more information: https://docs.aws.amazon.com/STS/latest/APIReference/ API_GetFederationToken.html.

Creating an STS Profile

Later in this chapter you'll rely upon AWS-supplied policies for CloudShell (AWS CloudShellFullAccess) and for EC2 (AmazonEC2FullAccess). However, there is no AWS supplied policy for STS, so you must therefore create your own.

Begin by searching within AWS console for **IAM**, select **Policies**, and then click **Create Policy**, as shown in Figure 2-5.

Figure 2-5. *Creating an IAM policy*

Figure 2-6 shows the options selected when creating a new policy. Select the **JSON** tab.

Figure 2-6. *JSON content*

Replace the highlighted content from Figure 2-6 with

```
{
    "Version": "2012-10-17",
    "Statement": [
        {
            "Sid": "STS",
            "Effect": "Allow",
            "Action": [
                "sts:AssumeRole",
                "sts:GetFederationToken"
            ],
            "Resource": "*"
        }

    ]
}
```

Then click **Next: Tags**. For the Create policy screen (not shown), click **Next: Review**. Enter the policy name of snowflake_private_link_policy and check the Service lists **STS**, as shown in Figure 2-7, and then click the **Create policy** button to conclude the policy configuration.

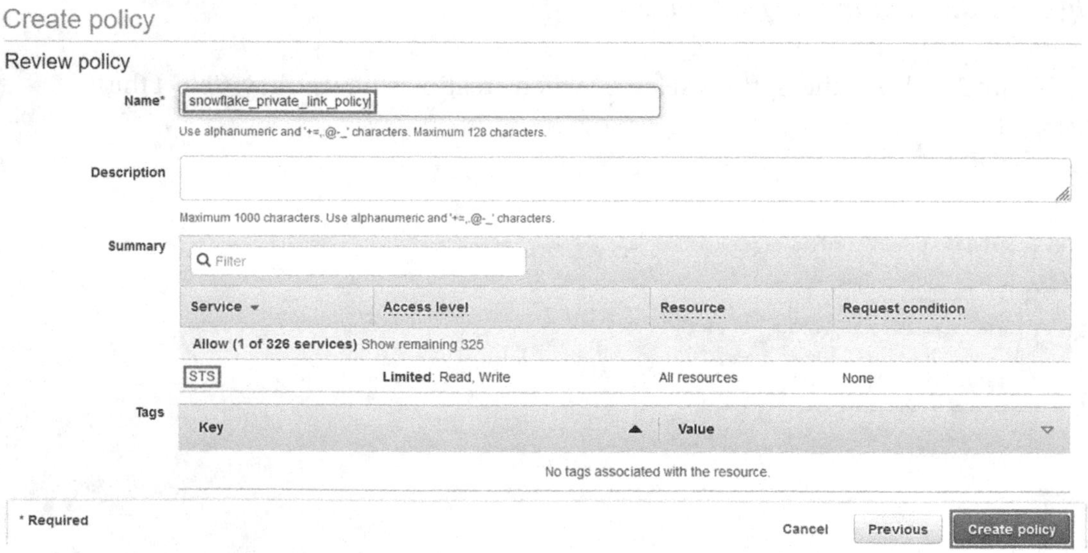

Create policy

Review policy

Name*	snowflake_private_link_policy	
	Use alphanumeric and '+=,.@-_' characters. Maximum 128 characters.	
Description		
	Maximum 1000 characters. Use alphanumeric and '+=,.@-_' characters.	
Summary	Q Filter	

Service ▾	Access level	Resource	Request condition
Allow (1 of 326 services) Show remaining 325			
STS	**Limited**: Read, Write	All resources	None

Tags	Key ▲	Value ▽
	No tags associated with the resource.	

* Required   Cancel   Previous   Create policy

Figure 2-7. *Policy creation*

If you later require additional entitlements for managing the PrivateLink configuration, you can edit snowflake_private_link_policy and add the new JSON code.

With your new policy available, you can move on to creating an IAM user and attaching required policies.

Creating an IAM User

According to best practices, you must create an AWS IAM user. Within the AWS console, search for **IAM**, select **Users**, and click **Add users**, as shown in Figure 2-8.

Figure 2-8. *Creating an IAM user*

Figure 2-9 shows the options selected when creating a new user named snowflake_private_link_user.

Ensure that the **Access key – Programmatic access** option is checked!

The whole purpose of creating a new IAM user is to generate the access key ID and secret access key. Without them you will not be able to proceed with configuring PrivateLink.

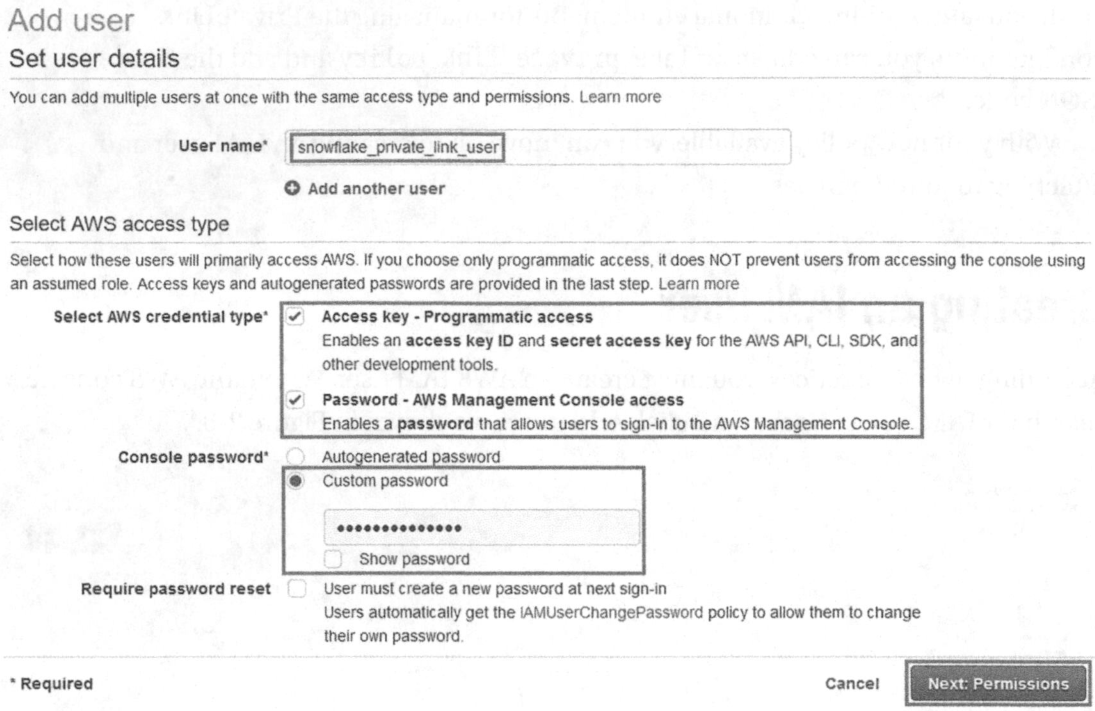

Add user

Set user details

You can add multiple users at once with the same access type and permissions. Learn more

User name*   `snowflake_private_link_user`

⊕ **Add another user**

Select AWS access type

Select how these users will primarily access AWS. If you choose only programmatic access, it does NOT prevent users from accessing the console using an assumed role. Access keys and autogenerated passwords are provided in the last step. Learn more

Select AWS credential type*   ☑ **Access key - Programmatic access**
Enables an **access key ID** and **secret access key** for the AWS API, CLI, SDK, and other development tools.

☑ **Password - AWS Management Console access**
Enables a **password** that allows users to sign-in to the AWS Management Console.

Console password*   ○ Autogenerated password
● Custom password

●●●●●●●●●●●●●●●
☐ Show password

Require password reset   ☐ User must create a new password at next sign-in
Users automatically get the IAMUserChangePassword policy to allow them to change their own password.

* Required   Cancel | Next: Permissions |

Figure 2-9. *IAM user credentials*

When complete, click **Next: Permissions**, which navigates to the next screen where an existing policy can be associated with the user. In this example, you assign snowflake_private_link_policy, as shown in Figure 2-10, noting that the Filter policies search criteria interactively selects defined policies. In this case, the partial text search returns AWS-defined AWS CloudShellFullAccess. You should also search for and add AmazonEC2FullAccess and your user-defined policy named snowflake_private_link_ policy for later use within this chapter.

Set the checkbox for each policy before clicking the **Next: Tags** button.

Figure 2-10. *IAM user permissions*

Check the box next to the desired policy and then click **Next: Tags** (screen not shown) and then click **Next: Review**, which navigates to the review screen as shown in Figure 2-11.

Add user

Review

Review your choices. After you create the user, you can view and download the autogenerated password and access key.

User details

User name	snowflake_private_link_user
AWS access type	Programmatic access and AWS Management Console access
Console password type	Custom
Require password reset	No
Permissions boundary	Permissions boundary is not set

Permissions summary

The following policies will be attached to the user shown above.

Type	Name
Managed policy	AWSCloudShellFullAccess
Managed policy	AmazonEC2FullAccess
Managed policy	snowflake_private_link_policy

Tags

No tags were added.

Cancel   Previous   Create user

Figure 2-11. *IAM user review*

Check that the policies are correct and then click the **Create user** button, which displays the final dialog shown in Figure 2-12, not forgetting to download the user security credentials.

After logging into the AWS console using snowflake_private_link_user, and for testing purposes only, you may wish to record the new password into the downloaded credentials.

The access key ID and secret access key will be required later. You **MUST** save this information now. This is the only time the secret access key is visible. It cannot be retrieved later.

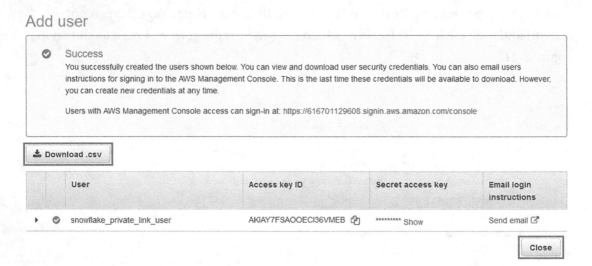

Figure 2-12. *IAM user download credentials*

With your access key ID and secret access key available, please store them in your checklist (see below). You may choose to store them in AWS Secrets Manager (out of scope for this chapter) for later automation of PrivateLink management.

Then click **Close**.

Getting a Federation Token

You must now create a federation token used to configure PrivateLink, noting tokens are only valid for short time periods and therefore require recreation when administering PrivateLink. This section walks through the necessary steps.

Using a different browser, navigate to the AWS console at `https://168745977517.signin.aws.amazon.com/console` and log in using IAM user snowflake_private_link_user and the password from the file downloaded above.

Search for **CloudShell**, which will take a minute or so to invoke, after which you should see a console, as shown in Figure 2-13.

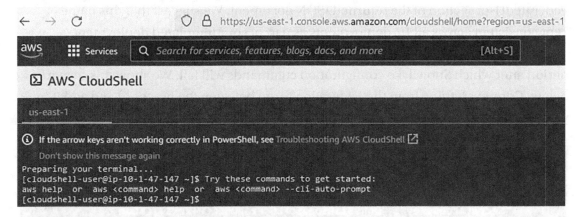

Figure 2-13. *AWS CloudShell console*

If AWS CloudShell is not available, or according to your preference, AWS provides a separate installable desktop download, refer to `https://docs.aws.amazon.com/cli/latest/userguide/getting-started-install.html`.

If AWS CloudShell fails to start, please refer to the Troubleshooting Guide below.

With your CloudShell console open, and assuming BASH (the default shell) is used, you must first set two environment variables using the access key ID and secret access key values downloaded when creating a new user:

```
export AWS_ACCESS_KEY_ID=<Your Access Key Here>
export AWS_SECRET_ACCESS_KEY=<Your Secret Access Key Here>
```

You now attempt to call the GetFederationToken operation:

```
aws sts get-federation-token --name snowflake
```

Your AWS environment variables determine whether the above command executes correctly or results in an error, as shown in Figure 2-14.

```
[cloudshell-user@ip-10-0-50-20 ~]$ aws sts get-federation-token --name snowflake
An error occurred (AccessDenied) when calling the GetFederationToken operation: Cannot call GetFederationToken with session credentials
```

Figure 2-14. *Access denied*

Assuming your environment variables are set correctly, you should see a response like Figure 2-15, noting the image has been obfuscated and your key and token will differ.

Another consideration is the —name attribute passed through when generating the federation token, as highlighted in Figure 2-15. This name becomes part of the FederatedUser section of the returned JSON document. We suggest that this name is kept consistent across all LZ deployments as part of your templated deployment.

Note the Expiration attribute indicating the token is only valid for a specific time period after which Snowflake configuration commands will fail. We show this scenario below. Copy everything from the Federation token between the braces { } and add it to your checklist.

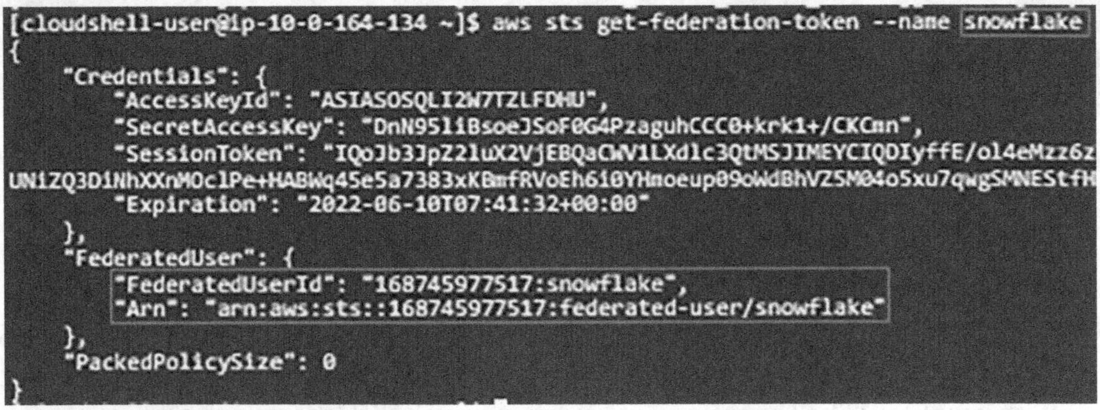

```
[cloudshell-user@ip-10-0-164-134 ~]$ aws sts get-federation-token --name snowflake
{
    "Credentials": {
        "AccessKeyId": "ASIASOSQLI2W7TZLFDHU",
        "SecretAccessKey": "DnN95liBsoeJSoF0G4PzaguhCCC0+krk1+/CKCmn",
        "SessionToken": "IQoJb3JpZ2luX2VjEBQaCWV1LXdlc3QtMSJIMEYCIQDIyffE/ol4eMzz6z
UNiZQ3DiNhXXnM0clPe+HABWq45e5a7383xKBmfRVoEh618YHmoeup09oWdBhVZSM04o5xu7qwgSMNEStfH
        "Expiration": "2022-06-10T07:41:32+00:00"
    },
    "FederatedUser": {
        "FederatedUserId": "168745977517:snowflake",
        "Arn": "arn:aws:sts::168745977517:federated-user/snowflake"
    },
    "PackedPolicySize": 0
}
```

Figure 2-15. *Federated token name*

Verifying PrivateLink Is Authorized

With your AWS account ID and federation token available, you now log into the Snowflake UI and switch to the classic interface. Use the classic interface because you can set your role on the command line, a sorely missed feature omitted from Snowsight. Alternatively, SnowSQL may also be used.

Ensure your AWS account ID is enclosed in quotes or PrivateLink commands fail with SQL execution internal error **Processing aborted due to error 300002:3397911273.**

PrivateLink commands are set at account level and therefore require the accountadmin role:

USE ROLE accountadmin;

Using both your ASW account ID and the returned federation token, retrieve the PrivateLink status. For the first invocation of PrivateLink commands, expect this command to return "Private link access not authorized."

Simply replace the AWS account ID with yours and then cut and paste the credentials from your CloudShell results, or from your checklist, before executing:

```
SELECT system$get_privatelink(
    '<Your AWS Account ID Here>',
    '{
        "Credentials": {
            "AccessKeyId": "<Your Access Key Here>",
            "SecretAccessKey": "<Your Secret Access Key Here>",
            "SessionToken": "<Your Session Token Here>",
            "Expiration": "<Your Expiry Time Here>"
        },
        "FederatedUser": {
            "FederatedUserId": "<Your AWS Account ID Here>:snowflake",
            "Arn": "arn:aws:sts:: <Your AWS Account ID Here>:federated-
            user/snowflake"
        },
        "PackedPolicySize": 0
    }' );
```

The expected response is "Private link access not authorized." or for later checks where system$authorise_privatelink() has been executed, "Private link access authorized."

If the expiration time is exceeded, this error message is displayed:

```
AWS API error code="ExpiredToken". Please contact your AWS Administrator
for further assistance.
```

To correct the error, a new token is required. Repeat the steps for getting a federation token and refactor the call to system$get_privatelink() and then retry.

Authorizing PrivateLink

Having successfully verified that your PrivateLink access is not authorized, you can now enable PrivateLink:

```
USE ROLE accountadmin;

SELECT system$authorize_privatelink(
      '<Your AWS Account ID Here>',
      '{
          "Credentials": {
              "AccessKeyId": "<Your Access Key Here>",
              "SecretAccessKey": "<Your Secret Access Key Here>",
              "SessionToken": "<Your Session Token Here>",
              "Expiration": "<Your Expiry Time Here>"
          },
          "FederatedUser": {
              "FederatedUserId": "<Your AWS Account ID Here>:snowflake",
              "Arn": "arn:aws:sts:: <Your AWS Account ID Here>:federated-
              user/snowflake"
          },
          "PackedPolicySize": 0
      }' );
```

Assuming you have not exceeded the expiration time, your command should return "Private link access authorized." See Figure 2-16.

Figure 2-16. *PrivateLink authorized confirmation*

Revoking PrivateLink

In the same manner as you authorize PrivateLink, you may also revoke PrivateLink:

```
USE ROLE accountadmin;

SELECT system$revoke_privatelink(
    '<Your AWS Account ID Here>',
    '{
        "Credentials": {
            "AccessKeyId": "<Your Access Key Here>",
            "SecretAccessKey": "<Your Secret Access Key Here>",
            "SessionToken": "<Your Session Token Here>",
            "Expiration": "<Your Expiry Time Here>"
        },
        "FederatedUser": {
            "FederatedUserId": "<Your AWS Account ID Here>:snowflake",
            "Arn": "arn:aws:sts:: <Your AWS Account ID Here>:federated-
            user/snowflake"
        },
        "PackedPolicySize": 0
    }' );
```

Assuming you have not exceeded the expiration time, your command should return "Private link access revoked." See Figure 2-17.

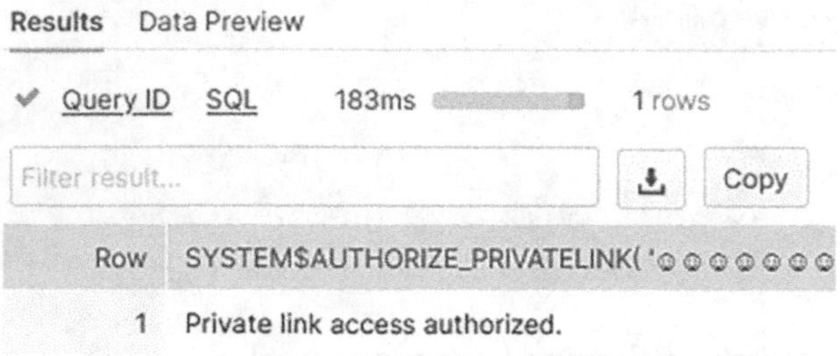

Results   Data Preview

✔ Query ID   SQL   183ms ▬▬▬▬▬   1 rows

Filter result...   ⬇   Copy

Row   SYSTEM$AUTHORIZE_PRIVATELINK('⊙ ⊙ ⊙ ⊙ ⊙ ⊙ ⊙

1   Private link access authorized.

Figure 2-17. *PrivateLink revocation confirmation*

Identifying PrivateLink Endpoints

Your final PrivateLink function allows you to identify all currently active PrivateLink endpoints; that is, all AWS accounts for which PrivateLink has been enabled as illustrated in Figure 2-18, noting self-service for Google Private Service Connect is not available at the time of writing.

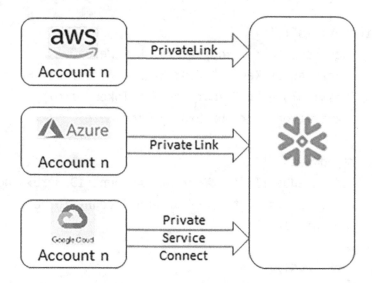

Figure 2-18. *Multiple PrivateLink connections to a single Snowflake account*

One legitimate use for implementing multiple PrivateLinks is to segregate products and networks for monitoring and cataloging tools onto different AWS accounts and establish PrivateLink to a single Snowflake account for each. We discuss this approach later within this book.

To identify endpoints, Snowflake provides `system$get_privatelink_authorized_endpoints()`:

```
USE ROLE accountadmin;

SELECT system$get_privatelink_authorized_endpoints();
```

When there are no authorized endpoints, the result set is empty, as shown in Figure 2-19.

Figure 2-19. *No PrivateLink endpoints enabled*

When one or more authorized endpoints exist, the result set is populated, as shown in Figure 2-20.

Figure 2-20. *PrivateLink endpoints enabled*

You might also consider adding PrivateLink endpoint identification to your security monitoring to ensure no unauthorized endpoints have been added to your system configuration, noting the endpoints can be either AWS or Azure with GCP to follow:

```
USE ROLE accountadmin;

SELECT REPLACE ( value:endpointId,      '"' ) AS AccountID,
       REPLACE ( value:endpointIdType, '"' ) AS CSP
FROM    TABLE (
          FLATTEN (
```

```
input => parse_json(system$get_privatelink_authorized_
endpoints())
        )
);
```

We leave the automation of any monitoring as an exercise for you to complete, noting the above SQL command must run as ACCOUNTADMIN; therefore, it is best suited to being wrapped in a stored procedure and called by task with result set stored into local table.

Configuring AWS VPC

You must now configure your VPC, and for this you may need the assistance of your engineering colleagues, particularly when operating inside your organization infrastructure.

Please also refer to documentation found here: https://docs.snowflake.com/ en/user-guide/admin-security-privatelink.html#configuring-your-aws-vpc- environment.

Log into your AWS console as administrator (root) and search for **VPC**, which takes you to the VPC Dashboard, part of which is shown in Figure 2-21, and select **Create VPC**.

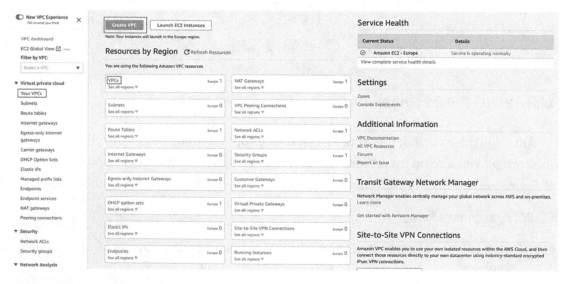

Figure 2-21. *Your VPCs*

Referring to Figure 2-22, select **VPC and more** (as highlighted) and then choose a name. In this example, we use Snowflake-VPC.

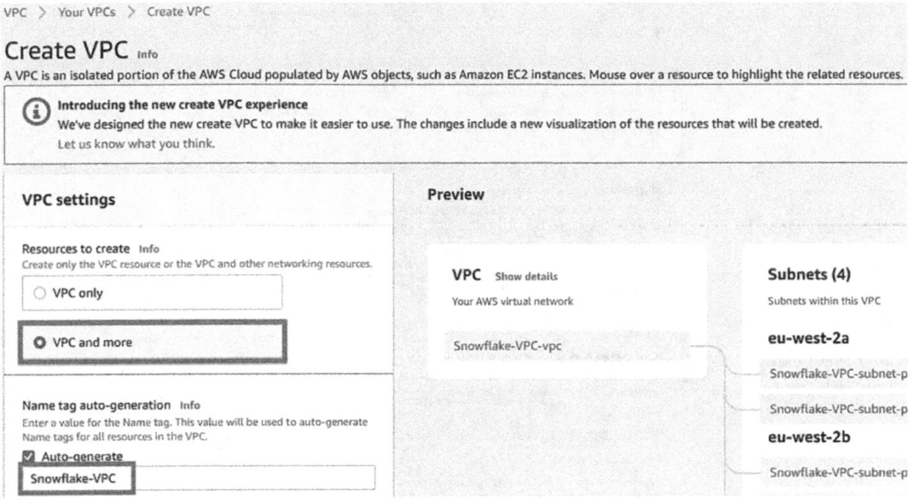

Figure 2-22. *Creating a VPC*

Scroll down and change the highlighted sections:

- **Number of Availability Zones (AZs)** to **3**

- **NAT gateways ($)** to **in 1 AZ**

- **VPC endpoints** to **None**

When complete, click **Create VPC**, as shown in Figure 2-23.

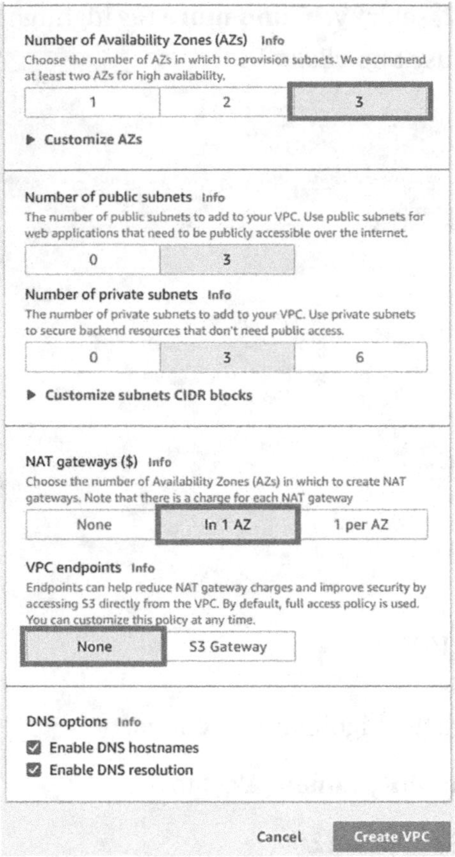

Figure 2-23. *Setting VPC availability zones*

The creation of a VPC takes a few minutes and you'll see the message "Wait NAT Gateways to activate" along with a percent complete, as shown in Figure 2-24 (noting some detail has been omitted).

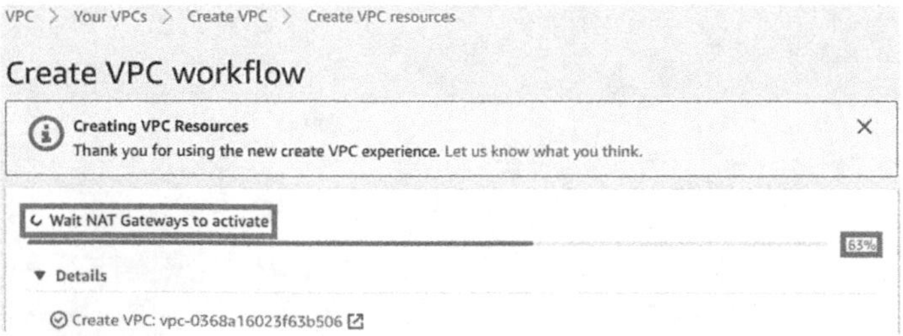

Figure 2-24. *VPC partial completion message*

Once complete, a Success message will appear, as shown in Figure 2-25 (noting some detail has been omitted).

Figure 2-25. *VPC Success message*

Select **View VPC** and refer to Figure 2-26 where State is shown as Available, confirming the VPC has been set up correctly. Please make a note of the new VPC, which in our example is vpc-0368a16023f63b506. Also note the IPv4 CIDR, which in our example is 10.0.0.0/16, and add them to your checklist as both are required later.

Figure 2-26. *VPC status and CIDR*

From the console's left-hand list, select **Security ➤ Security groups**, as shown in Figure 2-27.

Figure 2-27. *Security groups*

The screen will refresh and show available security groups. Select the checkbox for the new VPC, which in our case is vpc-0368a16023f63b506 (noting yours will differ). The Inbound rules count should state "1 Permission entry."

Make a note of the security group ID, which in our case is sg-0bbcea88372906e42 (noting yours will differ). Add this to your checklist.

Select the **Inbound rules** tab.

- **Type** should be **All traffic**.

- **Protocol** should be **All**.

- **Port Range** should be **All**.

Figure 2-28 shows the expected results.

Figure 2-28. *Inbound rules*

Then click **Edit inbound rules**, after which the screen will refresh to allow rules to be added, as shown in Figure 2-29.

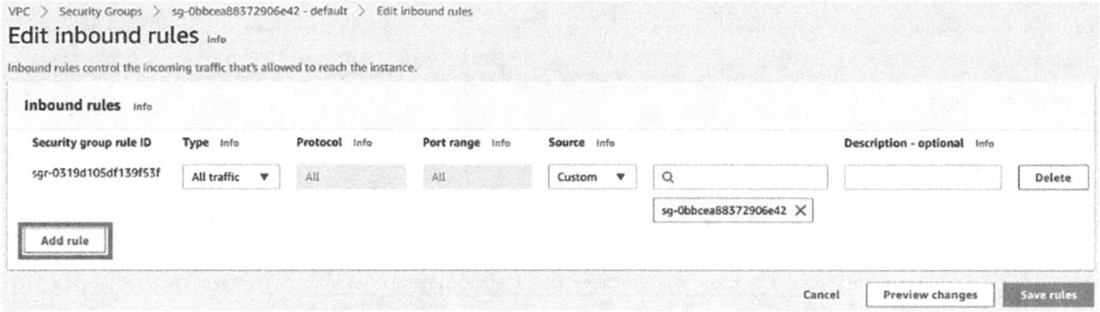

Figure 2-29. *Editing inbound rules*

Select **Add rule**, where the screen presents additional configuration options, as shown in Figure 2-30.

Delete the All traffic rule.

Add two rules for HTTP and HTTPS. Using the search dialog, ensure the same security group ID from your checklist is defaulted.

When both rules are configured, as shown in Figure 2-30, select **Save Rules**.

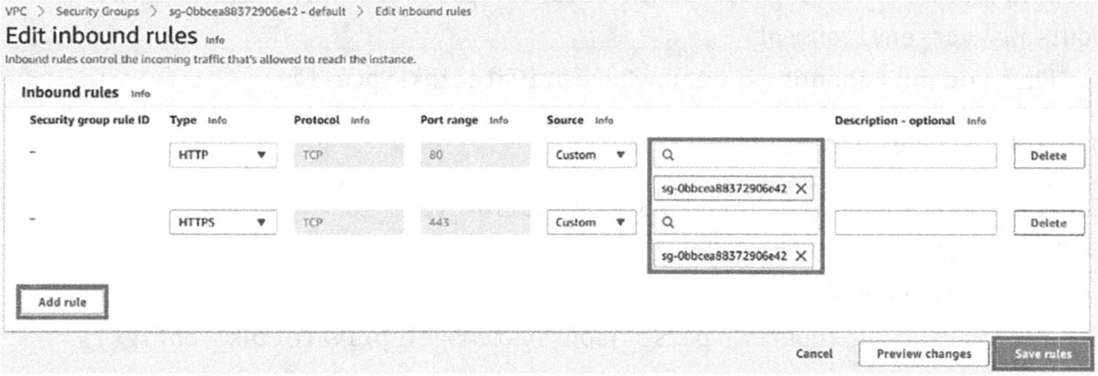

Figure 2-30. *Adding inbound rules*

Your screen will now refresh and two new inbound rules will be seen, as shown in Figure 2-31.

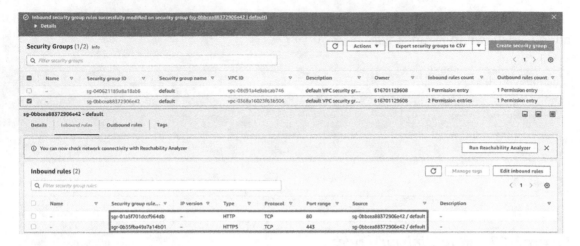

Figure 2-31. *Inbound rules summary*

Configuring an AWS VPC Endpoint (VPCE)

You must now configure your VPC endpoint and for this you may need the assistance of your engineering colleagues, particularly when operating inside your organization infrastructure. Please also refer to documentation found here: `https://docs. snowflake.com/en/user-guide/admin-security-privatelink.html#configuring-your-aws-vpc-environment`.

First, you must identify your Snowflake `privatelink-vpce-id`:

```
USE ROLE accountadmin;

SELECT REPLACE ( value, '"' ) AS privatelink_vpce_id
FROM   TABLE (
           FLATTEN (
                input => parse_json(system$get_privatelink_config())
                )
             )
WHERE  key = 'privatelink-vpce-id';
```

Our `privatelink-vpce_id` is `com.amazonaws.vpce.eu-west-2.vpce-svc-0839061a5300e5ac1`, noting your value may differ. Add this to your checklist.

While you can derive your Snowflake region (eu-west-2) from the logged in URL, you can also confirm via the user interface:

```
SELECT current_region();
```

Revert to your AWS console and set your AWS region to be the same as your Snowflake region. The drop-down can be found in top right-hand corner of the AWS console, as shown in Figure 2-32.

Figure 2-32. *Setting the correct AWS region*

Failure to set the AWS console region to the Snowflake region causes endpoint validation to fail.

You also need to extract the privatelink-account-url. You do this by running the following SQL in Snowflake. Please make a note of this in your checklist because it's required later when creating hosted ones.

```
USE ROLE accountadmin;
SELECT REPLACE ( value, '"' ) AS privatelink_vpce_id
FROM    TABLE (
            FLATTEN (
                    input => parse_json(system$get_privatelink_config())
                    )
                )
WHERE   key = 'privatelink-account-url';
```

Within the AWS console, search for **VPC**, which takes you to the VPC Dashboard, part of which is shown in Figure 2-33.

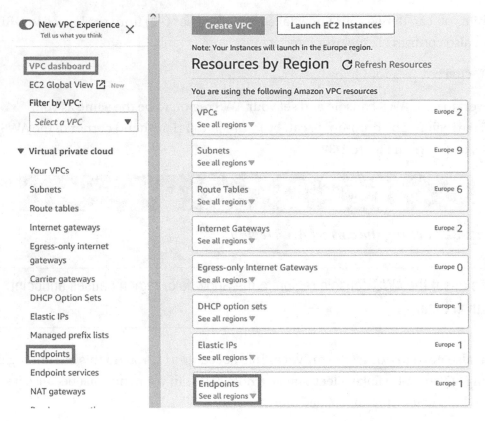

Figure 2-33. *VPC Dashboard*

Endpoints can be accessed from two places within the VPC dashboard. Click on either of the Endpoints options highlighted in Figure 2-33. Figure 2-34 shows the option to create a new endpoint.

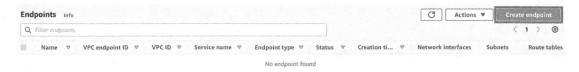

Figure 2-34. *Endpoints*

If you are reconfiguring an existing endpoint by deleting, please note that deleting endpoints may take a few minutes.

To create a new endpoint, click **Create endpoint**, which opens the screen shown in Figure 2-35.

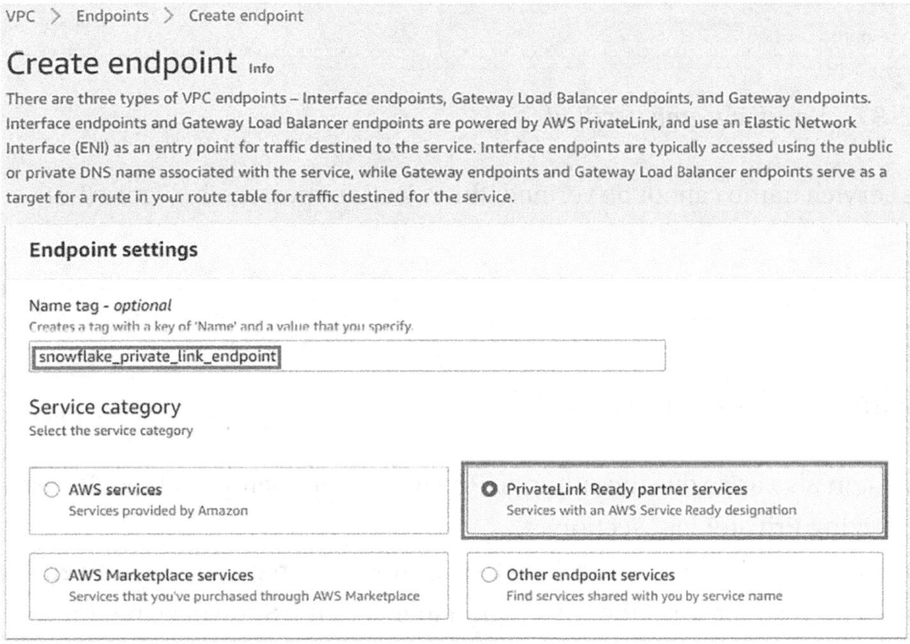

VPC > Endpoints > Create endpoint

Create endpoint Info

There are three types of VPC endpoints – Interface endpoints, Gateway Load Balancer endpoints, and Gateway endpoints. Interface endpoints and Gateway Load Balancer endpoints are powered by AWS PrivateLink, and use an Elastic Network Interface (ENI) as an entry point for traffic destined to the service. Interface endpoints are typically accessed using the public or private DNS name associated with the service, while Gateway endpoints and Gateway Load Balancer endpoints serve as a target for a route in your route table for traffic destined for the service.

Endpoint settings

Name tag - *optional*
Creates a tag with a key of 'Name' and a value that you specify.

snowflake_private_link_endpoint

Service category
Select the service category

○ AWS services
Services provided by Amazon

● PrivateLink Ready partner services
Services with an AWS Service Ready designation

○ AWS Marketplace services
Services that you've purchased through AWS Marketplace

○ Other endpoint services
Find services shared with you by service name

Figure 2-35. *Creating an endpoint*

Populate these fields:

- Set **Name tag** to **snowflake_private_link_endpoint**.

- Set **Service category** to **PrivateLink Ready partner service**.

As shown in Figure 2-36, set the Service name to the Snowflake value for the `privatelink-vpce_id` retrieved above, and in your checklist (`com.amazonaws.vpce.eu-west-2.vpce-svc-0839061a5300e5ac1`), noting your value will differ, and then click **Verify service**.

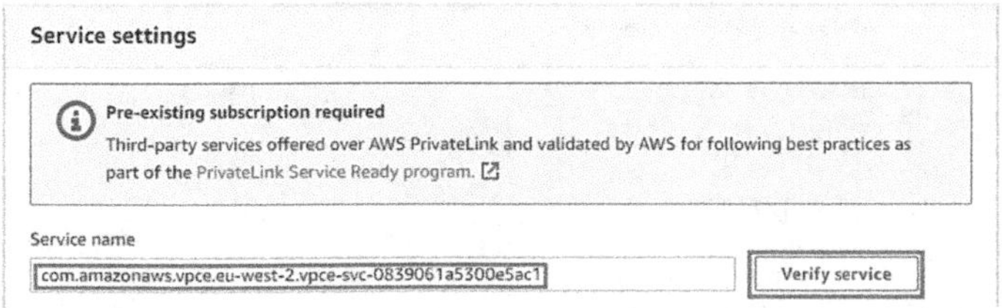

Service settings

ⓘ **Pre-existing subscription required**
Third-party services offered over AWS PrivateLink and validated by AWS for following best practices as part of the PrivateLink Service Ready program. ☐

Service name

com.amazonaws.vpce.eu-west-2.vpce-svc-0839061a5300e5ac1 | Verify service

Figure 2-36. *Verifying PrivateLink service*

Assuming the entered service name is valid for the AWS region where the endpoint is being created, the confirmation message in Figure 2-37 will appear.

> ⊘ Service name verified.

Figure 2-37. *Service name verified*

If the service name cannot be verified, the rejection message in Figure 2-38 will appear.

> ⊗ Service name could not be verified.

Figure 2-38. *Service name failure*

Failure can also arise due to Authorize PrivateLink not being enabled. Please refer to the "Authorizing PrivateLink" section.

Failure to set the AWS console region to the Snowflake region can cause endpoint validation to fail, so please set the AWS region to match the Snowflake region and retry.

The service name must be verified before VPC can be selected.

With the service name verified, you can select the VPC from the drop-down list you created in your example (see checklist), Snowflake-VPC-vpc, in which to create the endpoint, as shown in Figure 2-39.

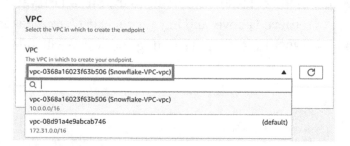

Figure 2-39. *Selecting the VPC*

Now configure subnets by selecting the checkbox for each availability zone (AZ). Select **Subnet ID** as the **private** option for all subnets from the corresponding drop-down box and set the **IP address type** to **IPV4**, as shown in Figure 2-40.

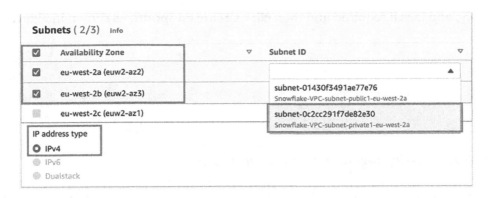

Figure 2-40. *Configuring a subnet for an AZ*

Now select the security group you edited earlier (see checklist), as shown in Figure 2-41, noting ports 80 and 443 must be open, which you set when creating the security group.

- 80: Required for the Snowflake OCSP cache server

- 443: Required for general Snowflake SSL traffic

Without digressing too far, and for those with curious minds, the Online Certificate Status Protocol (OCSP) is explained here: `https://en.wikipedia.org/wiki/Online_Certificate_Status_Protocol`.

Your organization may have defined several security groups according to local policy. Select the appropriate group, noting the example in Figure 2-41 uses the security group we created and is for our new VPC; see the checklist.

Figure 2-41. *Selecting security groups*

Security groups are stateful. If an inbound port is open, the same port will be opened for outbound. To check, right-click on the desired **Group ID** (in our example, sg-0bbcea88372906e42) and then **Open Link in a New Tab** and click the **Inbound rules** tab where the settings will appear (not shown here).

Finally, add tags if required and then click **Create endpoint**, as shown in Figure 2-42.

Figure 2-42. *Adding tags and creating an endpoint*

You should now have your endpoint displayed, as shown in Figure 2-43, noting the status is Pending. After a short while, the status will change to Available.

Figure 2-43. *Pending endpoint*

From the Endpoint page, you need to scroll down and make a note in your checklist of the **first** DNS names entry. You choose the **first** DNS name as every other sequential DNS name is AZ specific; they are called out in Figure 2-44. The DNS name is required in the next section when setting up CNAME, so add this to the checklist.

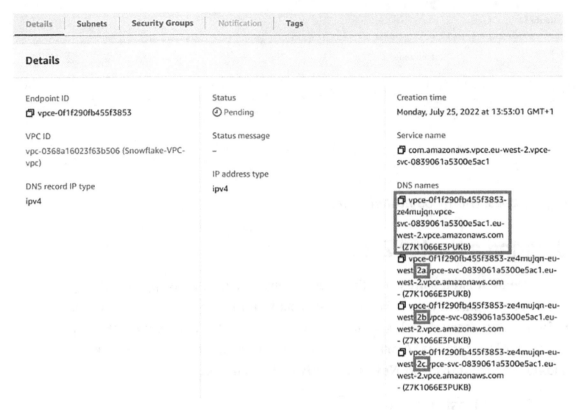

Figure 2-44. *Selecting a DNS name*

Creating CNAME Records

A canonical name, or CNAME, record is a type of DNS record that maps an alias name to a true or canonical domain name. For example, you might have two logical names that map to a single common name. Maintaining a single common name is preferrable to making changes in two or more places.

Within your Snowflake context, you need to create the relationship between your PrivateLink context and AWS VPCE. For this you must now log into the AWS console using the root user and select another AWS service called **Route 53** and select **Hosted zones**, as shown in Figure 2-45.

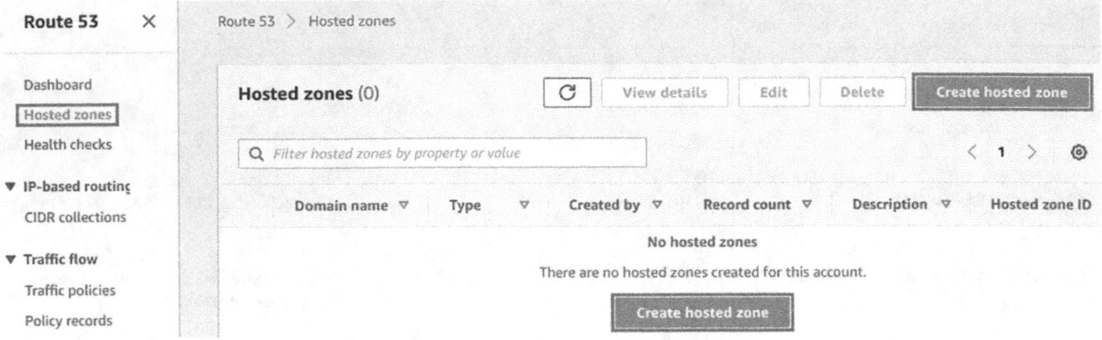

Figure 2-45. *Route 53 hosted zones*

Creating a Hosted Zone

Click **Create hosted zone** when the screen shown in Figure 2-46 appears. Here you need to enter part of your `privatelink-account-url` from your checklist. As our `privatelink-account-url` is `https://go51698.eu-west-2.aws.snowflakecomputing.com/console#/internal/worksheet`, we enter only `eu-west-2.privatelink.snowflakecomputing.com`. Next, select **Private hosted zone**.

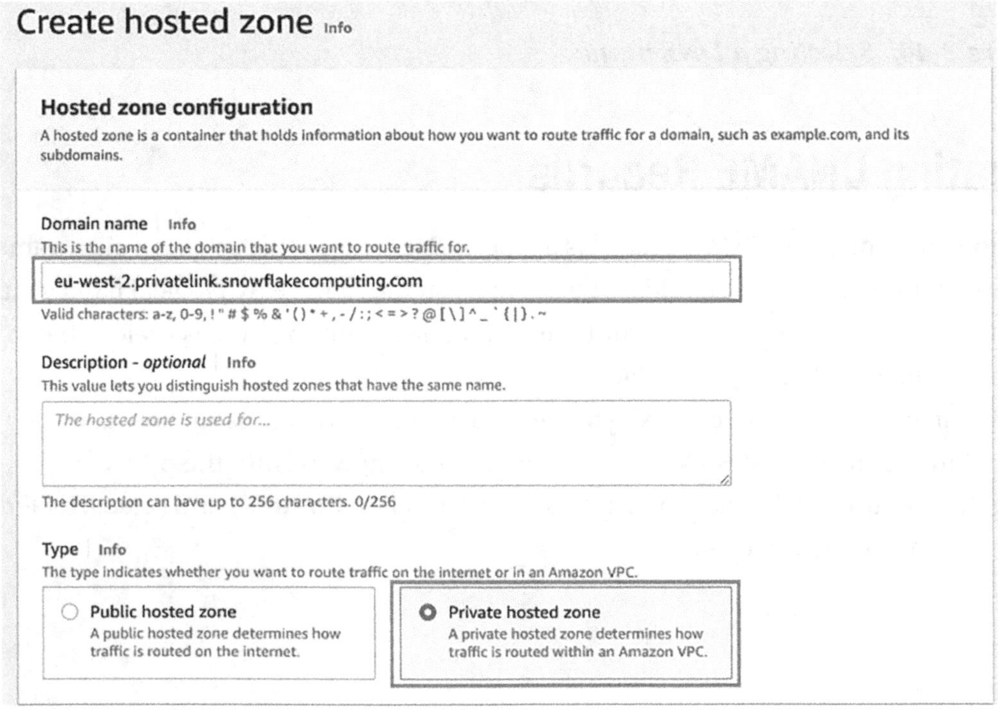

Figure 2-46. *Creating a Route 53 hosted zone*

Now select the region to match your Snowflake account region (eu-west-2) and then the VPC you created earlier from the drop-down lists, as shown in Figure 2-47.

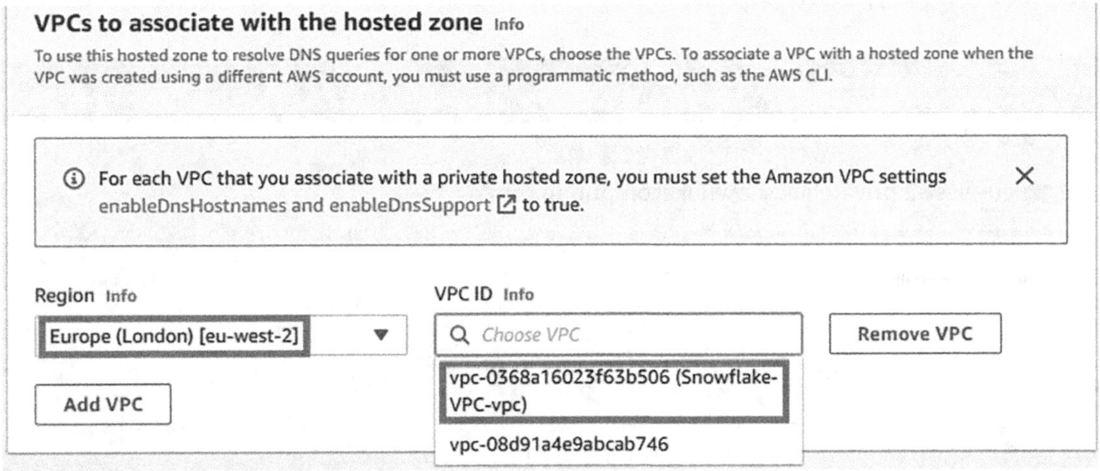

Figure 2-47. *Associating the VPC with a hosted zone*

Finally, add tags if required and then click **Create hosted zone**, as shown in Figure 2-48.

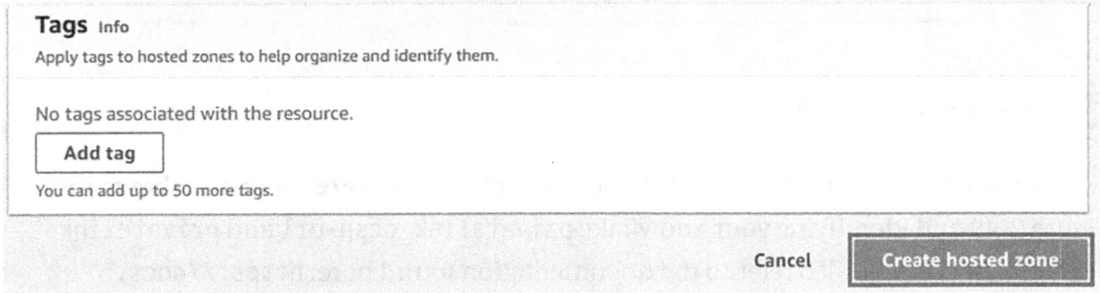

Figure 2-48. *Adding tags to a hosted zone*

Creating a Record

With your hosted zone created, the screen will refresh, as shown in Figure 2-49. Select **Create record**.

Within your hosted zone these two DNS record types are declared:

- NS: Identifies the name servers for the hosted zone

- SOA: Start of authority record provides information about a domain and the corresponding Amazon Route 53 hosted zone

For your reference, the list of supported DNS record types can be found here: `https://docs.aws.amazon.com/Route53/latest/DeveloperGuide/ResourceRecordTypes.html`.

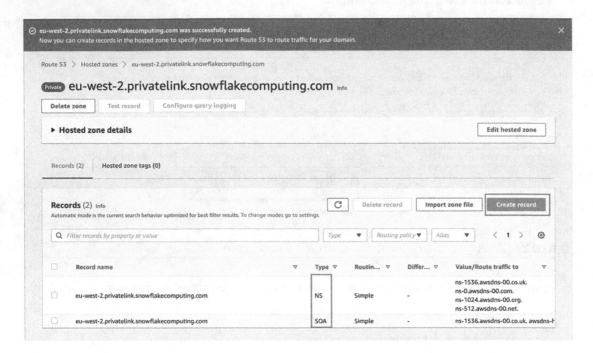

Figure 2-49. *Hosted zone details*

You must now create your CNAME record by clicking **Create record** as shown in Figure 2-49 and identifying your Snowflake `privatelink_ocsp-url` and `privatelink-account-url`. Please also refer to the documentation found here: `https://docs.snowflake.com/en/user-guide/admin-security-privatelink.html#creating-additional-cname-records`.

- Set **Record name** to the value returned from `SELECT current_account();` (in our example, go51698).

- Set **Record type** to **CNAME**.

- Set **Value** to the DNS name for the endpoint (in our example, vpce-0f1f290fb455f3853-ze4mujqn.vpce-svc-0839061a5300e5ac1.eu-west-2.vpce.amazonaws.com)

Then click **Create records**, as shown in Figure 2-50.

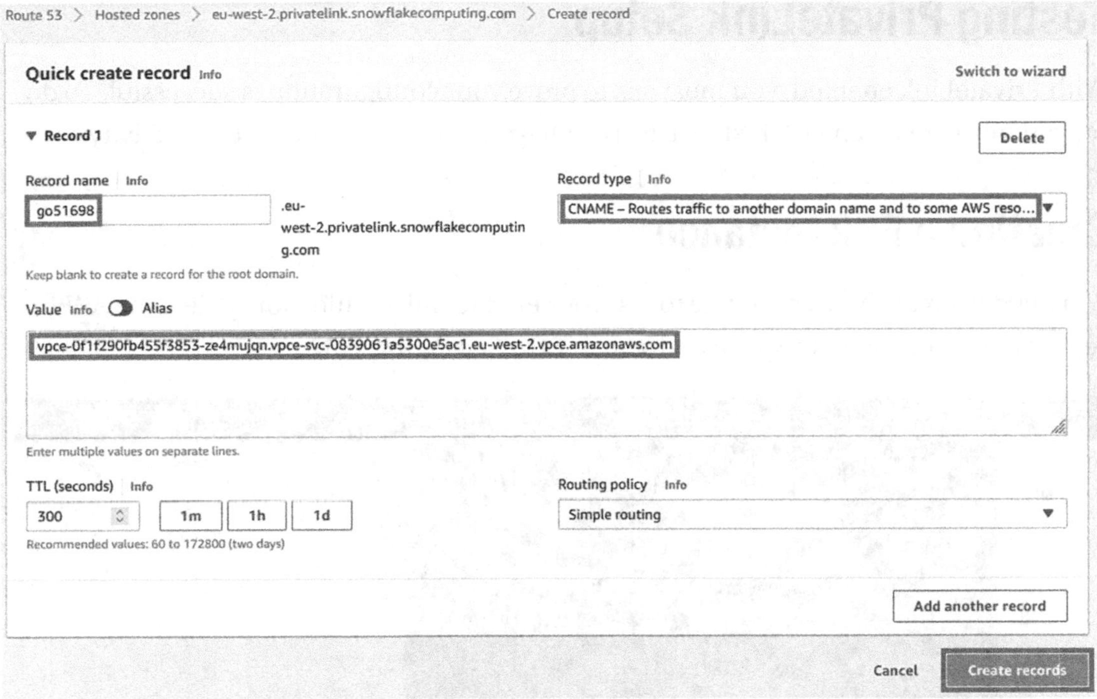

Figure 2-50. *Quick Create Record screen*

Figure 2-51 shows the new record added to the hosted zone.

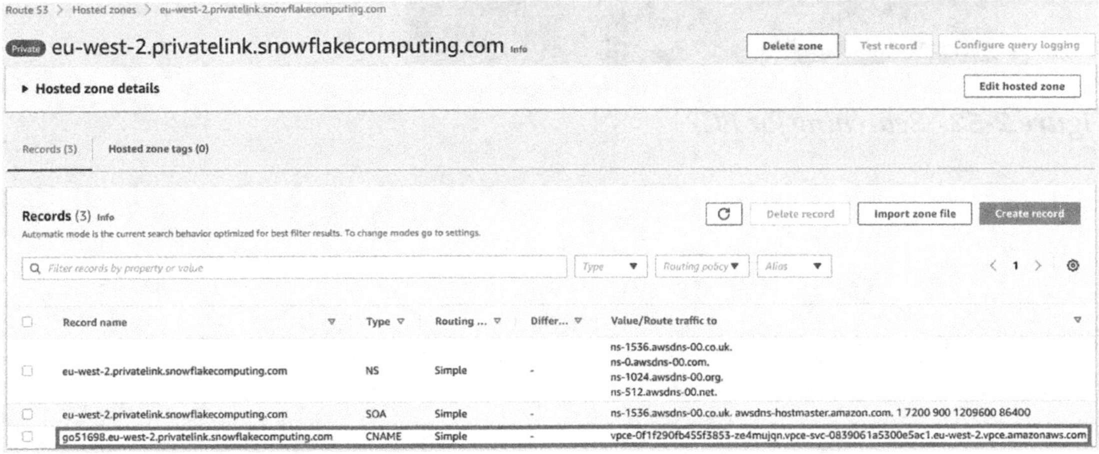

Figure 2-51. *Confirming a record*

Testing PrivateLink Setup

With PrivateLink enabled, you must test to prove your configuration is successful. To do so, you must create an EC2 instance to ssh into your VPC and validate connectivity.

Creating a EC2 Instance

You need to create a EC2 instance to test your PrivateLink connection. Select the AWS search bar at the top of the page and search **EC2**. See Figure 2-52.

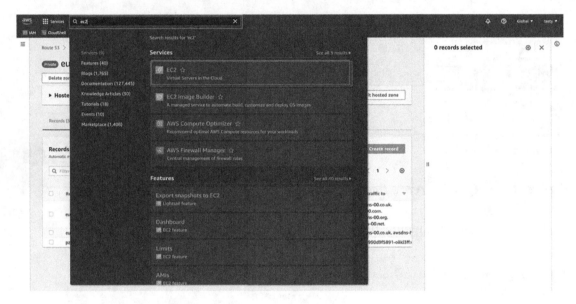

Figure 2-52. *Searching for EC2*

You will see the EC2 Dashboard, as shown in Figure 2-53.

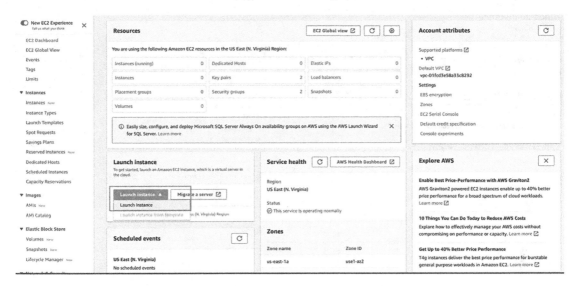

Figure 2-53. *Launching the EC2 instance*

Ensure the same AWS region as your Snowflake account is selected in the AWS command bar; otherwise, the VPC or subnets created previously cannot be linked to your EC2 instance.

Pick a name for the EC2 instance. In this example, we used Test-Snowflake-PrivateLink, leaving the Quick Start set to Amazon Linux for the purpose of the demo, as shown in Figure 2-54.

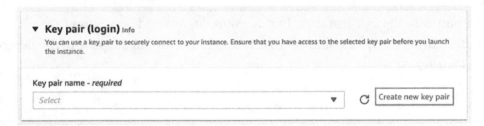

Figure 2-54. *EC2 instance options*

Scroll down to Key Pair (login) and select **Create new key pair**, as shown in Figure 2-55. This file is required to authenticate yourself to allow access to the instance.

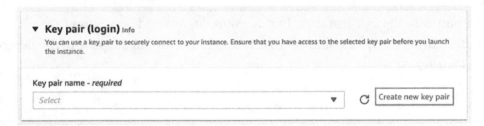

Figure 2-55. *Key pair (login)*

Click **Create new key pair** and the screen in Figure 2-56 appears.

- Set **Key pair name** to **Snowflake-Instance**.

- Leave **Key pair type** defaulted to **RSA**.

- Leave **Private key file format** defaulted to **.pem**.

Note the instruction to store the private key in a secure and accessible location.

Figure 2-56. *Creating a key pair*

Then click **Create key pair**, at which point a file named Snowflake-instance.pem is both generated and downloaded via your browser, Figure 2-57 shows the Microsoft Windows file location.

Figure 2-57. *.pem file location*

Associate the .pem file with your favorite editor. We use Vim.

With your key pair in hand, the AWS console screen will refresh and the next section to be updated is Network settings, as shown in Figure 2-58.

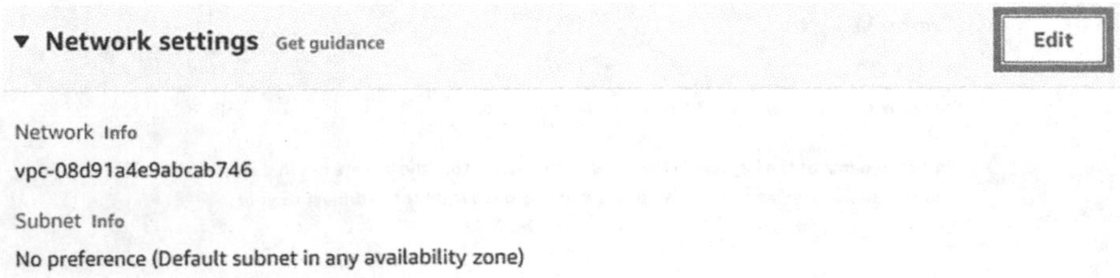

▼ **Network settings** Get guidance   Edit

Network Info
vpc-08d91a4e9abcab746

Subnet Info
No preference (Default subnet in any availability zone)

Figure 2-58. *Editing network settings*

Click **Edit** and make the following changes, as shown in Figure 2-59 (noting your values will differ):

- Set **VPC** to the VPC you created from your checklist (in our example, vpc-0368a16023f63b506).

- Set **Subnet** to the value for Snowflake-VPC-subnet-public1-eu-west-2a (note that this is typically **NOT** the default setting).

- Set **Auto-assign public IP** to **Enable**.

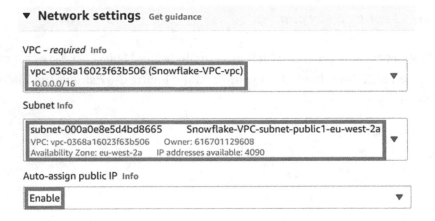

Figure 2-59. *Editing the VPC settings*

You must select the public subnet variant because you will SSH into your EC2 instance over the Internet from your personal machine.

Scroll down to Firewall (security groups), as shown in Figure 2-60:

- Leave **Key pair type** defaulted to **Create security group**.

- Set **Security group name** to **Snowflake-EC2-SG**.

- Leave **Type** defaulted to **SSH**.

- Set **Source type** to **My IP**, noting your local machine (or router) IP address will be displayed where indicated.

Firewall (security groups) Info

A security group is a set of firewall rules that control the traffic for your instance. Add rules to allow specific traffic to reach your instance.

● Create security group	○ Select existing security group

Security group name - *required*

Snowflake-EC2-SG

This security group will be added to all network interfaces. The name can't be edited after the security group is created. Max length is 255 characters. Valid characters: a-z, A-Z, 0-9, spaces, and ._-:/()#,@[]+=&;{}!$*

Description - *required* Info

launch-wizard created 2022-07-30T09:38:30.415Z

Inbound security groups rules

▼ Security group rule 1 (TCP, 22, 86.1.252.19/32)   Remove

Type Info	Protocol Info	Port range Info
ssh ▼	TCP	22

Source type Info	Source Info	Description - *optional* Info
My IP ▼	🔍 Add CIDR, prefix list or security	e.g. SSH for admin desktop
	Your IP Addr Here	

Figure 2-60. *Firewall configuration*

You do not need to make any further changes on the configuration panel.

To the right-hand side of the configuration panel you see the Summary, panel as illustrated in Figure 2-61.

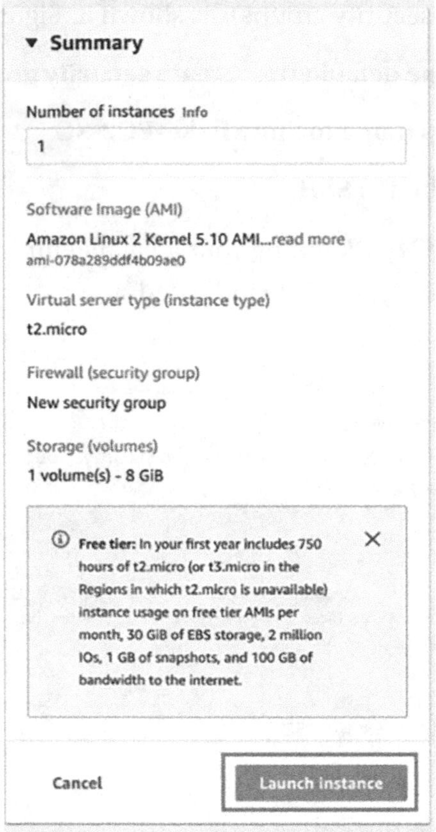

Figure 2-61. *Launch instance*

Select **Launch Instance**. After a few seconds the confirmation screen will appear, as shown in Figure 2-62.

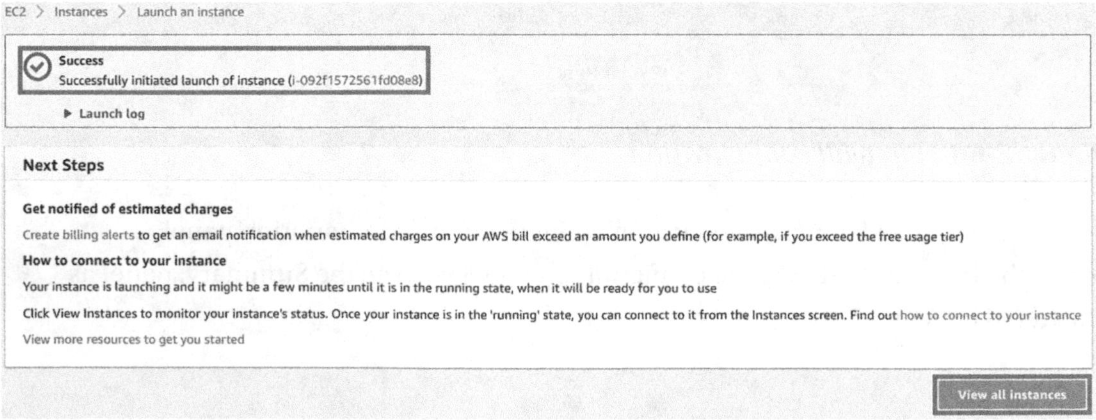

Figure 2-62. *Launch instance*

You may wish to make note of the EC2 instance handle, which in our case is i-092f1572561fd08e8.

Enabling VPC Endpoint Access

You must now enable your endpoint access to your VPC. Using the AWS console, search for **VPC** and then select **Endpoints**, as shown in Figure 2-63.

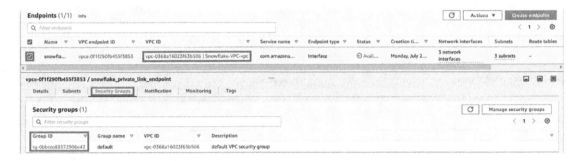

Figure 2-63. *Enabling an endpoint for VPC*

Select the **Group ID**, which in our example is sg-0bbcea88372906e42, to expand the security group, as shown in Figure 2-64.

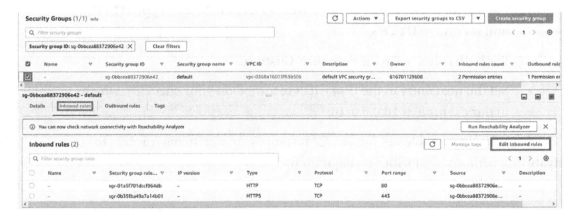

Figure 2-64. *Editing inbound rules*

Select the **Inbound rules** tab and then click **Edit inbound rules** to add two new rules, as shown in Figure 2-65.

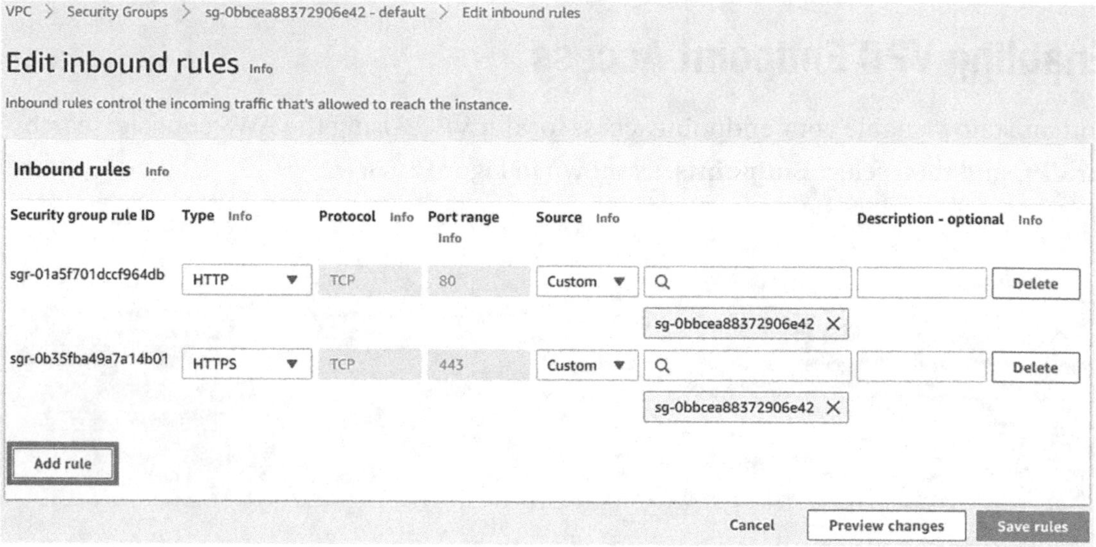

Figure 2-65. *Adding rules*

Select **Add Rule** twice, and referring to Figure 2-66,

- For the two new rules, change **Custom TCP**, the first to **Type HTTP** and second to **HTTPS**.

- For both rules, leave the **source** as **Custom**.

- Set the next entry to the IPv4 CIDR recorded on your checklist, which in our example is 10.0.0.0/16.

Using VPC IPv4 enables all future EC2 instances created to have access to your private link without any further configuration.

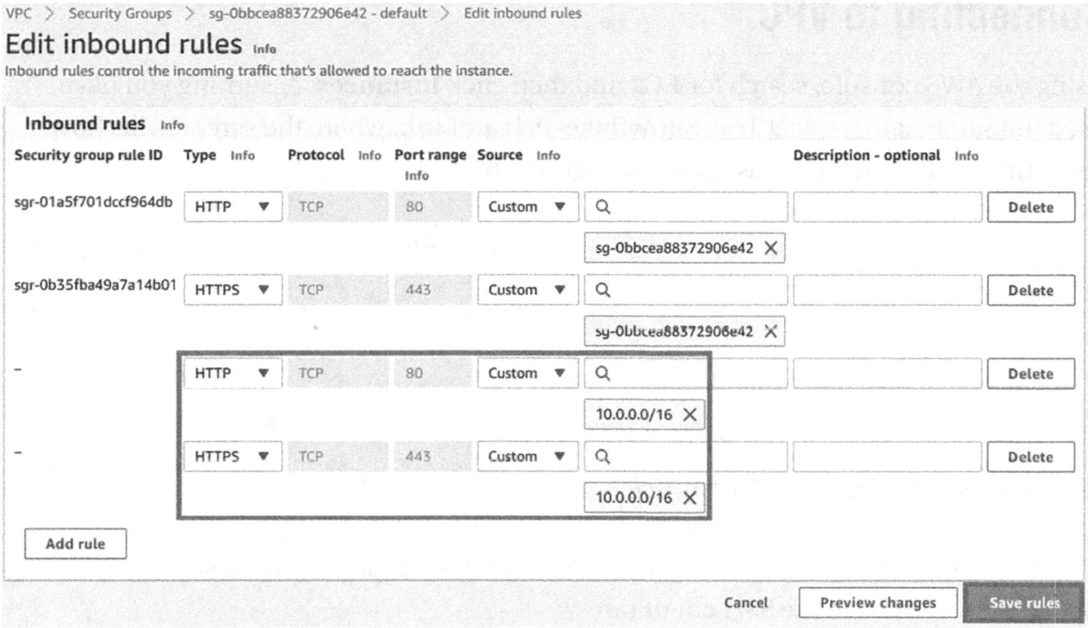

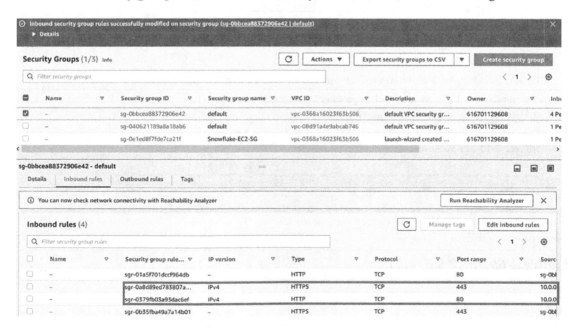

Figure 2-66. *Configuring new rules*

When the configuration is complete, click **Save rules**, after which the screen will refresh to show the security groups, which now includes your new rules, as shown in Figure 2-67.

Figure 2-67. *Security group confirmation*

Connecting to VPC

Using the AWS console, search for **EC2** and then click **Instances**. Assuming you have used the same name, select **Test-Snowflake-PrivateLink**, where the screen will show your running EC2 instance, as shown in Figure 2-68.

Figure 2-68. *Running EC2 instance*

Click **Connect** when the Connect to instance screen is displayed, as shown in Figure 2-69, and select the **SSH client** tab.

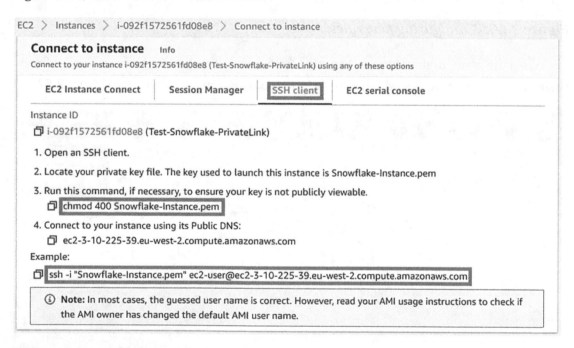

Figure 2-69. *EC2 SSH client*

Add the following two commands to your checklist:

- **The chmod command**: chmod 400 Snowflake-Instance.pem

- **SSH example**: ssh -i "Snowflake-Instance.pem" ec2-user@ ec2-3-10-225-39.eu-west-2.compute.amazonaws.com. Note that your target EC2 instance will differ.

Use your preferred SSH tool. For MS Windows, we use PuTTY and PuTTYgen; for Apple Mac, we use terminal.

Apple Mac Terminal

Using your Apple Mac, launch terminal and navigate to the folder where you saved the pem key downloaded earlier when creating the key-pair.

- Change the directory using cd to the path of the .pem file.

- Run chmod 400 Snowflake-Instance.pem.

Figure 2-70 shows the sequence of commands, noting the path has been obscured.

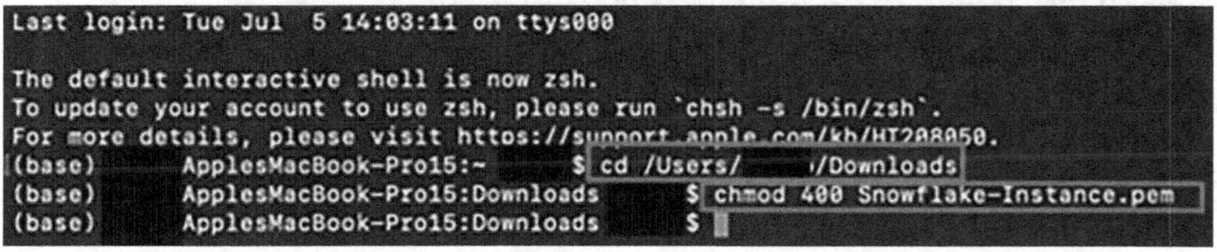

```
Last login: Tue Jul  5 14:03:11 on ttys000

The default interactive shell is now zsh.
To update your account to use zsh, please run `chsh -s /bin/zsh`.
For more details, please visit httos://support.apple.com/kb/HT208050.
(base)        ApplesMacBook-Pro15:~          $ cd /Users/      /Downloads
(base)        ApplesMacBook-Pro15:Downloads          $ chmod 400 Snowflake-Instance.pem
(base)        ApplesMacBook-Pro15:Downloads          $
```

Figure 2-70. *Terminal chmod PEM file*

As shown in Figure 2-71, copy and paste the example from AWS (which in our example is ssh -i "Snowflake-Instance.pem" ec2-user@ec2-3-10-225-39.eu-west-2.compute.amazonaws.com).

When asked "Are you sure you want to continue connecting?," type "yes" and press Enter.

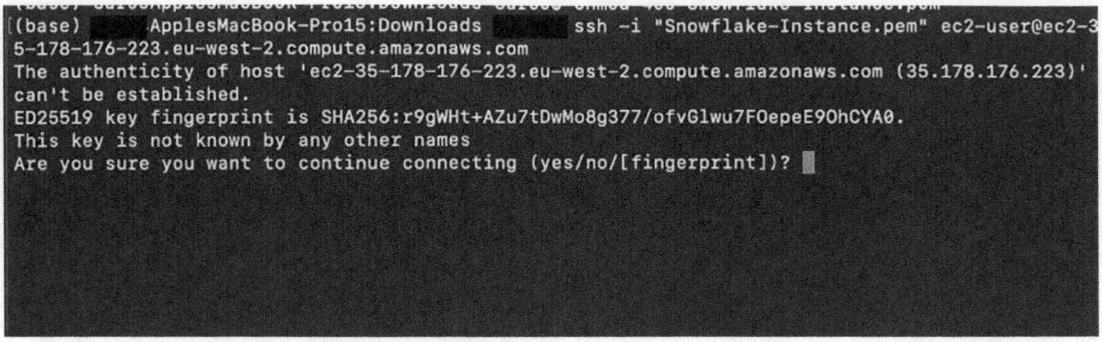

```
(base)        ApplesMacBook-Pro15:Downloads        ssh -i "Snowflake-Instance.pem" ec2-user@ec2-3
5-178-176-223.eu-west-2.compute.amazonaws.com
The authenticity of host 'ec2-35-178-176-223.eu-west-2.compute.amazonaws.com (35.178.176.223)'
can't be established.
ED25519 key fingerprint is SHA256:r9gWHt+AZu7tDwMo8g377/ofvGlwu7FOepeE9OhCYA0.
This key is not known by any other names
Are you sure you want to continue connecting (yes/no/[fingerprint])?
```

Figure 2-71. *Terminal ssh to EC2 instance*

A successful ssh into the EC2 instance is shown in Figure 2-72.

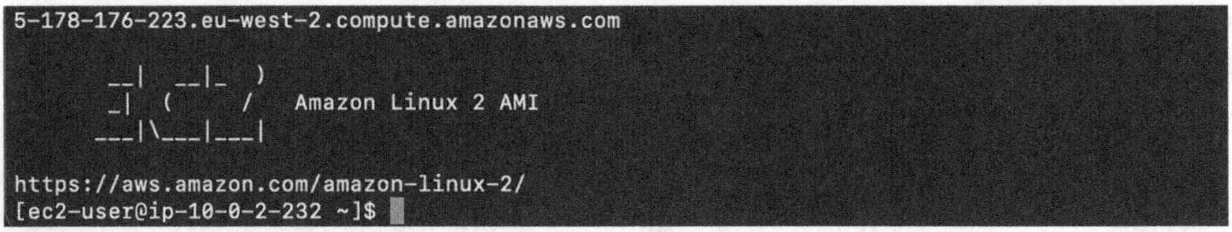

```
5-178-176-223.eu-west-2.compute.amazonaws.com

      __|  __|_  )
      _|  (     /    Amazon Linux 2 AMI
     ___|\___|___|

https://aws.amazon.com/amazon-linux-2/
[ec2-user@ip-10-0-2-232 ~]$
```

Figure 2-72. *Terminal successful ssh to EC2 instance*

Windows PuTTY/PuTTYgen

Before attempting to ssh into your EC2 instance, you must first convert the Snowflake-Instance.pem file using PuTTYgen. To do so using your MS Windows desktop machine, run PuTTYgen as shown in Figure 2-73.

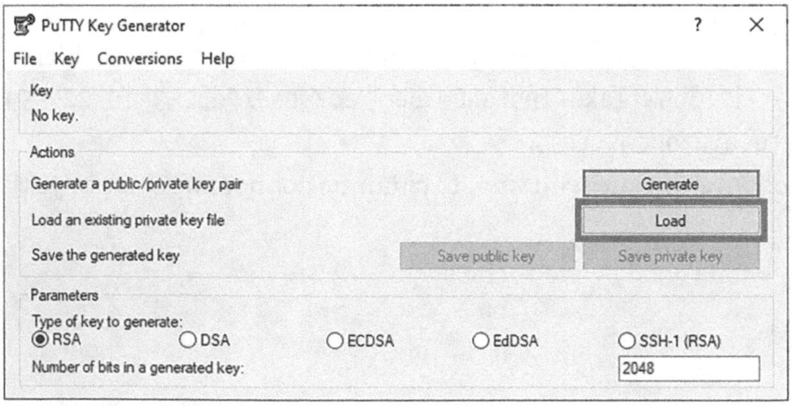

Figure 2-73. *PuTTYgen key generator*

Click **Load** when a standard Windows Explorer dialog opens. Navigate to the directory where **Snowflake-Instance.pem** is located, ensure **All Files (*.*)** in the bottom right corner is selected, and then pick the PEM file.

Key conversion happens automatically, resulting in a screen like Figure 2-74 (noting the key has been obscured).

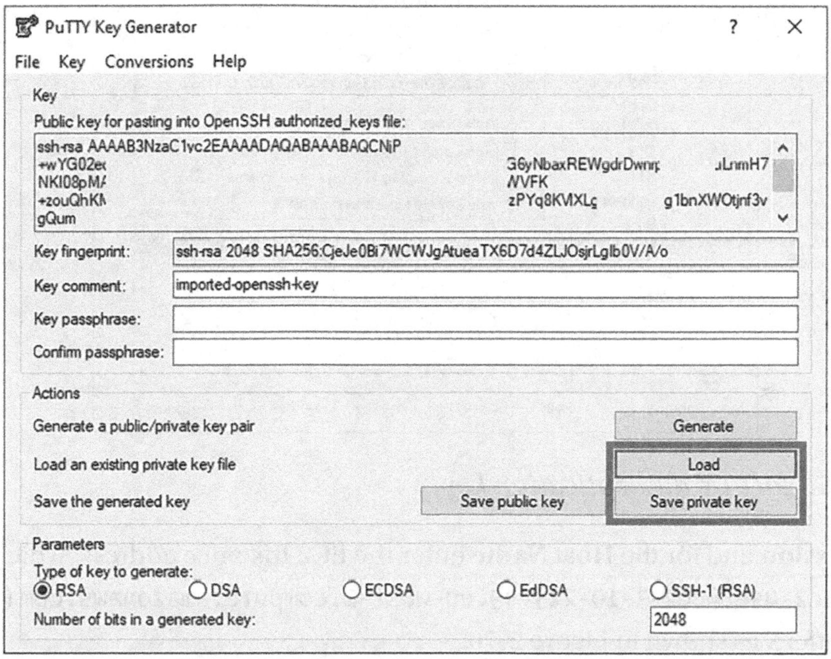

Figure 2-74. *PuTTYgen key conversion*

Click the **Save private key** button and store the .ppk file in a secure location, recording the saved location in your checklist before closing PuTTYgen.

Open PuTTY as shown in Figure 2-75 and navigate to Connection ➤ SSH ➤ Auth.

Click **Browse** and locate the private key generated by PuTTYgen from your secure location. By default the file extension will be .ppk.

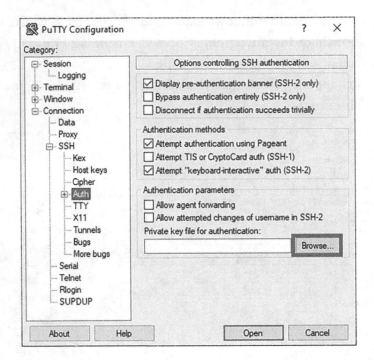

Figure 2-75. *PuTTY authentication key*

Click **Session** and for the **Host Name** enter the EC2 Instance address, which in our example is ec2-user@ec2-3-10-225-39.eu-west-2.compute.amazonaws.com (noting yours will differ), as shown in Figure 2-76.

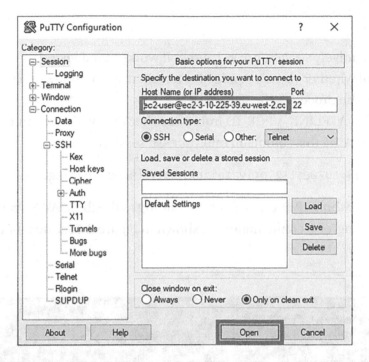

Figure 2-76. *PuTTY open session*

A successful ssh into the EC2 instance is shown in Figure 2-77.

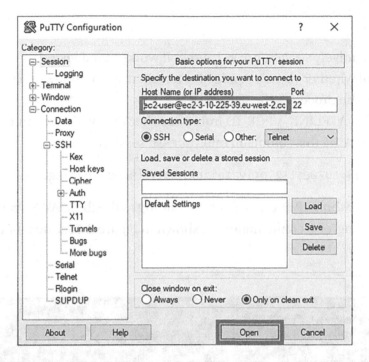

Figure 2-77. *PuTTY successful connection*

Test DNS

Testing DNS is relatively straightforward: either it works or it doesn't!

Regardless of whether you use Apple Mac terminal, MS Windows PuTTY, or your favorite ssh tool, the process is the same.

Using the PRIVATELINK_VPCE_ID from your checklist, which in our example will differ from yours, issue this command into your ssh session:

```
nslookup go51698.eu-west-2.privatelink.snowflakecomputing.com
```

You should see three IP addresses corresponding to the three AWS availability zones you specified during VPC configuration, as shown in Figure 2-78. Add all three to the checklist.

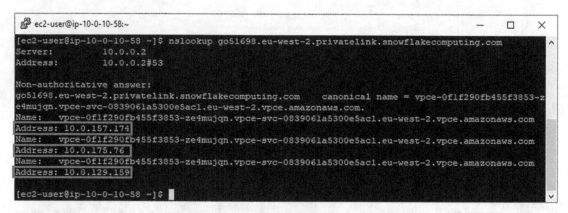

Figure 2-78. *nslookup Snowflake PrivateLink URL*

Using the AWS console, search for VPC ➤ Endpoints ➤ Subnets, as shown in Figure 2-79.

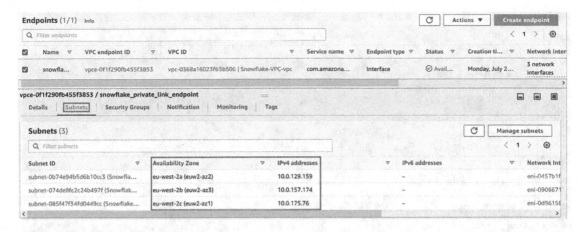

Figure 2-79. Endpoint subnets

Compare the IPv4 addresses with those returned by nslookup. All should match. In our example they are

- 10.0.175.76 - eu-west-2c (euw2-az1)

- 10.0.129.159 - eu-west-2a (euw2-az2)

- 10.0.157.174 - eu-west-2b (euw2-az3)

If the IP addresses do not match, PrivateLink is not configured correctly. Refer to the Troubleshooting Guide later within this chapter.

Connect via PrivateLink

You must now confirm your connection to the Snowflake PrivateLink URL within your EC2 instance. You must install Telnet, a client application for which more information can be found here: https://en.wikipedia.org/wiki/Telnet.

Using your ssh terminal, you first install Telnet on your EC2 instance by typing sudo yum -y install. Figure 2-80 shows successful installation confirmation.

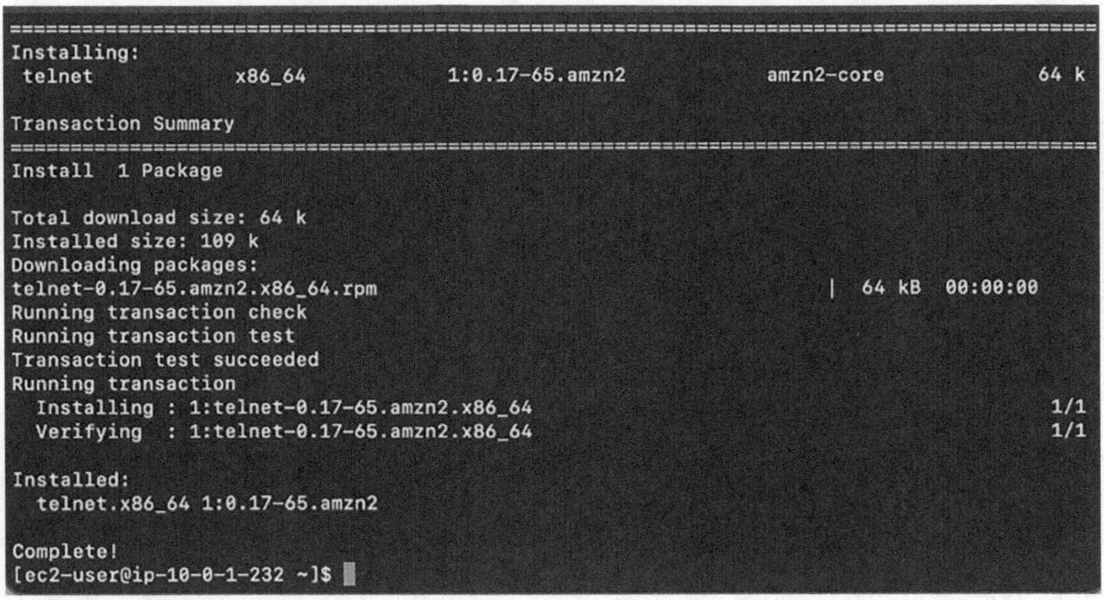

```
===============================================================================
Installing:
 telnet               x86_64          1:0.17-65.amzn2          amzn2-core              64 k

Transaction Summary
===============================================================================
Install  1 Package

Total download size: 64 k
Installed size: 109 k
Downloading packages:
telnet-0.17-65.amzn2.x86_64.rpm                              |  64 kB  00:00:00
Running transaction check
Running transaction test
Transaction test succeeded
Running transaction
  Installing : 1:telnet-0.17-65.amzn2.x86_64                                 1/1
  Verifying  : 1:telnet-0.17-65.amzn2.x86_64                                 1/1

Installed:
  telnet.x86_64 1:0.17-65.amzn2

Complete!
[ec2-user@ip-10-0-1-232 ~]$ ▊
```

Figure 2-80. *Telnet installation complete*

With Telnet installation complete, you can test connectivity to each of the three availability zones listed on your checklist, noting you must also state the port (80 for HTTP and 443 for HTTPS (SSL)):

```
telnet 10.0.175.76   443
telnet 10.0.129.159 443
telnet 10.0.157.174 443
```

Figure 2-81 shows the expected Telnet connected response.

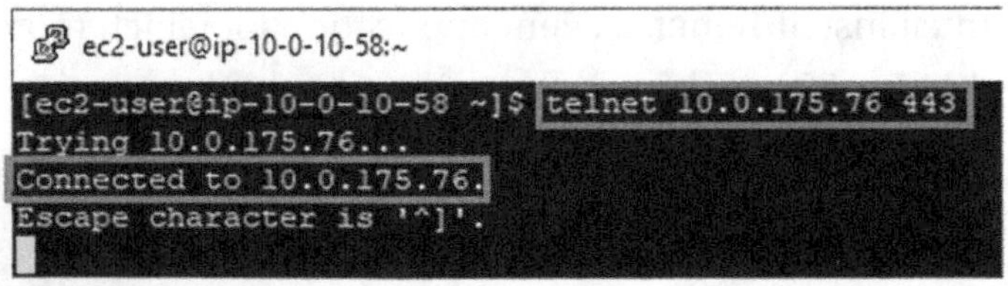

```
 ec2-user@ip-10-0-10-58:~
[ec2-user@ip-10-0-10-58 ~]$ telnet 10.0.175.76 443
Trying 10.0.175.76...
Connected to 10.0.175.76.
Escape character is '^]'.
```

Figure 2-81. *Telnet response*

If Telnet does not connect, check that the correct port (443) has been specified. Repeat for each of the remaining IP addresses and ports.

Endpoint to Reroute Snowflake Internal Stage Traffic

When a SQL command such as a SELECT statement is executed, Snowflake unloads
the data to an internal S3 bucket and returns the access URL to your VPC. The default
behavior is for your VPC to attempt to access the returned URL via HTTP over the public
Internet, resulting in the request being declined and no results returned.

To access the data in Snowflake's internal S3 bucket, you must create and configure
an endpoint to ensure the request will be redirected securely and thus return the
result for your query. We will cover how to do this in the next steps. Figure 2-82 shows
Snowflake traffic routing.

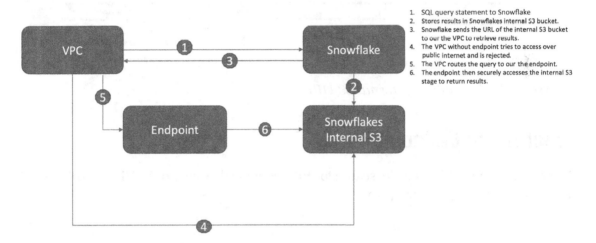

Figure 2-82. *Snowflake traffic route*

Retrieving Snowflake's Internal S3 Bucket Stage URL

First, you get the Snowflake internal S3 bucket stage URL for your snowflake account
by running the following SQL statement in your Snowflake account. Store the returned
value in your private link checklist under the Snowflake internal S3 bucket stage URL
section for later use.

```
USE ROLE accountadmin;
```

```
SELECT key, value FROM TABLE ( FLATTEN ( input=>parse_json ( system$get_
privatelink_config())));
```

Figure 2-83 shows an example result set, noting the value needed is privatelink-internal-stage.

```
1  use role accountadmin;
2  select key, value from table(flatten(input=>parse_json(system$get_privatelink_config())));
3
```

Row	KEY	VALUE
1	privatelink-account-name	"cm85487.eu-central-1.privatelink"
2	privatelink-account-url	"cm85487.eu-central-1.privatelink.snowflakecomputing.com"
3	privatelink-connection-urls	"[]"
4	privatelink-internal-stage	"sfc-eu-ds1-9-customer-stage.s3.eu-central-1.amazonaws.com"
5	privatelink-vpce-id	"com.amazonaws.vpce.eu-central-1.vpce-svc-0e506bf875de0062e"
6	privatelink_ocsp-url	"ocsp.cm85487.eu-central-1.privatelink.snowflakecomputing.com"
7	regionless-privatelink-account-url	"sfcsupport-pierre_aws_eucentral.privatelink.snowflakecomputing.com"

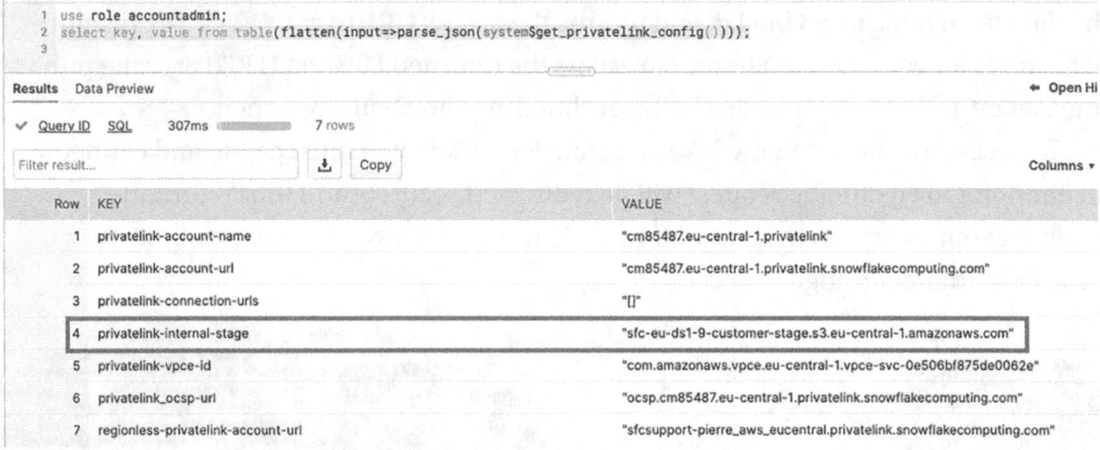

Figure 2-83. *Snowflake internal S3 URL*

Creating an Endpoint

Returning to your AWS console, search for **VPC**, which takes you to the VPC Dashboard, part of which is shown in Figure 2-84.

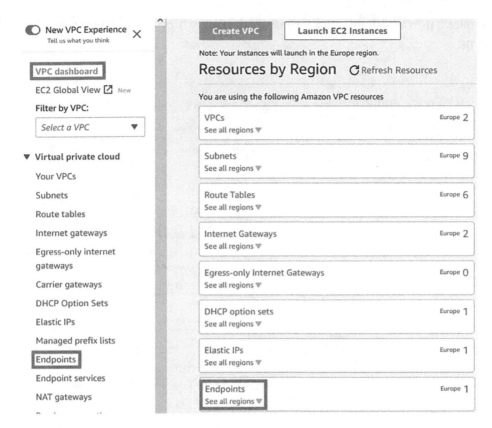

Figure 2-84. *VPC Dashboard*

Endpoints can be accessed from two places within the VPC Dashboard. Click one of the two **Endpoints** options highlighted in Figure 2-84.

The endpoint must be created within the same region as your Snowflake account.

After the screen refreshes, Figure 2-85 shows the option to create a new endpoint.

Figure 2-85. *Endpoints*

To create a new endpoint, click **Create endpoint**, which opens the screen shown in
Figure 2-86.

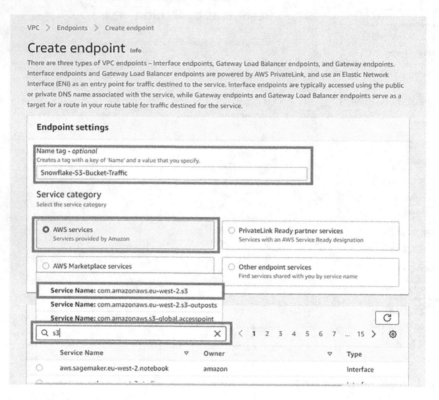

Figure 2-86. *Creating an endpoint*

Populate these fields:

- Set **Name** to **Snowflake_S3_Bucket_Traffic**.

- Set **Service category** to **AWS services**.

- Search **Service name** for **S3** and then select **com.amazon.eu-
  west-2.s3**.

Figure 2-87 shows the correct service name selected. The **Type** must be **Interface**.

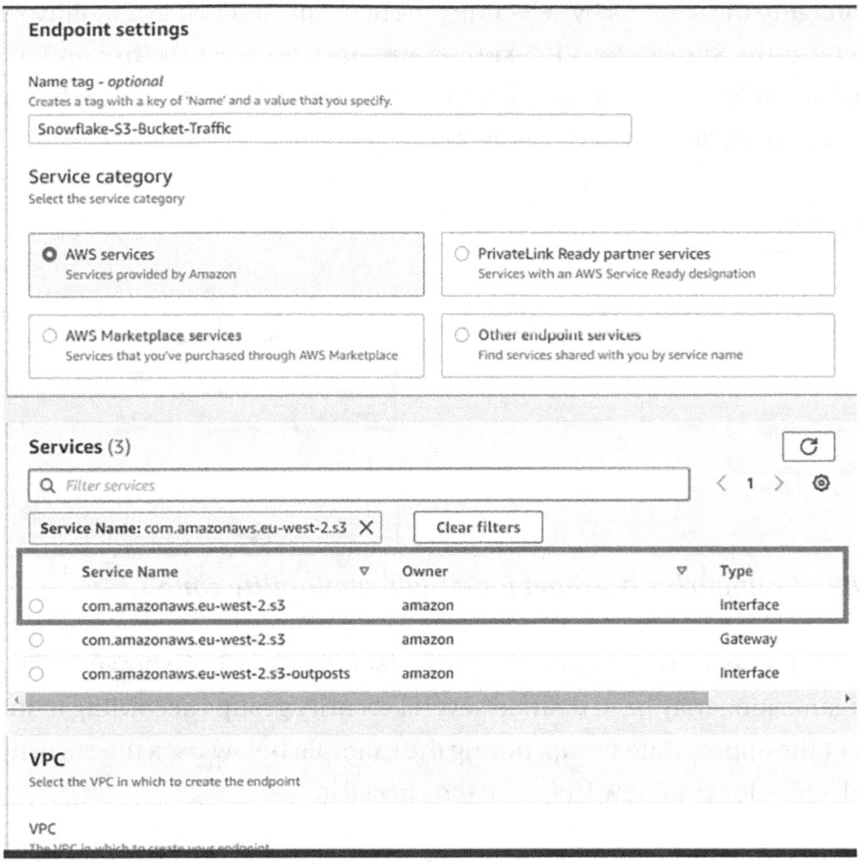

Endpoint settings

Name tag - *optional*
Creates a tag with a key of 'Name' and a value that you specify.

```
Snowflake-S3-Bucket-Traffic
```

Service category
Select the service category

- ○ **AWS services**
  Services provided by Amazon

- ○ **PrivateLink Ready partner services**
  Services with an AWS Service Ready designation

- ○ **AWS Marketplace services**
  Services that you've purchased through AWS Marketplace

- ○ **Other endpoint services**
  Find services shared with you by service name

Services (3)

🔍 Filter services

〈 1 〉 ⚙

Service Name: com.amazonaws.eu-west-2.s3 ✕   Clear filters

	Service Name ▽	Owner ▽	Type
○	com.amazonaws.eu-west-2.s3	amazon	Interface
○	com.amazonaws.eu-west-2.s3	amazon	Gateway
○	com.amazonaws.eu-west-2.s3-outposts	amazon	Interface

VPC
Select the VPC in which to create the endpoint

VPC
The VPC in which to create your endpoint

Figure 2-87. Service name set

Select the VPC from the drop-down list (see checklist) you created in your example, **Snowflake-VPC-vpc**. This is where you will create the endpoint, as shown in Figure 2-88.

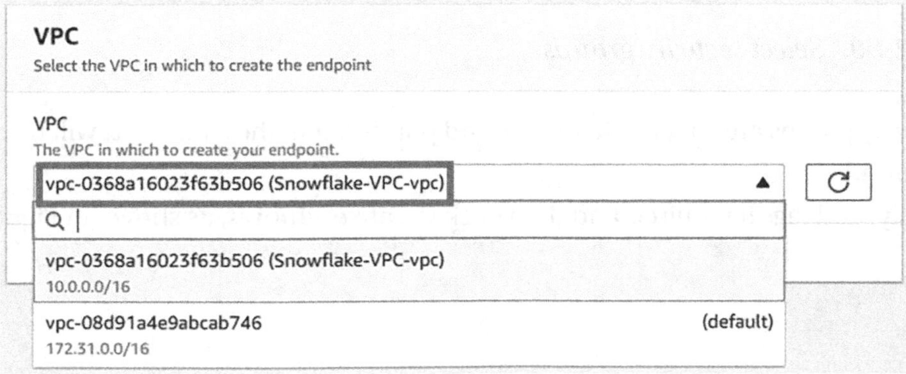

VPC
Select the VPC in which to create the endpoint

VPC
The VPC in which to create your endpoint.

```
vpc-0368a16023f63b506 (Snowflake-VPC-vpc)                        ▲    ⟳
```

🔍 |

vpc-0368a16023f63b506 (Snowflake-VPC-vpc)
10.0.0.0/16

vpc-08d91a4e9abcab746   (default)
172.31.0.0/16

Figure 2-88. Selecting a VPC

Now configure the subnets by selecting the checkbox for each availability zone (AZ). Set **Subnet ID** as the **Snowflake-VPC-subnet-private1-eu-west-2a/b/c** option in the example configuration for all subnets from the corresponding drop-down box and set **IP address type** to **IPV4**, as shown in Figure 2-89.

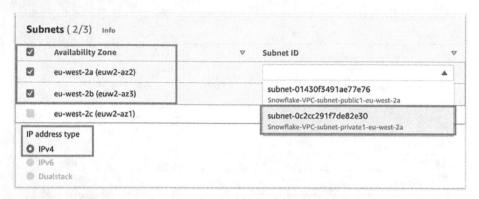

Figure 2-89. *Configuring a subnet for a high availability zone*

Now select the security group edited earlier (see checklist), as shown in Figure 2-90.

Your organization may have defined several security groups according to local policy. Select the appropriate group, noting the example below uses the security group you created and is for your new VPC. See the checklist.

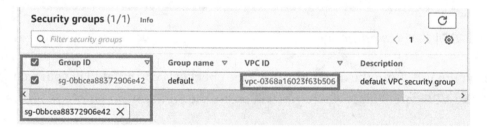

Figure 2-90. *Select security groups*

Security groups are stateful. If an inbound port is open, the same port will be opened for outbound.

Finally, add tags if required and then click **Create endpoint**, as shown in Figure 2-91.

Figure 2-91. *Adding tags and create endpoints*

You now have your endpoint displayed, noting the status is Pending. After a short while the status will change to Available.

Select the endpoint you just created and select the Subnets heading and record all the IPv4 addresses, as shown in Figure 2-92. Add them to your checklist under the S3-Traffic-Endpoint-IPs section.

Figure 2-92. *Endpoint subnet IP addresses*

Create Route 53/Hosted Zone Service

You now need to route your traffic securely through the endpoint you created to Snowflake's internal S3 buckets URL. For this you need to select another AWS service called **Route 53** and select **Hosted zones** and then select **Create hosted zone**, as shown in Figure 2-93.

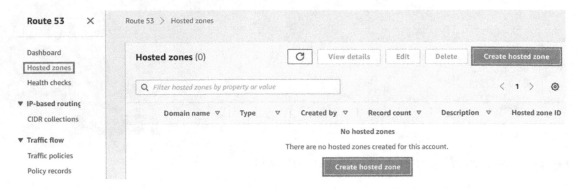

Figure 2-93. *Route 53 hosted zones*

Within the created hosted zones section, you need to enter your Snowflake account's internal stage URL from your checklist into the Domain name section. Under **Type**, select **Private Hosted Zone**, as shown in Figure 2-94.

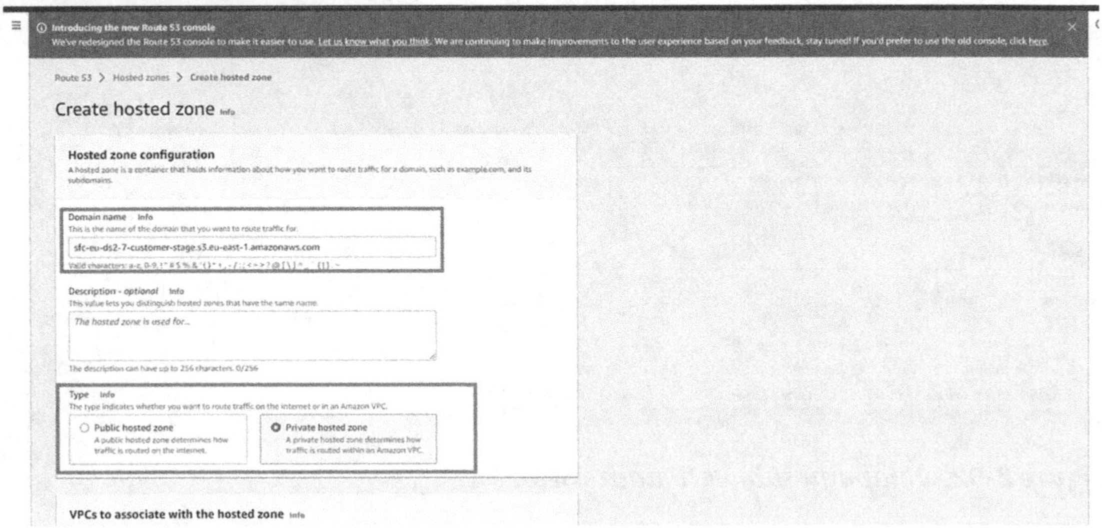

Figure 2-94. *Creating a Route 53 hosted zone*

Select the region to match your Snowflake account region (eu-west-2) and then the VPC you created earlier from the drop-down lists, as shown in Figure 2-95.

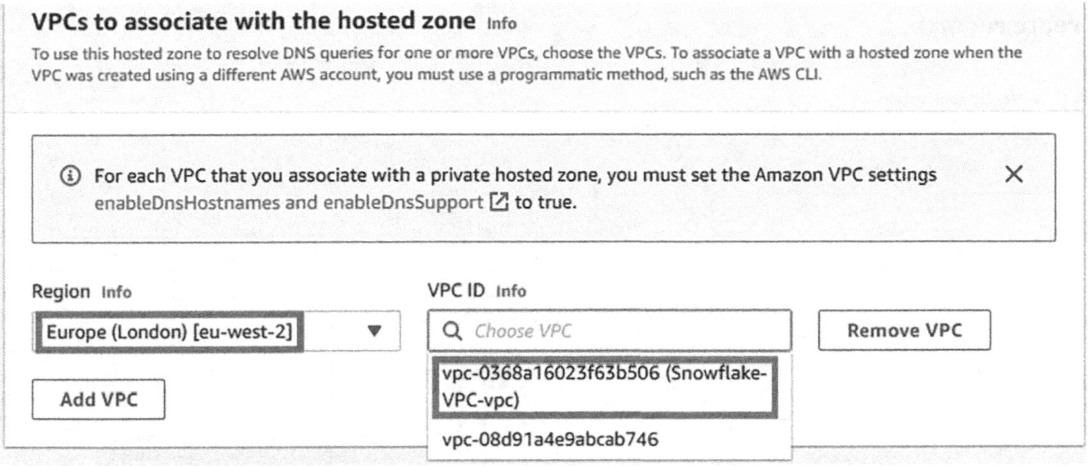

VPCs to associate with the hosted zone Info

To use this hosted zone to resolve DNS queries for one or more VPCs, choose the VPCs. To associate a VPC with a hosted zone when the VPC was created using a different AWS account, you must use a programmatic method, such as the AWS CLI.

ⓘ For each VPC that you associate with a private hosted zone, you must set the Amazon VPC settings   ✕
enableDnsHostnames and enableDnsSupport ⤢ to true.

Region Info   VPC ID Info

| Europe (London) [eu-west-2] | ▼ | 🔍 Choose VPC   Remove VPC

Add VPC   vpc-0368a16023f63b506 (Snowflake-
                                                VPC-vpc)

                                                vpc-08d91a4e9abcab746

Figure 2-95. *Associating the VPC with the hosted zone*

Finally, add tags if required and then click the **Create hosted zone** button, as shown in Figure 2-96.

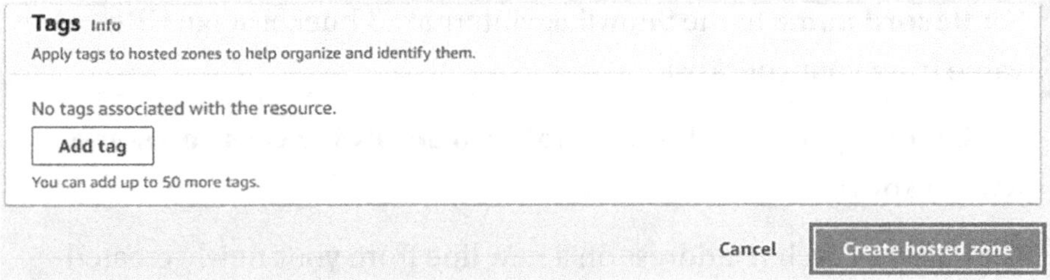

Tags Info

Apply tags to hosted zones to help organize and identify them.

No tags associated with the resource.

Add tag

You can add up to 50 more tags.

Cancel   Create hosted zone

Figure 2-96. *Adding tags to a hosted zone*

Creating a Record

With your hosted zone created, the screen will refresh, as shown in Figure 2-97. Select **Create record**.

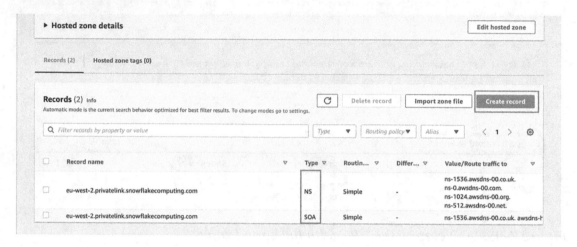

Figure 2-97. *Creating a record*

Populate these fields:

- Set **Record name** to the Snowflake internal S3 bucket stage URL value from your checklist.

- Set **Record type** to **A – Routes Traffic to an IPv4 access and some AWS resources**.

- Set **Value** to each IP address on a new line from your newly created endpoints stored in your checklist under S3-Traffic-Endpoint-IPs.

Then click **Create records**, as shown in Figure 2-98.

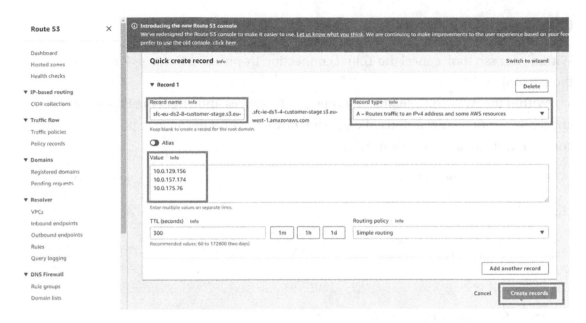

Figure 2-98. *Creating records*

Testing Your Endpoint

As in the section titled "Connecting to a VPC," you first connect to your EC2 Instance. Using terminal or PowerShell, run `nslookup <your snowflake internal stage URL>` to ensure that your Snowflake internal stage URL resolves to the S3 private endpoint. Figure 2-100 illustrates the steps using an example internal stage URL:

```
nslookup sfc-eu-ds2-7-customer-stage-s3.eu.east.1.amazonaws.com
```

You should see a response similar to this:

```
Server: ip-10-0-10-2, eu-central-1 compute, internal
Address: 10.0.10.3
```

```
Non-authorative answer:
Name: sfc-eu-ds2-7-customer-stage-s3.eu.east.1.amazonaws.com
Address: 10.0.10.11
```

Cleanup

Within your ssh tool, cancel the Telnet connection by pressing Control+Z and then close the ssh tool.

Within the AWS console, search for EC2 -> Instances and select the EC2 instance checkbox. From the drop-down dialog, shut down the EC2 instance to prevent unwanted consumption, as shown in Figure 2-99.

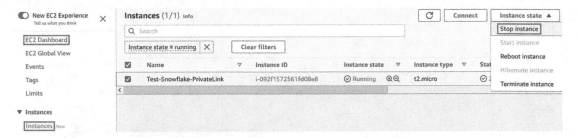

Figure 2-99. *Stopping an EC2 instance*

At the Stop instance confirmation, click **Stop**, as shown in Figure 2-100.

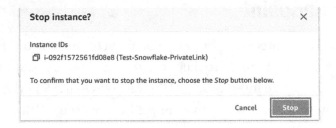

Figure 2-100. *EC2 instance confirmation*

The screen will refresh and the instance state will show Stopping before later changing to Stopped.

Troubleshooting Guide

This section contains some useful pointers for solving issues found during the writing of this chapter. The content is not exhaustive, and your local environment may offer alternative options to explore.

Confirm Configuration

As may be expected, there are plenty of opportunities to misconfigure one or more steps when enabling PrivateLink.

Assuming the checklist has been accurately maintained throughout the course of walking through this chapter, the first check is to ensure all entries are correct and match those example values provided herein.

If all values look correct, the second check is to go through this chapter from the beginning and review all AWS and Snowflake settings in the order given. For AWS, all configuration can either be edited or deleted and redone according to preference.

AWS continually updates the console. There may be differences between the screenshots in this chapter and your real-world experience.

Check Authorized Endpoints

Attempting to connect using a PrivateLink that has not been enabled results in a 403 Forbidden error.

If your session token has expired, regenerate using the AWS CloudShell commands above and then run `system$get_privatelink` to determine your PrivateLink status.

Confirm a PrivateLink entry using

```
USE ROLE accountadmin;

SELECT system$get_privatelink_authorized_endpoints();
```

Further troubleshooting information can be found here: `https://community.snowflake.com/s/article/AWS-PrivateLink-and-Snowflake-detailed-troubleshooting-Guide`.

Check That Endpoint Regions Align

When creating a VPC endpoint, if the AWS region and Snowflake region are not aligned, then the service name will not validate. Also, the VPC list will not match your expectations or will be greyed out.

To resolve the issue, set the AWS console region to the same as your Snowflake region and retry.

CloudShell Fails to Start

Occasionally CloudShell fails to start with this error: "Unable to start the environment. To retry, refresh the browser or restart by selecting Actions, Restart AWS CloudShell."

After a few minutes, the error message may change to "Unable to start the environment. To retry, refresh the browser or restart by selecting Actions, Restart AWS CloudShell. System error: Environment was in state: CREATION_FAILED. Expected environment to be in state: RUNNING. To retry, refresh the browser or restart by selecting Actions, Restart AWS CloudShell."

Try logging in with a different browser. Select a different region using the drop-down list next to the login username, as shown in Figure 2-101.

If using a different browser and region fails, in the top right of the browser is an **Actions** button, as shown in Figure 2-101.

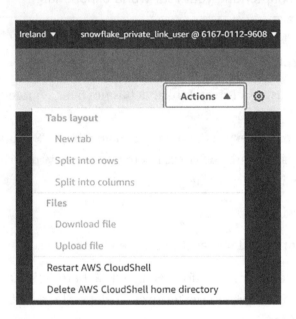

Figure 2-101. *CloudShell actions*

You should first try the **Restart AWS CloudShell** action, but if this fails, try the **Delete AWS CloudShell home directory** action, after which AWS attempts to restart CloudShell.

Deleting the CloudShell home directory removes all files.

Next, prove CloudShell works from your root account for the same region. If CloudShell can be successfully invoked using your root account, but not while using the IAM user, this may indicate an entitlement problem with the profile attached to your IAM user. See the IAM Policy Simulator (next) or this may indicate your account is On Hold (see below).

IAM Policy Simulator

Troubleshooting IAM policies can be performed using the IAM Policy Simulator found here:
`https://policysim.aws.amazon.com/home/index.jsp?#users/snowflake_private_link_user`. Then follow these steps:

- Log into the AWS console as root and then launch the simulator.

- Within the left had side drop-down, select **Users** and then **snowflake_private_link_user**.

- Under **Policy Simulator**, select **AWS CloudShell** and then **Select All**.

- Finally, click **Run Simulation** and examine the results for each service under the **Permission** column.

Figure 2-102 shows the expected results where all services are allowed.

Figure 2-102. *IAM Policy Simulator*

If any permissions fail, amend the snowflake_private_link_policy.

AWS Account on Hold

While writing this chapter and testing the AWS components, I received an email from Amazon:

> Case ID: 10166144851
>
> Subject: Action Required: Irregular activity in your AWS account
> <Your AWS Account ID Here>
>
> Severity: Urgent

I resolved the case, but my account was put "on hold," though there was no indication within AWS Console to indicate the status. Surprisingly, many features remained working, including both user and profile administration functions, but you will not be able to execute some processes using an IAM user though processes do execute as root.

Should you experience similar, raise a support ticket (see next) but don't hold your breath waiting for your account to be released. You may prefer to close your existing account and create another.

Raising a Support Ticket

If you are unable to solve the problem by yourself and other online resources, a support ticket may be raised here: https://us-east-1.console.aws.amazon.com/support/home?region=us-east-1&skipRegion=true#.

Within the ticket description put the following:

- Account ID

- IAM user

- Description of issue. Mine was: "IAM user with AWSCloudShellFullAccess cannot start CloudShell"

Use online chat to diagnose the issue with a service agent–a real human, not a chatbot.

According to your means, it may be possible to create another AWS account or use your organization support channel to reset your AWS account status.

SSO Over PrivateLink

SSO may be enabled for either public Internet access or for PrivateLink access. It is not possible to enable SSO for both public Internet access and PrivateLink access at the same time.

When implementing SSO for PrivateLink, the same setup and configuration steps must be completed as originally implemented for public Internet access. We revisit them in a later chapter.

Further information can be found here: `https://docs.snowflake.com/en/user-guide/admin-security-fed-auth-configure-idp.html`.

Checklists

This section collates all the information contained within this chapter into a template as a starting point for your later automation of the content. Feel free to add or amend the content in Table 2-2 according to your organization needs.

Table 2-2 contains information gathered within this chapter to confirm configuration steps have been completed, provide an electronic audit trail, and assist later automation.

Table 2-2. *PrivateLink Checklist*

Item	Value
AWS Account ID:	
Created Policy Name:	
Created AWS Username:	
Created AWS User Password:	
Access and Secret Keys Downloaded Filename:	
Access Key ID:	
Secret Access Key:	
Federation Token:	
VPC Name:	

(continued)

Table 2-2. (*continued*)

Item	Value
VPC IPv4 CIDRs	
Default Security Group ID:	
PrivateLink-vpce-id:	
Snowflake Region:	
privatelink-account-url:	
Endpoint Name:	
Endpoint DNS Name:	
EC2 Instance Name:	
Key-Pair Name:	
Key-Pair Location:	
EC2 Instance Security Group Name:	
Connect to Instance chmod Command:	
Connect to Instance Example Command:	
nslookup/Endpoint Subnet Address:	
Telnet Installation Command:	
Telnet IP/Ports	
S3-Traffic-Endpoint-Ips	

Table 2-3 contains information from `system$get_privatelink_authorized_endpoints()`. These are all the CSP accounts configured for PrivateLink access to your Snowflake account and require VPCE configuration.

```
USE ROLE accountadmin;

SELECT REPLACE ( value:endpointId,      '"' ) AS AccountID,
       REPLACE ( value:endpointIdType, '"' ) AS CSP
FROM   TABLE (
         FLATTEN (
```

```
            input => parse_json(system$get_privatelink_authorized_
            endpoints())
            )
    );
```

Table 2-3 shows indicative results.

Table 2-3. *system$get_privatelink_*
authorized_endpoints() Output

Item	Value
Account ID#1	168745977517
CSP#1	Aws Id

Table 2-4 contains information from system$get_privatelink_config() used for all VPCE configuration:

```
USE ROLE accountadmin;

SELECT REPLACE ( key,    '"' ) AS Key,
       REPLACE ( value, '"' ) AS Value
FROM    TABLE (
            FLATTEN (
                input => parse_json(system$get_privatelink_config())
                )
            )
ORDER BY value DESC;
```

Table 2-4 shows indicative results.

Table 2-4. *system$get_privatelink_config() Output*

Item	Value
regionless-privatelink-account-url	ypeuipd-fc49369.privatelink. snowflakecomputing.com
privatelink_ocsp-url	ocsp.go51698.eu-west-2.privatelink. snowflakecomputing.com
privatelink-account-url	go51698.eu-west-2.privatelink. snowflakecomputing.com
privatelink-account-name	go51698.eu-west-2.privatelink
privatelink-vpce-id	com.amazonaws.vpce.eu-west-2.vpce-svc-0839061a5300e5ac1
privatelink-connection-urls	[]
Snowflake internal S3 bucket stage URL	

Summary

You began this chapter by exploring changes released by Snowflake for users to self-serve PrivateLink. You discovered that self-service is not as simple as we might hope for, but instead requires significant AWS configuration.

After preparing your AWS profile and user, you were able to generate a security federated token and later use the same token within Snowflake to check PrivateLink status and then both enable and disable PrivateLink.

You saw how to determine which endpoints have been enabled for PrivateLink and an approach for monitoring.

You then moved on to creating your AWS VPC and VPCE, noting the requirement for both AWS and Snowflake regions to match before creating CNAME records.

The testing proved more complex than anticipated, requiring an EC2 instance and ssh from your local desktop, for which we provided guides for both Apple Mac terminal and MS Windows PuTTY use.

The high degree of manual configuration required for PrivateLink provides many opportunities for misconfiguration. We discussed some common problems and proposed troubleshooting solutions.

Finally, to aid automated maintenance and deployment, we proposed a checklist of all components and configuration items required before developing an automation framework which we leave for your further investigation.

CHAPTER 3

Security Monitoring

Security monitoring utilizes components that allow you to monitor, control, and manage your Snowflake estate. They are the internal and external alerting mechanisms defined and implemented to meet your organization's security profile, where generated events provide advanced warning of malicious activity and cause you to take remedial action.

You also use third-party tooling for security monitoring because they provide another layer of protection against unauthorized system access, malicious actors, and bad intent. Despite your best intent and building internal controls defined by the best and brightest people your organization can employ, there is always someone smarter or more devious with ill intent who may be able to evade the best-in-breed defenses you have designed.

The preferred method of protecting your systems is to implement multiple layers of defense. In this chapter, you'll explore how to implement a layered defense, one where a single compromised component does not automatically render all of your Snowflake vulnerable but instead may isolate the compromise to a single part of the system.

As you work through this chapter, you will identify and develop compensating controls for security monitoring. You start by imagining your LZ-delivered Snowflake platform account has a Production application implemented and ask yourself what controls you expect to have been deployed to ensure conformance to best practice.

For organizations working with Snowflake to deploy a LZ, your Snowflake Sales Engineer will be able to provide a template security features checklist. This list is generic in nature and asks many questions to prompt and direct your thinking to consider where to focus efforts and prevent security vulnerabilities from arising.

This chapter demonstrates a formal approach to defining policies, processes, and procedures with a deep dive into a few examples to start you on your journey. Within this chapter we use both Classic UI and SnowSQL. Refer to Chapter 1 for installation instructions and to get started.

© Andrew Carruthers and Sahir Ahmed 2023
A. Carruthers and S. Ahmed, *Maturing the Snowflake Data Cloud*, https://doi.org/10.1007/978-1-4842-9340-9_3

Control Definition

In this section we describe a limited-scope suite of controls expected to be defined by your organization in the form of several approved policies. For each policy document implemented, there will be processes and then procedures, as shown in Figure 3-1.

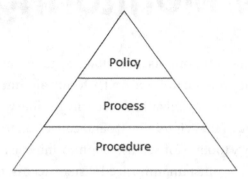

Figure 3-1. *Policy, process, procedure*

What Is a Policy?

A policy specifies in general terms the objective or standard to be achieved without mandating a particular approach or implementation. ISO 9001 provides the criteria for a quality management system. More information can be found here: `www.iso.org/iso-9001-quality-management.html`.

You'll develop sample policies for your expected controls as you progress through this chapter.

What Is a Process?

A process is a suite of activities to be completed to deliver the policy objective. You may better articulate each step for your process by using a flowchart, checklist, or other mechanism designed to ensure the completion of your process.

An example of an effective process may be found within your organization for making changes to your systems where you are required to work through a series of checks and balances, attestations, and proofs before being authorized to implement change.

You will develop sample processes for your expected controls as you progress through this chapter.

What Is a Procedure?

A procedure implements an individual step within your process and itself may consist of several constituent parts. Only when all constituent parts are finished and correct can the procedure be considered complete.

You will develop sample procedures for your expected controls as you progress through this chapter.

Sample Expected Controls

Within this section we identify some commonly implemented account-level controls that you might reasonably expect to be found within your organization. Naturally we cannot provide an exhaustive list of controls as your organization's security posture and operating environment may differ, so the controls presented here are merely a starting point for your further consideration.

For a fuller suite of controls, you may also consider National Institute of Standards and Technology (NIST), a U.S. Department of Commerce organization that publishes standards for cyber security; go to www.nist.gov/cyberframework. I discuss NIST in my earlier book, *Building the Snowflake Data Cloud*, Chapter 4, which also outlines some account security measures.

In this section we adopt a different approach. We assume the expected controls have been implemented, so we are more concerned with the end-to-end integration to your organization's infrastructure and monitoring tooling.

Automation

Depending upon the agreed frequency, and the agreed mechanism for delivery, we offer some pointers to aid your automation of control event information to consuming parties, as outlined in Figure 3-2. The distribution channels are not exhaustive, merely indicative, and all require some degree of integration with third-party tooling.

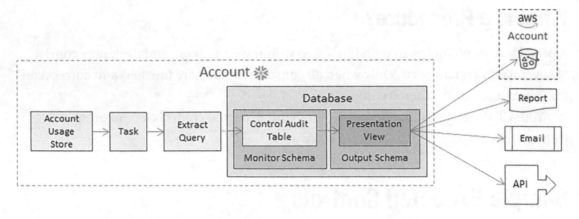

Figure 3-2. *Event detection and automated distribution*

The general pattern for event detection and distribution to consuming parties is the following:

- Identify the data source for the event.

- Extract and save the events to a local table for audit purposes.

- Present the events for consumption.

- Publish the events.

Depending upon the approach, the following constructs will be required.

Event Extraction

You must provide a means to extract all relevant events for each control. In general terms, this means a specific extract capability and scheduling mechanism for each control.

Within Snowflake you typically use streams and tasks to detect the presence of data, useful for triggering outbound event data for consumption but not possible when accessing the Account Usage Store views. However, you may use a scheduled event with an extract query that repeats periodically to create your event audit trail into a control table.

Writing to an S3 Bucket

Writing to an S3 bucket (Azure Blob Storage, Google Cloud Storage) requires a CSP account with a Snowflake storage integration mapped to the S3 location and a stage defined for SQL to write the events to a file.

Report

This may be a PowerBI, Tableau, ThoughtSpot, or similar report with capability to export data to flat file. A connector or gateway will require configuring with an agreed authentication mechanism, preferably token-based but a fallback option is to create a service user with (at least) 16-digit random mixed character/number combination generated password.

We also expect a third-party reporting tool-specific network policy to be created. We discuss this below.

Email

By default, Snowflake does not provide in-built email capability, therefore integration with CSP email capability is required. Email integration may be accomplished by AWS lambda wrapping simple email service (SES) functionality via API calls or an SMTP interface. Using MIME it is possible to send attachments, noting some file type extensions are prohibited.

External API

Modern data interchange may also be implemented using Extract, Load, and Transform (ELT) tooling such as Rivery, Fivetran, and Boomi; the ELT market is saturated with vendors. For simplicity's sake, we assume an API will be the target to offload event information, although many alternative endpoints are supported.

We also expect a third-party ELT tool-specific network policy to be created. We discuss this below.

Restricted Use of ACCOUNTADMIN

When a Snowflake account is first provisioned, there are usually two or three nominated users given the ACCOUNTADMIN role in order to begin the account delivery within your organization. We draw a distinction between Snowflake as a platform and the use of Snowflake to develop products, applications, or components.

In accordance with Snowflake best practice, ACCOUNTADMIN must only be used for platform-specific activities and not under any circumstances used for application development.

Policy

To comply with Snowflake best practices, the use of ACCOUNTADMIN must be restricted to account maintenance activities only.

The ACCOUNTADMIN role can only be used where another Snowflake role cannot be used, and then only by approved users.

For emergency use by suitably skilled staff, the ACCOUNTADMIN role must be controlled via a privileged access management (PAM) or "break glass" system.

All uses of the ACCOUNTADMIN role must be audited and notified to one or more nominated support teams. You should also specify the maximum permitted elapsed time before events are notified.

Process

Disable Snowflake Sales Engineering Users

With the exception of Snowflake trial accounts, the provision of Snowflake accounts by Snowflake Sales Engineering (SE) usually results in one or two users with the ACCOUNTADMIN role. For those new to Snowflake, and whose organizations have just engaged Snowflake, it is perfectly acceptable for the SE users to remain enabled while the staff are learning Snowflake. For organizations using Snowflake that are more mature and established with experienced staff, any SE provisioned users must be first disabled and then removed as soon as possible.

Create the "Break Glass" User

Regardless of the method of account provisioning, and in accordance with Snowflake best practices, there should be a minimum of two users with the ACCOUNTADMIN role and we recommend a third generic user be created under the control of PAM for "break glass" use in an emergency.

For all "break glass" usage, you must identify, authenticate, and record the users' information as the "break glass" user will not belong to a specific person.

Engage the PAM Team

Your PAM team will have its own process for managing the Snowflake "break glass" user usage, which you must weave into your control process.

Identify the Monitoring Recipients

Most organizations have a team dedicated to monitoring cyber security threats. You must identify the internal consumers of all events raised by the use of the ACCOUNTADMIN role and agree how they will receive event messages.

You should also send the same event notifications to your Operational Support teams for an independent audit trail recording and as a check and balance to ensure actions carried out are in accordance with expected activities.

Pre-Use Notification

Your process should make provision to notify all event notification recipients prior to the ACCOUNTADMIN role usage. You don't want any surprises! However, there are times when unexpected usage can occur, but they should be notified as soon as the activity completes, along with an explanation of why the ACCOUNTADMIN role was used.

Develop Notifications

With all requirements from your interested parties in place, you can then design and develop your notification dataset and delivery mechanisms.

Test and Sign Off

Naturally, you want to ensure compliance with policy. In an ideal world, the process steps will translate directly into a suite of deliverables typically raised as requirements and formally recorded in your change tracking tool. Regardless your delivery methodology, you must manage your technical estate and provide an audit trail of all activity.

Procedure

With your process definitions available, you can now define your detailed implementation procedures.

Disable or Remove Snowflake SE Users

Assuming your Snowflake account was delivered by Snowflake SE, you must disable their users, but first, ensure that Snowflake SE users have not created objects because ownership will revert to the role that created the user.

When you're satisfied any object ownership issues have been resolved, proceed to either disable or remove Snowflake SE users, first setting your role:

```
USE ROLE securityadmin;
```

Check for Snowflake SE users. They are usually suffixed with _SFC but do check with your SE first:

```
SHOW users;
```

For trial accounts, the SNOWFLAKE user cannot be disabled.

Figure 3-3 shows the Snowflake users for a trial account including a user SNOWFLAKE along with others that may have been created (in this example, ANDYC is my default user created on account initialization).

```
AndyC#(no warehouse)@(no database).(no schema)>SHOW users;
+-----------+-------------------------------+-----------------+----------------+
| name      | created_on                    | login_name      | display_name   |
+-----------+-------------------------------+-----------------+----------------+
| ANDYC     | 2022-09-30 11:12:47.262 -0700 | ANDYC           | ANDYC          |
| SNOWFLAKE | 2022-09-30 11:12:47.315 -0700 | SNOWFLAKE       | SNOWFLAKE      |
+-----------+-------------------------------+-----------------+----------------+
```

Figure 3-3. *Trial account users*

Assuming you have users to disable, including Snowflake SE users:

```
ALTER USER <Snowflake SE User>
SET disabled = TRUE;
```

Alternatively, you may remove users:

```
DROP USER <Snowflake SE User>;
```

Create the "Break Glass" ACCOUNTADMIN User

Precise implementation steps for integrating PAM into your Snowflake account are highly dependent upon the tooling chosen; therefore. it is not possible to deliver a single series of steps that satisfy all scenarios. Instead, please refer to third-party tooling integration guides.

Extract Event Information

In Chapter 6 of *Building the Snowflake Data Cloud*, we developed monitoring capability to detect activity conducted via role the ACCOUNTADMIN, which we recommend as a fully worked example.

The core query returning relevant information is partially repeated here, noting Account Usage Store latency. Further details on the query_history view can be found here: https://docs.snowflake.com/en/sql-reference/account-usage/query_history.html.

You start with declarations that enable later automation (if you wish):

```
SET monitor_owner_role = 'monitor_owner_role';
SET monitor_warehouse  = 'monitor_wh';
```

Change the role to sysadmin and create a warehouse:

```
USE ROLE sysadmin;

CREATE OR REPLACE WAREHOUSE IDENTIFIER ( $monitor_warehouse ) WITH
WAREHOUSE_SIZE       = 'X-SMALL'
AUTO_SUSPEND         = 60
AUTO_RESUME          = TRUE
MIN_CLUSTER_COUNT    = 1
MAX_CLUSTER_COUNT    = 4
SCALING_POLICY       = 'STANDARD'
INITIALLY_SUSPENDED  = TRUE;
```

Change the role to securityadmin, create a new role called monitor_owner_role, and grant entitlement:

```
USE ROLE securityadmin;

CREATE OR REPLACE ROLE IDENTIFIER ( $monitor_owner_role );
GRANT IMPORTED PRIVILEGES ON DATABASE snowflake TO ROLE IDENTIFIER (
$monitor_owner_role );

GRANT USAGE   ON WAREHOUSE IDENTIFIER ( $monitor_warehouse ) TO ROLE
IDENTIFIER ( $monitor_owner_role  );
GRANT OPERATE ON WAREHOUSE IDENTIFIER ( $monitor_warehouse ) TO ROLE
IDENTIFIER ( $monitor_owner_role  );
```

Now grant the role monitor_owner_role to yourself:

```
GRANT ROLE IDENTIFIER ( $monitor_owner_role )
TO USER <Your User Here>;
```

Change the role and warehouse:

```
USE ROLE       IDENTIFIER ( $monitor_owner_role );
USE WAREHOUSE IDENTIFIER ( $monitor_warehouse  );
```

You can now identify and extract event information from the Account Usage Store, noting that the quoted latency of up to 45 minutes may affect your results:

```
SELECT start_time,
       role_name,
       database_name,
       schema_name,
       user_name,
       query_text,
       query_id
FROM   snowflake.account_usage.query_history
WHERE  role_name = 'ACCOUNTADMIN'
ORDER BY start_time DESC;
```

Table functions remove latency but are not available for the Account Usage Store query_history view. Instead, you must use the corresponding information_schema view, which you know exists for every database.

Here is how to access a database information_schema using a table function:

```
SELECT start_time,
       role_name,
       database_name,
       schema_name,
       user_name,
       query_text,
       query_id
FROM   TABLE ( snowflake.information_schema.query_history())
WHERE  role_name = 'ACCOUNTADMIN'
ORDER BY start_time DESC;
```

Simply replace the Snowflake database with any existing database in your account to which your role has entitlement to read.

The challenge is to bring all real-time query_history data from all desired databases together into a single object, while recognizing that your production Snowflake account has periodic releases where new databases can be either removed or deployed. You must avoid all hard-coded dependencies within your monitoring tooling as would be the case with a single view that spans all required databases. These constraints lead you to consider developing a JavaScript stored procedure to dynamically consolidate any real-time monitoring requirements by accessing each database metadata. But first a note of caution: Accessing metadata can be a resource- and time-intensive operation.

Assuming you have captured event information, your next step is to deliver the event information to your consumers.

Deliver Event Information

With your event information available, you have several options for delivery:

- Unloading event information to a file within an S3 bucket for collection by interested parties

- Making event information accessible via a user interface for interactive reporting

- Sending event information via email to all interested parties

- Presenting information for an external process to call a remote API to unload event information

You may define other interfaces, which we leave for your further consideration.

Checklist

The checklist for this control is not exhaustive but includes the items shown in Table 3-1.

Table 3-1. *Restricted Use of ACCOUNTADMIN Checklist*

Item	Value
Policy signed off?	
Process signed off?	
Snowflake Sales Engineering users disabled?	
"Break glass" user managed by PAM?	
Event source identified?	
All monitoring recipients identified?	
All monitoring recipients engaged and trained?	
Event payload documented and agreed?	
Event response plan agreed?	
Event payload delivery mechanism implemented?	
Event audit trail implemented?	
Event preplanned activity notification process agreed?	
Procedures implemented?	
Operations Manual updated?	
Last review date:	

Network Policy

When a Snowflake account is first provisioned, the account is accessible over the public Internet. Your organization may mandate restricting access to on-prem locations and known, approved Internet locations. To restrict access to known IP ranges, you implement network policies.

It is not possible to SET a network policy that prevents access from the current session. We prove this assertion later in this chapter.

You can only use IPv4 ranges; IPv6 are not currently supported.

Policy

End users may only access a Snowflake account from on-prem IP ranges. There must be no open Internet access to Snowflake.

Third-party tooling and service users may only connect from approved IP ranges and each third-party tool or service user must have its own network policy. Sharing of network policies across third-party tool connections and/or service users is prohibited.

All network policy declarations must be both audited and notified to one or more nominated support teams in the event of changes taking place. You should also specify the maximum permitted elapsed time before events are notified.

Process

Implement Account-Level Network Policy

Your Cyber Security and/or Network Management colleagues will maintain a list of allowable IP ranges that must remain secure to prevent unauthorized access.

Identify all the allowable IP ranges along with any explicit or mandated disallowed ranges too.

Implement a Third-Party-Specific Network Policy

Identify approved third-party tools and service users along with their allowable IPv4 ranges.

Identify the Monitoring Recipients

You should also send the same event notifications to your Operational Support teams for independent audit trail recording and as a check and balance to ensure actions carried out are in accordance with expected activities.

Network Policy Change Notification

Your process should make a provision to notify all event notification recipients prior to making a network policy change.

You may be required to make a network policy change whenever either internal IP ranges change, third-party tools make changes to their IP ranges, or service users migrate to new environments.

Develop Notifications

With all requirements from your interested parties in place, you can then design and develop your notification dataset and delivery mechanisms.

For network policies, you may choose to implement a comparison against known "good" network policies to ensure your perimeter remains as designed and as a safeguard against attack.

We suggest that checking the Account Usage Store `query_history` view for changes to network policies is not an adequate guardrail for two reasons:

- The `query_history` view has up to 45 minute latency so changes will not be immediately visible.

- The `query_history` ages out after a year and it is entirely possible for a network policy to not be amended during this time

Test and Sign Off

Naturally, you want to ensure compliance with policy. In an ideal world, the process steps will translate directly into a suite of deliverables typically raised as requirements and formally recorded in your change tracking tool. Regardless your delivery methodology, you must manage your technical estate and provide an audit trail of all activity.

Procedure

With your process definitions available, you can now define your detailed implementation procedures.

Create Account-Level Network Policy

Using the ACCOUNTADMIN role, create an account-level network policy containing all the allowable IP ranges along with any explicit or mandated disallowed ranges too. Note the below example is just that, an example not intended for real-world use:

```
USE ROLE accountadmin;
```

```
CREATE OR REPLACE NETWORK POLICY my_network_policy ALLOWED_IP_LIST = (
'192.168.0.0/22', '192.168.0.1/24' );
```

To enable a network policy after declaration:

```
ALTER ACCOUNT SET network_policy = my_network_policy;
```

Attempting to set the account network policy from an IP range not included within the allowed_ip_list results in an error:

> *Network policy MY_NETWORK_POLICY cannot be activated.*
> *Requestor IP address, <Your IP Address here>, must be included in the allowed_ip_list.*
>
> *To add the specific IP, execute command "ALTER NETWORK POLICY MY_NETWORK_POLICY SET ALLOWED_IP_LIST= ('192.168.0.0/22','192.168.0.1/24','<Your IP Address here>');".*
> *Similarly, a CIDR block of IP addresses can be added instead of the specific IP address.*

Note the error message contains an ALTER statement that includes your current machine IP address.

If your Internet connection uses DHCP, your IP may change between sessions. Ensure you UNSET your network policy before disconnecting!

Correct the network policy and retry.

Please also refer to Snowflake documentation found here: https://docs.snowflake.com/en/user-guide/network-policies.html#network-policies.

Create a Third-Party-Specific Network Policy

For each approved third-party tool or service user, create a network policy for the specific use of the third-party tool or service user containing all the allowable IP ranges. Note the below example is just that, an example not intended for real-world use:

```
USE ROLE accountadmin;

CREATE OR REPLACE NETWORK POLICY powerbi_gateway_policy ALLOWED_IP_LIST=(
'192.168.0.2' );
```

You assign a product-specific network policy to specific users. In this example we imagine a PowerBI service user for use within the PowerBI Gateway hosted on a remote Internet service:

```
USE ROLE securityadmin;

CREATE OR REPLACE USER test
PASSWORD             = 'test'
DISPLAY_NAME         = 'Test User'
EMAIL                = 'test@test.xyz'
DEFAULT_ROLE         = 'monitor_owner_role'
DEFAULT_NAMESPACE    = 'SNOWFLAKE.reader_owner'
DEFAULT_WAREHOUSE    = 'monitor_wh'
COMMENT              = 'Test user'
NETWORK_POLICY       = powerbi_gateway_policy
MUST_CHANGE_PASSWORD = FALSE;
```

Extract Event Information

The question you must answer is whether any network policy declared remains as specified, and whether declared network policies remain active. The distinction is important: both aspects must be fulfilled as changes to one aspect or to both aspects may invalidate your control.

Within this control you are not concerned with the assignment of a network policy to an individual service user. You expect user administration to fall under a separate control for your further consideration.

With two aspects to consider, you start with ensuring network policy declarations remain as specified. In Chapter 6 of *Building the Snowflake Data Cloud*, we developed monitoring capability to compare the current network policy settings match those contained within a reference schema, and suggest the same codebase provides sufficient details to meet this control requirement.

Your second check requires a little more thought. First, establish the network policies available:

```
USE ROLE accountadmin;

SHOW NETWORK POLICIES IN ACCOUNT;
```

You should see two network policies:

- `MY_NETWORK_POLICY`

- `POWERBI_GATEWAY_POLICY`

You can identify the IP ranges for each network policy:

```
DESCRIBE NETWORK POLICY my_network_policy;
DESCRIBE NETWORK POLICY powerbi_gateway_policy;
```

To identify whether an account-level network policy is active or not:

```
SHOW PARAMETERS LIKE 'NETWORK_POLICY' IN ACCOUNT;
```

Figure 3-4 shows both states where firstly a network policy is active for the account and secondly where the network policy has been disabled for the account.

Row	key	value	default	level	description	type
1	NETWORK_POLICY	MY_NETWORK_POLICY		ACCOUNT	Network policy assigned for the given target.	STRING

Row	key	value	default	level	description	type
1	NETWORK_POLICY				Network policy assigned for the given target.	STRING

Figure 3-4. *Active and inactive account network policies*

The SHOW command accesses the Snowflake Global Services Layer, which itself is based upon FoundationDB, they key-pair store that underpins all Snowflake metadata operations. This is where you encounter a problem: SHOW is not a true SQL command insofar as you can't use SHOW within a SQL statement. For example, you can't create a VIEW based upon SHOW or use SHOW to drive query content.

Before you can query results, you must set your warehouse. For ease, reuse an earlier declaration:

```
USE WAREHOUSE monitor_wh;
```

You use RESULT_SCAN to convert the SHOW output to a resultset, but you must understand the interaction. When RESULT_SCAN is used with last_query_id, you can't run commands in between as the query_id changes:

```
SELECT "key", "value", "level"
FROM TABLE ( RESULT_SCAN ( last_query_id()));
```

The presence of data within "value" and "level" (set to 'ACCOUNT') indicates an account-level network policy has been set, and the absence of data within these attributes indicates either no account-level network has been SET or the account-level network policy has been UNSET.

Unfortunately, Snowflake has not exposed network policy information within the Account Usage Store Users view, so you are left with the SHOW command to check whether a service user has a network policy:

```
SHOW PARAMETERS LIKE 'NETWORK_POLICY' IN USER test;
```

Figure 3-5 shows both states where firstly a network policy is active for the user and secondly where the network policy has been disabled for the user.

Row	key	value	default	level	description	type
1	NETWORK_POLICY	POWERBI_GATEWAY_POLICY		USER	Network policy assigned for the given target.	STRING

Row	key	value	default	level	description	type
1	NETWORK_POLICY				Network policy assigned for the given target.	STRING

Figure 3-5. *Active and inactive user network policies*

As you saw for the account network policy, you use RESULT_SCAN to convert the SHOW output to a resultset:

```
SELECT "key", "value", "level"
FROM TABLE ( RESULT_SCAN ( last_query_id()));
```

As shown in Chapter 6 of *Building the Snowflake Data Cloud*, you may choose to write a JavaScript stored procedure to wrap both statements and return data for onward propagation for event creation. We leave this for your further investigation.

Deliver Event Information

With your event information available, you reuse your delivery mechanism of choice, as articulated above.

Checklist

The checklist for this control is not exhaustive but includes the items shown in Table 3-2.

Table 3-2. *Network Policy Checklist*

Item	Value
Policy signed off?	
Process signed off?	
Network policy change identified?	
Network policy assignment/deassignment identified?	
Event source identified?	
All monitoring recipients identified?	
All monitoring recipients engaged and trained?	
Event payload documented and agreed?	
Event response plan agreed?	
Event payload delivery mechanism implemented?	
Event audit trail implemented?	
Event preplanned activity notification process agreed?	
Procedures implemented?	
Operations Manual updated?	
Last review date:	

Additional Controls

Having provided two fully worked examples, and to assist your further investigation into developing more controls, we offer these samples and leave the process and procedures open for your interpretation. Note that the list is not exhaustive; your list should be more complete and developed in accordance with your Cyber Security colleagues' requirements.

Unauthorized Data Download Policy

To prevent data loss, and to protect the reputation of your organization, by default all end users must not be able to unload data to external cloud storage locations. By default, Snowflake accounts permit ad-hoc data unloads to external cloud locations which must be disabled.

Recognizing there are valid business and technical reasons for enabling access to specific CSP storage locations, all such operations must utilize a predefined storage integration. Snowflake provides the means to enable specific CSP storage locations.

Snowflake Custom Roles

In a multi-tenant environment, such as an account where more than a single application team may conduct deployments, there may be a requirement to create tenant-specific roles enabling administration of a subset of Snowflake features.

All LZ-defined administrative roles and any application-specific defined roles must adhere to the least privilege access requirement. In other words, non-Snowflake supplied roles must only provision entitlement according to the minimum required to fulfill the role purpose.

Application-specific roles must segregate object owner entitlement from object usage entitlement. Roles must ensure objects can't be created dynamically in order to preserve schema integrity and facilitate troubleshooting.

Continuous Monitoring

All administrative events must be continually monitored for compliance. As a minimum, these roles must be monitored: ORGADMIN, ACCOUNTADMIN, SYSADMIN, SECURITYADMIN, and all LZ-defined administrative roles.

Share Monitoring

When data shares have been configured, ensure all external consumer access and usage is monitored to the greatest extend Snowflake permits. Note that external share monitoring is an expanding capability so we suggest a monthly review of features for inclusion and reporting.

Tri-Secret Secure

Implement Bring Your Own Key (BYOK)/Customer Managed Key (CMK) for Tri-Secret Secure encryption at rest for each Snowflake account, thereby providing a master "kill switch" to disable each account under your organization's control.

Ensure periodic rekeying is implemented for BYOK/CMK.

Data Masking

Ensure all sensitive data is masked according to your organization's policy. Report any unplanned changes to data masking policies or assignment of data masking policies.

Multiple Failed Logins

Monitor and report when multiple login attempts fail. The number of attempts may vary according to your organization's policy. We suggest three to five attempts as an optimal number of failed login attempts in the absence of an existing policy.

Cleanup

Having created various objects within this chapter, you now reset your account by removing each object in turn:

```
USE ROLE accountadmin;

ALTER ACCOUNT UNSET network_policy;

DROP NETWORK POLICY my_network_policy;
DROP NETWORK POLICY powerbi_gateway_policy;

USE ROLE securityadmin;
```

```
ALTER USER test UNSET network_policy;

DROP USER       test;
DROP ROLE       monitor_owner_role;
DROP WAREHOUSE  monitor_wh;
```

Periodic Access Recertification

Periodic access recertification is the process by which all Snowflake consumers are reviewed to ensure their access is both appropriate and currently approved for each Snowflake account.

Typically, organizations enable end user creation and apply user roles via some form of federated tooling such as Microsoft Azure Active Directory or Okta. We do not discuss how to use these tools within this chapter but instead focus on the Snowflake capabilities. Just be aware other options exist, and they should be used in preference if available. Ideally, Snowflake manual user creation should not be implemented, but not every organization is sufficiently mature in its approach to managing users and entitlements with automated deployment and revocation.

As a general principle, and your organization may adopt a tighter or more relaxed schedule than we discuss here, we suggest periodic access recertification should be conducted at a minimum of every three months. Our preference is to embed periodic user recertification as a monthly operational procedure within each team's Operational Support team manual.

Regardless of the accepted recertification period, there are several aspects to consider:

- Administrative entitlement to users

- Service user entitlement and object entitlement scope

- End user entitlement and object entitlement scope

- Data share object entitlement and consumer account

- Reader account entitlement and entitlement scope

- Private data exchange membership

- Snowflake Marketplace producer and consumer recertification

Recognizing some of the above points relate to individual account usage and are not related to the centralizcd provisioning of an account, we do not cover each aspect. Specifically, we do not cover private data exchange and Snowflake Marketplace for which administration is available via Snowsight.

Our intent is to highlight areas for your consideration aside from providing a starting point for further investigation.

User Entitlement

Let's first address how to identify the active roles in your Snowflake account. Roles are the foundation of your entitlement model for role-based access control (RBAC).

To identify the active roles in your account, you must first change the role to ACCOUNTADMIN in order to access the Account Usage Store:

```
USE ROLE accountadmin;
```

You can now identify all the Snowflake roles:

```
SELECT name,
       owner,
       comment
FROM   snowflake.account_usage.roles
WHERE  deleted_on IS NULL
ORDER BY name ASC;
```

Within the returned resultset, all roles with "owner" set to NULL are Snowflake reserved roles, providing a quick way to differentiate from LZ- or application-specific roles. Note the comment attribute is optional and therefore may not be populated.

Having identified all roles of interest, you have the starting point to begin your recertification process. Joining to grants_to_users allows you to view the roles granted to a user. Your first review should be to check role assignments to each user:

```
SELECT ur.role,
       ur.grantee_name,
       r.owner
FROM   snowflake.account_usage.roles            r,
       snowflake.account_usage.grants_to_users ur
WHERE  r.name          = ur.role
```

```
AND     r.deleted_on  IS NULL
AND     ur.deleted_on IS NULL
ORDER BY r.name ASC;
```

After satisfying yourself that the roles have correctly been assigned to users, your second review is to check all object entitlements are correct for each role:

```
SELECT r.name,
       gr.privilege,
       gr.table_catalog||'.'||gr.table_schema||'.'||gr.name AS object_name
FROM   snowflake.account_usage.roles             r,
       snowflake.account_usage.grants_to_roles gr
WHERE  r.name        = gr.name
AND    r.deleted_on  IS NULL
AND    gr.deleted_on IS NULL
ORDER BY r.name ASC;
```

These queries satisfy the first three requirements: identifying roles for administration users, service users, and end users and then validating the entitlements granted to each role. You do not expect to report the findings to a central monitoring team, but instead to work collaboratively with the role owners to disable users who have either left your organization or moved role internally within your organization. For those users who remain within your organization, entitlement and access must be confirmed according to their organization role profile and job description.

You now turn your attention to considering data shares and the consuming Snowflake accounts.

Data Shares

Let's first address how to identify the data shares in your Snowflake account and the corresponding consumers of each data share. Shares are a first-order object whose declaration and management are typically performed using the ACCOUNTADMIN role, the use of which should be tightly restricted. Data sharing is only supported between Snowflake accounts. For further information, see the documentation found here: https://docs.snowflake.com/en/user-guide/data-sharing-intro.html.

You must first change the role to ACCOUNTADMIN in order to access the Account Usage Store:

```
USE ROLE accountadmin;
```

Before you can query results, you must set your warehouse. For ease, reuse an earlier declaration:

```
USE WAREHOUSE monitor_wh;
```

You can now identify all the data shares along with their consumers:

```
SHOW shares;
```

As you saw earlier, you use RESULT_SCAN to convert the SHOW output to a resultset:

```
SELECT "kind",
       "name",
       "database_name",
       "owner",
       "to",
       "listing_global_name"
FROM TABLE ( RESULT_SCAN ( last_query_id()))
ORDER BY "kind", "name";
```

You may also be interested to see the objects contained within a share:

```
DESCRIBE SHARE <share name here>;
```

And in similar fashion to above, you use RESULT_SCAN to convert the SHOW output to a resultset:

```
SELECT "kind",
       "name",
       "shared_on"
FROM TABLE ( RESULT_SCAN ( last_query_id()))
ORDER BY "kind", "name";
```

The query output will list the constituent objects for the data share.

As an interesting exercise, try this:

```
DESCRIBE SHARE snowflake.account_usage;
```

You should see a partial screenshot similar to Figure 3-6, which illustrates the object types in the kind column. Note there are many more objects than those shown. Of particular interest are the SCHEMA as you scroll down the list.

kind	name	shared_on
DATABASE	SNOWFLAKE	2021-01-25 17:57:07.345 -0800
SCHEMA	SNOWFLAKE.ACCOUNT_USAGE	2021-01-25 17:57:06.037 -0800
FUNCTION	SNOWFLAKE.ACCOUNT_USAGE."TAG_REFERENCES_WITH_LINEAGE$V1(TAG_NAME_INPUT V...	2022-04-13 17:27:48.227 -0700
FUNCTION	SNOWFLAKE.ACCOUNT_USAGE."TAG_REFERENCES_WITH_LINEAGE(TAG_NAME_INPUT VAR...	2021-05-27 17:49:17.803 -0700
VIEW	SNOWFLAKE.ACCOUNT_USAGE.ACCESS_HISTORY	2022-10-06 20:01:00.183 -0700

Figure 3-6. *Account Usage Store inbound share*

With all data share information to hand, you can now determine the consuming Snowflake accounts along with the objects available within each data share to validate the correctness of data distribution. Note that we do not consider the data content (i.e., the actual data values shared), simply the ability for a consumer to read data from the shared objects; we discuss data share consumption in a later chapter.

Reader Accounts

Unfortunately, Snowflake uses the term "reader account" to provide an access path into a Snowflake account for use by consumers who are not Snowflake customers. A better term would be "reader user" as the syntax is synonymous with user creation. For further information, see the documentation found here: https://docs.snowflake.com/en/ user-guide/data-sharing-reader-create.html.

We do not discuss reader accounts managing access to data shares or configuring the reader account within this section. Our objective is to allow reader accounts to be identified for revalidation purposes only.

Creating a Reader Account

To test this approach, you must first create a reader account. Reader accounts are managed via the ACCOUNTADMIN role, so you must switch to it first:

```
USE ROLE accountadmin;
```

You may prefer to check for existing reader accounts first:

```
SHOW MANAGED ACCOUNTS;
```

Now create a reader account using the example from the Snowflake documentation, noting you may need to wait up to 5 minutes for provisioning:

```
CREATE MANAGED ACCOUNT reader_acct1
ADMIN_NAME     = user1,
ADMIN_PASSWORD = 'Sdfed43da!44',
TYPE           = READER;
```

You should see a status response similar to this, noting your account information will differ:

```
{"accountName":"OJ69034","loginUrl":"https://oj69034.eu-west-2.aws.
snowflakecomputing.com"}
```

Identifying Managed Accounts

```
SHOW MANAGED ACCOUNTS;
```

And in similar fashion to before, you use RESULT_SCAN to convert the SHOW output to a resultset:

```
SELECT "name",
       "cloud",
       "region",
       "locator",
       "url"
FROM TABLE ( RESULT_SCAN ( last_query_id()))
ORDER BY "name" ASC;
```

With all reader account information to hand, you can now validate whether each reader account should be retained or removed.

Additional Information

Assuming your reader accounts are in use and fully configured to access your Snowflake data shares, further information can be found within the Account Usage Store. You may investigate the views within `snowflake.reader_account_usage`.

Cleanup

Having created various objects within this chapter, you now reset your account by removing each object in turn:

```
DROP MANAGED ACCOUNT reader_acct1;
```

Security Information and Event Management (SIEM)

Your organization often requires security-relevant events from disparate sources to be collated into a single tool for analysis and long-term storage. Snowflake is no exception to this information gathering exercise and, as you will discover, Snowflake offers a comprehensive suite of monitoring capabilities to satisfy SIEM requirements.

As Snowflake is constantly evolving, we recommend a periodic review of Snowflake SIEM monitoring capabilities to include your Cyber Security colleagues, who may also have evolving SIEM event detection requirements as the threat space changes. An example of Snowflake evolution is the introduction of the `access_history` view, which appeared in February 2021.

Our approach is to highlight the various Snowflake information sources to fulfill demands from our Cyber Security colleagues for reporting via their SIEM tooling without specifying a particular delivery mechanism or recommending specific SIEM tooling. We also show a simple example for each SIEM aspect with the expectation that you will investigate further to satisfy your organizations SIEM needs.

For each of the Account Usage Store views there are equivalent `information_schema` views and table functions. We focus on deriving SIEM event detection from the Account Usage Store as this is the common location for all activity within the account.

Login History

The starting point is the Account Usage Store, firstly focusing on logins within the last 365 days, for which documentation can be found here: `https://docs.snowflake.com/ en/sql-reference/account-usage/login_history.html`. Note the latency period for `login_history` is up to 2 hours.

As is common throughout this book when accessing the Account Usage Store, you first set your role to ACCOUNTADMIN:

```
USE ROLE accountadmin;
```

Your first `login_history` SIEM event is to detect all failed login events:

```
SELECT event_id,
       event_type,
       user_name,
       error_code,
       error_message
FROM   snowflake.account_usage.login_history
WHERE  error_message IS NOT NULL
ORDER BY event_timestamp DESC;
```

Other SIEM events to report include

- User logins without multi-factor authentication

- Successful logins with IP address

- Successful logins from unapproved tooling

Along with any other organization-specific SIEM event detection requirements, we leave you to investigate `login_history` further.

Query History

Any user successfully logged in to Snowflake will conduct activity, some of which will be of interest to your Cyber Security colleagues.

You may wish to identify queries issued within the last 365 days, for which documentation can be found here: `https://docs.snowflake.com/en/sql-reference/ account-usage/query_history.html`. Note the latency period for `query_history` is up to 45 minutes.

```
USE ROLE accountadmin;
```

Your first `query_history` SIEM event is to detect all ACCOUNTADMIN role usage:

```
SELECT user_name,
       query_id,
       query_text,
       start_time
FROM   snowflake.account_usage.query_history
WHERE  query_text ILIKE '%ACCOUNTADMIN%'
ORDER BY start_time DESC;
```

Other SIEM events to report include

- Administrative events such as changing network policies and updates to account parameters

- User creation, updates, and deletes

- Role assignments to users

- Changes to storage integrations

- Changes to periodic data rekeying

Along with any other organization-specific SIEM event detection requirements, we leave you to investigate `query_history` further.

Access History

Lastly, you may also consider the access history of Snowflake objects within the last 365 days, for which documentation can be found here: `https://docs.snowflake.com/en/sql-reference/account-usage/access_history.html`. Latency period for `access_history` is 3 hours.

```
USE ROLE accountadmin;
```

The sample SIEM event is to return 10 `access_history` records to expose the JSON records within ARRAY attributes:

```
SELECT user_name,
       query_id,
```

```
        direct_objects_accessed,
        base_objects_accessed,
        objects_modified,
        query_start_time
FROM    snowflake.account_usage.access_history
ORDER BY query_start_time DESC
LIMIT 10;
```

We leave you to determine how to use the data returned by access_history view according to your organization's SIEM event detection requirements.

We offer these two SQL statements to expand upon the embedded JSON attributes within the access_history view.

This SQL statement shows how to access nested JSON attributes within access_history.direct_objects_accessed and joins with query_history to facilitate time-banded queries:

```
SELECT ah.user_name,
       ah.query_id,
       ah.direct_objects_accessed,
       doa.value:"objectDomain"::string              AS doa_object_domain,
       doa.value:"objectName"::string                AS doa_base_object,
       doa2.value:"columnName"::string               AS doa_column_name,
       ah.base_objects_accessed,
       boa.value:"objectDomain"::string              AS boa_object_domain,
       boa.value:"objectName"::string                AS boa_base_object,
       ah.objects_modified,
       om.value:"objectDomain"::string               AS om_object_domain,
       om.value:"objectName"::string                 AS om_base_object,
       ah.query_start_time
FROM    snowflake.account_usage.access_history                  ah,
        snowflake.account_usage.query_history                   qh,
        LATERAL FLATTEN ( direct_objects_accessed         ) doa,
        LATERAL FLATTEN ( doa.value:columns, OUTER => TRUE ) doa2,
        LATERAL FLATTEN ( base_objects_accessed           ) boa,
        LATERAL FLATTEN ( objects_modified                ) om
WHERE   ah.query_id                              = qh.query_id
AND     doa.value:"objectDomain"::string         = 'Table'
```

```
AND     doa_base_object                                  = '<Your Table Here>'
AND     qh.start_time                                    > DATEADD ( month, -1,
                                                           current_timestamp())
ORDER BY ah.query_start_time DESC;
```

This SQL statement further joins to account_usage.tables to provide the capability to filter by database, schema, and table. Note that this query can run for many minutes without filter criteria applied.

```
SELECT  ah.user_name,
        ah.query_id,
        ah.direct_objects_accessed,
        doa.value:"objectDomain"::string               AS doa_object_domain,
        doa.value:"objectName"::string                 AS doa_base_object,
        ah.base_objects_accessed,
        boa.value:"objectDomain"::string               AS boa_object_domain,
        boa.value:"objectName"::string                 AS boa_base_object,
        ah.objects_modified,
        om.value:"objectDomain"::string                AS om_object_domain,
        om.value:"objectName"::string                  AS om_base_object,
        ah.query_start_time
FROM    snowflake.account_usage.access_history      ah,
        snowflake.account_usage.query_history       qh,
        snowflake.account_usage.tables              t,
        LATERAL FLATTEN ( direct_objects_accessed ) doa,
        LATERAL FLATTEN ( base_objects_accessed   ) boa,
        LATERAL FLATTEN ( objects_modified        ) om
WHERE   ah.query_id                                    = qh.query_id
AND     doa.value:"objectId"::int                      = t.table_id
AND     doa.value:"objectDomain"::string               = 'Table'
AND     t.table_catalog                                = '<Your Database Here>'
AND     t.table_schema                                 = '<Your Schema Here>'
AND     t.table_name                                   = '<Your Table Here>'
AND     t.deleted IS NULL
AND     qh.start_time                                  > DATEADD ( month, -1,
                                                         current_timestamp())
ORDER BY ah.query_start_time DESC;
```

SaaS Security Posture Management (SSPM)

SSPM tooling provisions automated monitoring of security risks in Software as a Service (SaaS) products focused around misconfiguration, entitlement, user provisioning, and other cloud security risks.

We include this brief section to make you aware of the emergence of SSPM tooling. We expect the Information Technology industry will mature in this direction as products and platforms move to the cloud. With this perspective in mind, your security monitoring profile will need to evolve.

You should also be mindful of Snowflake partners either currently established or moving their products into the SSPM space, offering you the opportunity to leverage their skill, knowledge, and expertise.

Archive Data Retention Period

Account Usage Store views data persists for 365 days before ageing out. You must therefore protect your audit trail for the period specified by your organization's data retention policy by ensuring that all required Account Usage Store data is periodically copied into local tables.

I discuss Account Usage Store view history retention in Chapter 6 of my book *Building the Snowflake Data Cloud*, which lists several views available at the time of writing. As Snowflake matures as a product, we observe the Account Usage Store views change relatively frequently. Our original advice of reviewing views every six months (in March and September) may no longer hold true when viewed from your LZ perspective.

The LZ must provide a common implementation pattern and a common deployment mechanism for your organization to historize Account Usage Store views as part of the core platform delivery. Additionally, the common historization mechanism must be both regularly maintained and contain a migration path as the underlying Account Usage Store views are enhanced over time. As a consequence of maturing the archive retention process, we now propose a monthly review of your view data retention procedures.

Working on the assumption that you are required to hold the full history of every action within your account from inception to current date, you need to periodically copy data over to local tables before information ages out.

Preserving login_history

Taking a single Account Usage Store view `login_history` as an example, we will work through the steps required to provide a fully historized audit trail from within your account from inception to current date. Figure 3-7 illustrates our approach to extracting and persisting the `login_history` into an audit table. The challenge is that Snowflake evolves constantly.

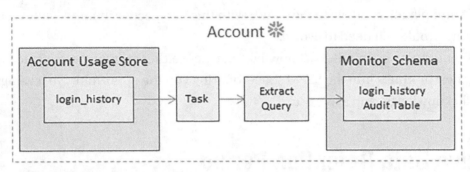

Figure 3-7. *Account Usage Store historized pattern*

Given the complexity of managing archive data in a federated environment where you may not have complete control of the target Snowflake accounts, your process must be fault-tolerant, easily extensible, and operationally supported.

But let's make a start and address issues as you encounter them. In common with earlier sections in this chapter, you reuse the same declarations and entitlements but this time create a database and schema to hold your audit trail:

```
SET monitor_owner_role    = 'monitor_owner_role';
SET monitor_warehouse     = 'monitor_wh';
SET monitor_database      = 'MONITOR';
SET monitor_owner_schema  = 'MONITOR.monitor_owner';

USE ROLE sysadmin;

CREATE OR REPLACE DATABASE IDENTIFIER ( $monitor_database ) DATA_RETENTION_
TIME_IN_DAYS = 90;

CREATE OR REPLACE SCHEMA IDENTIFIER ( $monitor_owner_schema );

USE ROLE securityadmin;
```

```
GRANT USAGE    ON DATABASE  IDENTIFIER ( $monitor_database        ) TO ROLE
IDENTIFIER ( $monitor_owner_role  );
GRANT USAGE   ON SCHEMA    IDENTIFIER ( $monitor_owner_schema   ) TO ROLE
IDENTIFIER ( $monitor_owner_role  );

GRANT USAGE                       ON SCHEMA IDENTIFIER ( $monitor_owner_
schema    ) TO ROLE IDENTIFIER ( $monitor_owner_role );
GRANT MONITOR                     ON SCHEMA IDENTIFIER ( $monitor_owner_
schema    ) TO ROLE IDENTIFIER ( $monitor_owner_role );
GRANT MODIFY                      ON SCHEMA IDENTIFIER ( $monitor_owner_
schema    ) TO ROLE IDENTIFIER ( $monitor_owner_role );
GRANT CREATE TABLE                ON SCHEMA IDENTIFIER ( $monitor_owner_
schema    ) TO ROLE IDENTIFIER ( $monitor_owner_role );
GRANT CREATE VIEW                 ON SCHEMA IDENTIFIER ( $monitor_owner_
schema    ) TO ROLE IDENTIFIER ( $monitor_owner_role );
GRANT CREATE SEQUENCE             ON SCHEMA IDENTIFIER ( $monitor_owner_
schema    ) TO ROLE IDENTIFIER ( $monitor_owner_role );
GRANT CREATE FUNCTION             ON SCHEMA IDENTIFIER ( $monitor_owner_
schema    ) TO ROLE IDENTIFIER ( $monitor_owner_role );
GRANT CREATE PROCEDURE            ON SCHEMA IDENTIFIER ( $monitor_owner_
schema    ) TO ROLE IDENTIFIER ( $monitor_owner_role );
GRANT CREATE STREAM               ON SCHEMA IDENTIFIER ( $monitor_owner_
schema    ) TO ROLE IDENTIFIER ( $monitor_owner_role );
GRANT CREATE MATERIALIZED VIEW  ON SCHEMA IDENTIFIER ( $monitor_owner_
schema    ) TO ROLE IDENTIFIER ( $monitor_owner_role );
GRANT CREATE FILE FORMAT          ON SCHEMA IDENTIFIER ( $monitor_owner_
schema    ) TO ROLE IDENTIFIER ( $monitor_owner_role );
GRANT CREATE TAG                  ON SCHEMA IDENTIFIER ( $monitor_owner_
schema    ) TO ROLE IDENTIFIER ( $monitor_owner_role );
```

With your database created, you must first set your context:

```
USE ROLE      monitor_owner_role;
USE WAREHOUSE monitor_wh;
USE DATABASE  MONITOR;
USE SCHEMA    monitor_owner;
```

Then create your audit table, noting the use of SELECT *. This is one of the very few valid use cases for SELECT *, which in general we view as unacceptable practice. With a simple statement you create a table and populate with available data up to the point in time the SQL statement ran.

As a side note, this form of SQL query is often referred to as CTAS, which stands for Create Table AS.

Note that we do not use the CREATE OR REPLACE syntax as we wish to preserve the audit table contents from the point of initial creation

---

```
CREATE TABLE aud_login_history
AS
SELECT *
FROM    snowflake.account_usage.login_history;
```

With your audit table created, you can manually add records as part of your periodic operational procedures. Alternatively, you can automate the process using a task, and for this you must grant appropriate entitlement to the monitor_owner_role:

```
USE ROLE securityadmin;
```

```
GRANT CREATE TASK ON SCHEMA monitor.monitor_owner TO ROLE monitor_
owner_role;
```

Revert to monitor_owner_role:

```
USE ROLE monitor_owner_role;
```

With your audit table created, let's check the most recent timestamp as it will be the starting point for subsequent data loads into the audit table:

```
SELECT MAX ( event_timestamp )
FROM    monitor.monitor_owner.aud_login_history;
```

You may encounter latency of up to 2 hours when testing the next SQL statements.

Now prove there are records ready for consumption:

```
SELECT *
FROM    snowflake.account_usage.login_history
WHERE   event_timestamp >
        (
        SELECT MAX ( event_timestamp )
        FROM    monitor.monitor_owner.aud_login_history
        );
```

If no records are available, you may need to create a separate login from the Classic UI, SnowSight, SnowSQL, or other connected tooling and then repeat the above SQL statement.

With records available, your INSERT statement can be tested, noting the records will auto-commit:

```
INSERT INTO monitor.monitor_owner.aud_login_history
SELECT *
FROM    snowflake.account_usage.login_history
WHERE   event_timestamp >
        (
        SELECT MAX ( event_timestamp )
        FROM    monitor.monitor_owner.aud_login_history
        );
```

When creating your task you must be mindful of scheduling the task execution. Using the minute option limits execution cycles to a maximum of 11520 minutes, which equates to every 8 days. In order to set tasks to run on longer periods, you must use the CRON option, for which more details can be found here: https://docs.snowflake.com/en/sql-reference/sql/create-task.html#optional-parameters.

In this example you set the execution frequency to be on the last day of the month at 02:00 GMT. Your testing should use a more convenient CRON setting:

```
CREATE OR REPLACE TASK task_aud_login_history_load
WAREHOUSE = monitor_wh
SCHEDULE = 'USING CRON 0 2 L 1-12 * GMT'
AS
INSERT INTO monitor.monitor_owner.aud_login_history
```

```
SELECT *
FROM   snowflake.account_usage.login_history
WHERE  event_timestamp >
       (
       SELECT MAX ( event_timestamp )
       FROM   monitor.monitor_owner.aud_login_history
       );
```

However, your task will fail if the underlying Account Usage Store view definition changes and the audit table is not maintained to reflect the source view definition.

Maintaining login_history

While this section addresses `login_history` as a worked example, you need a comprehensive solution catering for all Account Usage Store view changes and adopt this perspective for the remainder of this section.

While not an exhaustive list, you need to consider the following scenarios:

- Addition or removal of view attributes

- Changes to attribute datatype and precision

- Addition of new views or removal of old views

- Schema restructuring where views migrate to new schemas

- Rollout to disparate Snowflake targets within your organization

- Maintenance of audit versions across Snowflake accounts

Having identified the challenges, you have a few options available for maintaining audit table changes, which you now explore.

Automated Audit Maintenance

You could build an automated tool to cater for all foreseen future changes to Account Usage Store views and automatically deploy all changes into each Snowflake account.

We caution against creating an automated deployment process as there are many potential pitfalls as highlighted above, rendering this option unattractive. What you must not do is burden your Operations staff with unnecessary maintenance activities. Whatever you deliver must be robust and fault tolerant.

Semi-Automated Audit Maintenance

An alternative approach most likely to be successful is to implement tooling to identify the difference between the current audit tables and the source views. With the differences identified, the same tooling can propose changes by generating corresponding scripts. The sharp eyed among you will identify a pattern used in *Building the Snowflake Data Cloud*. We use code generators to great effect, and the same approach can be adapted here.

In outline, our suggested approach takes each of the bullet points above and delivers a JavaScript stored procedure focused on one objective. We caution against creating a monolithic procedure that "does everything;" we prefer discrete procedures that "do one thing well." You may also create a wrapper procedure to call each child procedure sequentially in order to collate all functionality.

With the discrete script in hand, you can proceed to integrate the changes within your Continuous Integration process along with updated test scripts ready for deployment, which we consider next.

Deployment

Your organization may prefer to decentralize its Snowflake accounts, in which case central maintenance is much more difficult.

Imagine a scenario where two companies have merged, with differing configuration management and deployment practices for their Snowflake estate or where Snowflake accounts are federated. In either scenario, your audit update deployment is not likely to be automated and a degree of manual intervention or handoff will be required.

Considering the complexities deployment brings, the semi-automated audit maintenance approach works best and, while more manual in approach, provides checks and balances that will prevent costly mistakes.

Cleanup

```
USE ROLE        accountadmin;

DROP ROLE       monitor_owner_role;
DROP WAREHOUSE monitor_wh;
DROP DATABASE   monitor;

DROP TASK        task_aud_login_history_load;
```

Summary

The scope of this chapter was to identify how to monitor, control, and manage your Snowflake estate. You saw how to deliver control output to a variety of interfaces to facilitate event consumption.

You then defined control definitions in terms of policy, process, and procedure before identifying and working through some sample controls. You also identified a sample suite of additional controls for your consideration, recognizing organizational requirements may differ.

Later you explored end user access recertification and considered the wider Snowflake infrastructure to identify other forms of consumption.

In support of SIEM you identified and explored worked examples of how to expose SIEM-relevant information. A brief mention of SSPM followed to expose the subject as an emerging theme.

Lastly, you learned how to preserve the full audit trail of Account Usage Store history with a single worked example for `login_history`. You saw the pitfalls of attempting to automate audit trail maintenance and you saw a workable, pragmatic solution.

You now move onto a subject of interest to your Financial Operations (FinOps) colleagues: cost monitoring.

CHAPTER 4

Cost Reporting

Every organization is rightly obsessed with costs. Snowflake, with its consumption-based model, is a prime candidate for cost monitoring and reallocation of consumption costs to the consuming team, line of business, or operating division. Your Financial Operations (FinOps) colleagues will be very keen to analyze current consumption and project future consumption using any and all metrics you expose to them.

You may face other internal and regulatory requirements for cost monitoring. The centralized provisioning of cost monitoring lends itself to wider utilization including

- Future Snowflake consumption trend projections

- Identification of consumption spikes and anomalies

- Automated inclusion of new Snowflake consumption

Organizations that provision Snowflake accounts via a centralized landing zone (LZ) have an ideal opportunity to capture all consumption metrics through the creation of both a Centralized Monitoring Store (CMS) and cost monitoring data sharing from each provisioned account. Alternatively, each Snowflake account needs to deliver its metrics individually. In a decentralized, federated organization without a single reference architecture or consistent approach, considerable time would be spent reconciling disparate views of data. It's far better to have a single, consistent, and centrally maintained approach.

Within this chapter, the objective is to deliver a "plug-and-play" cost monitoring capability where all future LZ-provisioned accounts deliver their consumption metrics into a single hub for central reporting. You could extend the cost monitoring capability beyond Snowflake into the cloud service provider (CSP) tooling; we discuss extending the monitoring capability later. Our approach is to deliver a step-by-step walk through of the components required to create the CMS and capabilities to transport metrics to the CMS.

© Andrew Carruthers and Sahir Ahmed 2023
A. Carruthers and S. Ahmed, *Maturing the Snowflake Data Cloud*, https://doi.org/10.1007/978-1-4842-9340-9_4

Please check the actual credit cost, which varies according to region and discount. Applying a simple additive approach will result in errors. The sample code does not take multiple costs into consideration and is left for your further consideration. The same is true for egress costs, which may differ according to CSP region.

Figure 4-1 illustrates the conceptual approach to delivering a CMS for the common collation of consumption metrics.

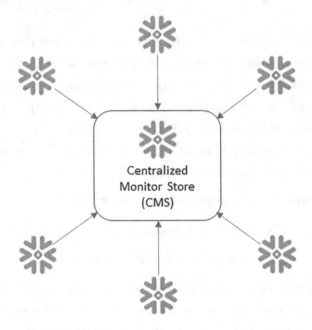

Figure 4-1. *Conceptual Centralized Monitoring Store (CMS)*

As you will read later within this chapter, the concept of a CMS is readily extended to capturing other categories of information where centralized data access is required.

The CMS has several primary use cases:

- Capturing all Snowflake consumption metrics for your organization for consolidated reporting purposes

- Capturing federated metrics. Some examples are

  - Inform Marketing and Sales teams, as explained in Chapter 5

  - Inform Disaster Recovery decision making, as explained in Chapter 12.

- Capturing externally generated (i.e., non-Snowflake) consumption metrics for consolidated reporting purposes

Provisioning cost monitoring through your LZ enables a centralized maintenance team to build out and test enhancements. Changes can be deployed in a controlled manner to all accounts, thus ensuring all metrics are consistent and correct at the point of delivery.

Snowflake Cost Monitoring

We assume two Snowflake accounts are available to implement the cost monitoring code. Chapter 8 explains how to create Snowflake trial accounts, and for the purposes of this chapter walk through, we are not concerned whether the accounts are in the same CSP region or federated across different CSPs and regions. What is important is that you have two accounts available, and to guide the walk through for this chapter, we have prefixed some headings with either "Spoke" or "CMS" to indicate where code should be executed.

Cost Drivers

Within this section, you'll examine the Snowflake components that drive costs. They are split into three key themes and each theme is examined individually next.

Compute Costs

Compute costs are the variable, on-demand costs incurred when you consume Snowflake resources and where performance tuning has the most impact.

All compute costs are Snowflake warehouse-related, either by the explicit declaration and use of a named warehouse or by serverless compute, which we discuss in detail below.

We categorize the following as compute costs:

- **Virtual warehouses:** Compute costs directly attributable to a named warehouse by a local user

- **Pipes:** Snowpipe serverless compute data ingestion

- **Materialized views:** Serverless compute relating to micro-partition maintenance activities

- **Table clustering**: Serverless compute relating to micro-partition maintenance activities

- **Search optimization**: Serverless compute consumed by the query optimizer for query predicates

- **Reader account usage**: Compute costs directly attributable to a named warehouse by a reader account

As you will see, the compute costs are not expressed by a single Account Usage Store view but are instead spread across several views.

Storage Costs

Storage costs are the physical cost of holding data within Snowflake and are a pass-through cost from the underlying CSP.

We categorize the following as storage costs:

- **Database tables**: Physical CSP micro-partition storage assigned to an individual table

- **Materialized views**: Physical CSP micro-partition storage assigned to an individual materialized view

- **Internal stages**: Physical CSP micro-partition storage assigned to a named stage, table stage, or user stage

Figure 4-2 illustrates the different types of stages.

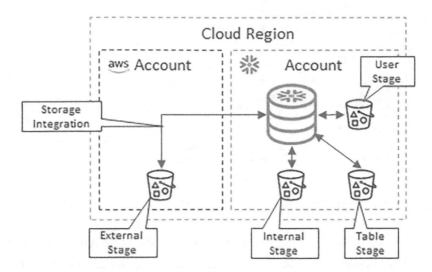

Figure 4-2. *Different types of stages*

Data Transfer Costs

Not all external stage costs are covered within cost metrics exposed by Snowflake. Data egress costs from external stages to other CSPs lie within the CSP cost reporting and not Snowflake.

We categorize the following as data transfer costs:

- **External stages**: Data unload costs to CSP storage

- **Data replication**: Intra-CSP data transfer costs for replication

Core Components

In order to build out a cost monitoring solution, let's identify the core components and discuss their purpose. Figure 4-3 illustrates a single spoke account sourcing statistics from the Snowflake Account Usage Store and persisting data sets into a cost reporting database. The cost reporting database is replicated to the single CMS hub account where consumption metrics are consolidated into the common hub repository.

You should expect your CMS to generate costing information too, which must also be both captured and exposed. Your approach must address consumption metrics delivered from one or more external spoke accounts, along with the internally generated CMS consumption metrics.

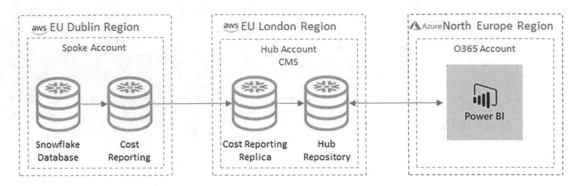

Figure 4-3. *Centralized monitoring store detail*

The provisioning of PowerBI reporting for your FinOps colleagues is beyond the scope of this book; however, we do demonstrate how to expose common metrics into views for reporting. Our example illustrates the use of a standard reporting tool, in this case PowerBI, to interactively "dice and slice" the hub data according to your FinOps colleagues' requirements.

A hub-and-spoke model provides a pattern-based approach to implementing cost monitoring. Both Secure Direct Data Sharing (SDDS) and database replication provide sufficient capability and are available out of the box.

We also acknowledge that Private Listings will achieve the same objective but note that they require the use of Snowsight to implement, and Private Listings cannot be delivered via scripting, which is a key requirement for LZ automated capability provisioning. We therefore exclude Private Listings as a suitable candidate for further discussion within this chapter.

Developing two differing integration patterns for SDDS and database replication is wasteful of both time and resources, and it leads to the maintenance of two code bases. We therefore focus our efforts on developing a single pattern for database replication, as this single pattern will satisfy both single CSP region and cross-region CSP Snowflake implementations. Some components developed for database replication will be reusable should you wish to implement SDDS instead, an exercise we leave for your investigation.

Identifying Consumption Metrics

In this section you investigate where consumption metrics are stored within each Snowflake account and how you can access them. You use a single source object to walk through this approach. Later you'll see a list of all other sources of metrics.

You start your investigation with the spoke account, which in our case is called PC52900. Yours will differ.

```
USE ROLE accountadmin;

SELECT current_account();
```

Metrics can be sourced from three locations:

- Every Snowflake account has a single imported database, centrally populated by Snowflake and holding 365 days of history.

- Every Snowflake database has an `information_schema` holding 14 days of history.

- Externally provisioned metrics, which we touch upon later.

We refer to the imported database as the Snowflake Account Usage Store and you will use it to source your cost monitoring data, noting the latency varies according to the view accessed. Further details can be found here: https://docs.snowflake.com/en/sql-reference/account-usage.html#account-usage.

Figure 4-4 shows how to access the Snowflake Account Usage Store using the Snowsight user interface (UI), noting a subset of views are shown. There are many more, but not all are useful for cost monitoring.

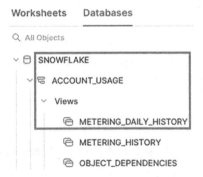

Figure 4-4. *Accessing the Snowflake Account Usage Store views*

Using the spoke account (PC52900 for us), choose `metering_daily_history` to identify suitable metrics that contain data for the past year, noting a latency of up to 3 hours.

This view provides consolidated metrics across a suite of Snowflake-supplied serverless compute services as listed here:

- Automatic clustering

- Materialized view refresh

- Snowpipe

- Query acceleration

- Replication

- Search optimization

- Serverless task

- Warehouse metering

- Reader accounts

A full explanation of `metering_daily_history` attributes is available here: `https://docs.snowflake.com/en/sql-reference/account-usage/metering_daily_history.html#metering-daily-history-view`.

Before executing any query, you must declare the warehouse to use. In this example, we use `compute_wh`; substitute any other declared warehouse available:

```
USE WAREHOUSE compute_wh;
```

Now add attributes to each extract in order to uniquely identify the data source of each dataset. These attributes should be added to every wrapper view to enrich the data.

- `current_region()`

- `current_account()`

- `current_timestamp()` aliased to `extract_timestamp`

To determine daily credits billed:

```
SELECT current_region()  AS current_region,
       current_account() AS current_account,
       service_type,
       usage_date,
       credits_billed,
```

```
        current_timestamp() AS extract_timestamp
FROM    snowflake.account_usage.metering_daily_history
ORDER BY service_type, usage_date DESC;
```

If no results are returned, this is due to serverless compute capability not being automatically invoked.

You may wish to return to this chapter after completing Chapter 12 but you must ensure replication is disabled first.

Noting the spoke account was used for testing replication, your results may be similar to those shown in Figure 4-5.

CURRENT_REGION	CURRENT_ACCOUNT	SERVICE_TYPE	USAGE_DATE	CREDITS_BILLED	EXTRACT_TIMESTAMP
AWS_EU_WEST_1	PC52900	REPLICATION	2023-02-06	0.000288699	2023-02-14 02:42:35.049 -0800
AWS_EU_WEST_1	PC52900	WAREHOUSE_METERING	2023-02-13	0.0175	2023-02-14 02:42:35.049 -0800
AWS_EU_WEST_1	PC52900	WAREHOUSE_METERING	2023-02-12	0.723055555	2023-02-14 02:42:35.049 -0800
AWS_EU_WEST_1	PC52900	WAREHOUSE_METERING	2023-02-10	0.000032222	2023-02-14 02:42:35.049 -0800
AWS_EU_WEST_1	PC52900	WAREHOUSE_METERING	2023-02-09	0.000149166	2023-02-14 02:42:35.049 -0800
AWS_EU_WEST_1	PC52900	WAREHOUSE_METERING	2023-02-07	0.729722222	2023-02-14 02:42:35.049 -0800
AWS_EU_WEST_1	PC52900	WAREHOUSE_METERING	2023-02-06	0.037222222	2023-02-14 02:42:35.049 -0800

Figure 4-5. *Example metering daily history results*

As you see from Figure 4-5, within the spoke account credits have been consumed for both replication and warehouse metering serverless compute consumption.

We encourage you to experiment with other Account Usage Store views and become familiar with capturing consumption metrics. We provide a list of views for your further consideration within the "Maintenance" section below.

Capturing Consumption Metrics

Having identified example metrics, let's now consider how to generate and store cost metrics whilst ensuring latency is taken into consideration. You must be consistent in your approach when generating metrics of any sort, and in this example, you know from Snowflake documentation that the maximum latency period for Account Usage Store views is 3 hours, so let's use this period below.

Spoke – Creating a Consumption Metrics Repository

You must establish a new database named cost_reporting and then create a schema named cost_owner. Finally, create a role named cost_owner_role to create objects that contain your consumption metrics.

```
USE ROLE sysadmin;

CREATE OR REPLACE DATABASE cost_reporting
DATA_RETENTION_TIME_IN_DAYS = 90;

CREATE OR REPLACE SCHEMA cost_reporting.cost_owner;

CREATE OR REPLACE WAREHOUSE cost_wh WITH
WAREHOUSE_SIZE       = 'X-SMALL'
AUTO_SUSPEND         = 60
AUTO_RESUME          = TRUE
MIN_CLUSTER_COUNT    = 1
MAX_CLUSTER_COUNT    = 4
SCALING_POLICY       = 'STANDARD'
INITIALLY_SUSPENDED = TRUE;

USE ROLE securityadmin;

CREATE OR REPLACE ROLE cost_owner_role;

GRANT USAGE ON DATABASE cost_reporting            TO ROLE cost_owner_role;
GRANT USAGE ON SCHEMA    cost_reporting.cost_owner TO ROLE cost_owner_role;

GRANT USAGE             ON SCHEMA cost_reporting.cost_owner TO ROLE cost_
owner_role;
GRANT MONITOR           ON SCHEMA cost_reporting.cost_owner TO ROLE cost_
owner_role;
GRANT MODIFY            ON SCHEMA cost_reporting.cost_owner TO ROLE cost_
owner_role;
GRANT CREATE TABLE      ON SCHEMA cost_reporting.cost_owner TO ROLE cost_
owner_role;
GRANT CREATE VIEW       ON SCHEMA cost_reporting.cost_owner TO ROLE cost_
owner_role;
```

```
GRANT CREATE SEQUENCE    ON SCHEMA cost_reporting.cost_owner TO ROLE cost_
owner_role;
GRANT CREATE FUNCTION    ON SCHEMA cost_reporting.cost_owner TO ROLE cost_
owner_role;
GRANT CREATE PROCEDURE   ON SCHEMA cost_reporting.cost_owner TO ROLE cost_
owner_role;
GRANT CREATE STREAM      ON SCHEMA cost_reporting.cost_owner TO ROLE cost_
owner_role;
GRANT CREATE TAG         ON SCHEMA cost_reporting.cost_owner TO ROLE cost_
owner_role;

GRANT USAGE   ON WAREHOUSE cost_wh TO ROLE cost_owner_role;
GRANT OPERATE ON WAREHOUSE cost_wh TO ROLE cost_owner_role;
```

Enable access to the Account Usage Store:

```
GRANT IMPORTED PRIVILEGES ON DATABASE snowflake TO ROLE cost_owner_role;
```

Grant the cost owner role to ACCOUNTADMIN:

```
GRANT ROLE cost_owner_role TO ROLE accountadmin;
```

Grant the cost owner role to yourself:

```
GRANT ROLE cost_owner_role TO USER <Your User Here>;
```

Spoke - Creating Consumption Metrics Objects

Within your spoke account, Figure 4-6 illustrates the database and objects to be created.

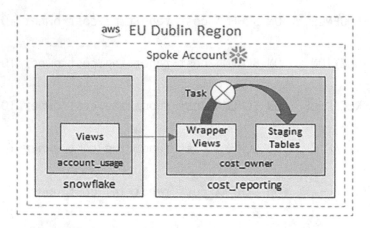

Figure 4-6. *Spoke data capture and persistence*

Creating Source Views

Using the Account Usage Store SQL query for metering_daily_history identified above, create a view named v_metering_daily_history to "wrap" the underlying view. Wrapping views allows you to implement bespoke logic. We now illustrate the addition of attributes for region, account, and the time data was extracted.

Create your wrapper view in the cost_reporting.cost_owner schema and first set your context:

```
USE ROLE      cost_owner_role;
USE WAREHOUSE cost_wh;
USE DATABASE  cost_reporting;
USE SCHEMA    cost_reporting.cost_owner;
```

Then create your wrapper view:

```
CREATE OR REPLACE VIEW v_metering_daily_history COPY GRANTS
AS
SELECT current_region()  AS current_region,
       current_account() AS current_account,
       service_type,
       usage_date,
       credits_billed,
       current_timestamp() AS extract_timestamp
FROM   snowflake.account_usage.metering_daily_history;
```

And check to see the expected results, which should match those from Figure 4-5:

```
SELECT current_region,
       current_account,
       service_type,
       usage_date,
       credits_billed,
       extract_timestamp
FROM   v_metering_daily_history;
```

Creating a Staging Table

Now create a staging table named stg_metering_daily_history to persist your metrics:

```
CREATE OR REPLACE TABLE stg_metering_daily_history
(
current_region      VARCHAR ( 30 ) NOT NULL,
current_account     VARCHAR ( 30 ) NOT NULL,
service_type        VARCHAR ( 30 ) NOT NULL,
usage_date          DATE,
credits_billed      NUMBER ( 38, 10 ),
extract_timestamp   TIMESTAMP_LTZ
);
```

Enable change tracking on stg_metering_daily_history because you will later create a stream on the same table within the replicated database:

```
ALTER TABLE stg_metering_daily_history
SET change_tracking = TRUE;
```

Let's assume only the previous day's data is required, and to implement this approach, you will "flush and fill" your staging table. Using the OVERWRITE keyword ensures your staging table is first truncated and then loaded with fresh data from the INSERT statement, noting the data will be for the previous day:

```
INSERT OVERWRITE INTO cost_reporting.cost_owner.stg_metering_daily_history
SELECT current_region,
       current_account,
       service_type,
```

```
        usage_date,
        credits_billed,
        extract_timestamp
FROM    cost_reporting.cost_owner.v_metering_daily_history
WHERE   usage_date = DATEADD ( 'day', -1,
                        TO_DATE (
                            DATE_PART ( 'year', extract_timestamp )||'-'||
                            DATE_PART ( 'mm',   extract_timestamp )||'-'||
                            DATE_PART ( 'day',  extract_timestamp )));
```

Unfortunately, Snowflake does not currently support Julian dates, which would simplify the SQL.

Now prove your INSERT statement has loaded data into your staging table:

```
SELECT *
FROM    cost_reporting.cost_owner.stg_metering_daily_history;
```

Creating a Task

You have two options to automate the generation of daily statistics. Your first option is to create a task for each source object and call each INSERT statement individually. Alternatively, when you have several source views, you may prefer to create a stored procedure to encapsulate all similar SQL statements and execute them serially.

Switch to the ACCOUNTADMIN role:

```
USE ROLE accountadmin;
```

Set the schema for the task to reside in:

```
USE DATABASE   cost_reporting;
USE SCHEMA     cost_reporting.cost_owner;
```

Prove the ACCOUNTADMIN role can insert data into stg_metering_daily_history:

```
INSERT OVERWRITE INTO cost_reporting.cost_owner.stg_metering_daily_history
SELECT current_region,
       current_account,
       service_type,
       usage_date,
```

```
        credits_billed,
        extract_timestamp
FROM    cost_reporting.cost_owner.v_metering_daily_history
WHERE   usage_date = DATEADD ( 'day', -1,
                        TO_DATE (
                            DATE_PART ( 'year', extract_timestamp )||'-'||
                            DATE_PART ( 'mm',   extract_timestamp )||'-'||
                            DATE_PART ( 'day',  extract_timestamp )));
```

Create a new task named task_metering_daily_history, noting the use of serverless compute for this task.

For testing purposes, schedule the task to run every 5 minutes:

```
CREATE OR REPLACE TASK cost_reporting.cost_owner.task_metering_
daily_history
USER_TASK_MANAGED_INITIAL_WAREHOUSE_SIZE = XSMALL
SCHEDULE = '5 MINUTE'
AS
INSERT OVERWRITE INTO cost_reporting.cost_owner.stg_metering_daily_history
SELECT current_region,
       current_account,
       service_type,
       usage_date,
       credits_billed,
       extract_timestamp
FROM    cost_reporting.cost_owner.v_metering_daily_history
WHERE   usage_date = DATEADD ( 'day', -1,
                        TO_DATE (
                            DATE_PART ( 'year', extract_timestamp )||'-'||
                            DATE_PART ( 'mm',   extract_timestamp )||'-'||
                            DATE_PART ( 'day',  extract_timestamp )));
```

Now resume the task:

```
ALTER TASK cost_reporting.cost_owner.task_metering_daily_history RESUME;
```

And prove the task is started:

```
SHOW tasks;
```

The `state` attribute should be `started`.

You can also check the `information_schema.task_history` table, which is easier to read:

```
SELECT timestampdiff ( second, current_timestamp, scheduled_time ) as
next_run,
        scheduled_time,
        current_timestamp,
        name,
        state
FROM    TABLE ( information_schema.task_history())
ORDER BY completed_time DESC;
```

Figure 4-7 shows the expected response, noting NEXT_RUN is 280 seconds in the future where the STATE is SCHEDULED and historical runs are shown as the number of seconds since the last run; in this example, 20 seconds have elapsed.

NEXT_RUN	SCHEDULED_TIME	CURRENT_TIMESTAMP	NAME	STATE
280	2023-02-17 13:37:07.079 -0800	2023-02-17 13:32:27.246 -0800	TASK_METERING_DAILY_HISTORY	SCHEDULED
-20	2023-02-17 13:32:07.059 -0800	2023-02-17 13:32:27.246 -0800	TASK_METERING_DAILY_HISTORY	SUCCEEDED

Figure 4-7. Task schedule confirmation

When STATE is FAILED, you should investigate the reason for the failure. This explains why you rechecked if your SQL statement would execute from the ACCOUNTADMIN role.

When your task executes, results are stored in the `cost_reporting.cost_owner. stg_metering_daily_history` table, noting the `extract_timestamp` attribute indicates when the data was created. Ensure the timestamp correlates to the task called SCHEDULED_TIME:

```
SELECT *
FROM    cost_reporting.cost_owner.stg_metering_daily_history;
```

Rescheduling a Task

When you are satisfied your task is working as expected, you can change the SCHEDULE to 05:00 GMT daily to allow for Account Usage Store view latency of up to 3 hours to expire. Setting the SCHEDULE to 05:00 GMT also takes British Summer Time (BST) time changes into consideration; refactor for your own timezone.

First, you must SUSPEND your task:

```
ALTER TASK cost_reporting.cost_owner.task_metering_daily_history SUSPEND;
```

With your task suspended, you can reschedule:

```
ALTER TASK IF EXISTS cost_reporting.cost_owner.task_metering_daily_
history SET
SCHEDULE = 'USING CRON 0 5 * * * GMT';
```

Then resume your task:

```
ALTER TASK cost_reporting.cost_owner.task_metering_daily_history RESUME;
```

Recheck your next task submission time:

```
SELECT timestampdiff ( second, current_timestamp, scheduled_time ) as
next_run,
        scheduled_time,
        current_timestamp,
        name,
        state
FROM    TABLE ( information_schema.task_history())
ORDER BY completed_time DESC;
```

You should see that your SCHEDULED_TIME matches your expectation.

Suspending a Task

When your testing is complete, you should SUSPEND your task to prevent excessive resource consumption:

```
ALTER TASK cost_reporting.cost_owner.task_metering_daily_history SUSPEND;
```

Propagating Consumption Metrics

With your daily snapshot established, you must propagate your spoke account data to the CMS using database replication. The center of Figure 4-8 shows two spoke databases feeding data into a CMS. Each spoke is assumed to capture cost reporting metrics.

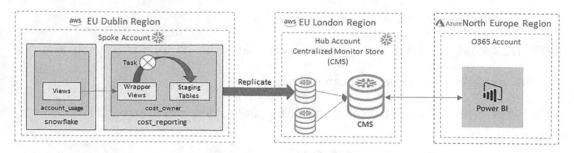

Figure 4-8. *Spoke metrics propagation to CMS*

Spoke – Replicating a Database to CMS

Consolidating spoke metrics into a single source is dependent upon the ingestion of replicated databases into the CMS. You must therefore ensure all spoke accounts are enabled for database replication. We leave this for your further investigation, noting we provide a full walkthrough of organizations in Chapter 8 and database replication is explained in Chapter 12.

You must also know your CMS account, which in this example is NP62160, noting the organization prefix is NUYMCLU.

By default, `ACCOUNTADMIN` is the only role entitled to implement database replication:

```
USE ROLE accountadmin;
```

Check that your database is available:

```
SHOW DATABASES;
```

Then enable the `cost_reporting` database replication to your CMS account:

```
ALTER DATABASE cost_reporting ENABLE REPLICATION
TO ACCOUNTS NUYMCLU.NP62160;
```

You should see a response stating "Statement executed successfully."
Replication provides read-only access for the replicated database.
Confirm that the database `cost_reporting` is enabled for replication:

```
SHOW REPLICATION DATABASES;
```

You should see a response similar to Figure 4-9.

snowflake_region	account_name	name	primary	replication_allowed_to_accounts
AWS_EU_WEST_1	REPLICATION_TE	COST_REPO	NUYMCLU.REPLICATION_TERTIARY.CO	NUYMCLU.NP62160, NUYMCLU.REP

Figure 4-9. *Exported replication database*

CMS – Ingesting a Replicated Database

You can now ingest the database exported from your spoke account. Create a database called NUYMCLU_REPLICATION_TERTIARY_COST_REPORTING as exported from the primary account database. Our naming convention embeds the organization and account identifiers into the imported database name but yours may differ.

```
USE ROLE accountadmin;

CREATE DATABASE nuymclu_replication_tertiary_cost_reporting
AS REPLICA OF nuymclu.replication_tertiary.cost_reporting
DATA_RETENTION_TIME_IN_DAYS = 90;
```

You should see a response stating "Database NUYMCLU_REPLICATION_TERTIARY_COST_REPORTING" successfully created."

You must refresh your imported database:

```
ALTER DATABASE nuymclu_replication_tertiary_cost_reporting REFRESH;
```

Your newly imported database will appear in your list of databases, as shown in Figure 4-10.

Figure 4-10. *Imported replication database*

Prove that your imported database contains the same data as its source:

```
SELECT *
FROM    nuymclu_replication_tertiary_cost_reporting.cost_owner.stg_metering_
daily_history;
```

Having successfully imported a spoke database, you must provision the CMS to ingest the contents for centralized reporting.

CMS – Building a Repository

You must establish a new database named cms and then create a schema named cms_ owner. Finally, create a role named cms_owner_role to create objects that contain your consumption metrics.

```
USE ROLE sysadmin;

CREATE OR REPLACE DATABASE cms
DATA_RETENTION_TIME_IN_DAYS = 90;

CREATE OR REPLACE SCHEMA cms.cms_owner;

CREATE OR REPLACE WAREHOUSE cms_wh WITH
WAREHOUSE_SIZE      = 'X-SMALL'
AUTO_SUSPEND        = 60
AUTO_RESUME         = TRUE
MIN_CLUSTER_COUNT   = 1
MAX_CLUSTER_COUNT   = 4
SCALING_POLICY      = 'STANDARD'
INITIALLY_SUSPENDED = TRUE;

USE ROLE securityadmin;

CREATE OR REPLACE ROLE cms_owner_role;

GRANT USAGE ON DATABASE cms            TO ROLE cms_owner_role;
GRANT USAGE ON SCHEMA    cms.cms_owner TO ROLE cms_owner_role;

GRANT USAGE             ON SCHEMA cms.cms_owner TO ROLE cms_owner_role;
GRANT MONITOR           ON SCHEMA cms.cms_owner TO ROLE cms_owner_role;
GRANT MODIFY            ON SCHEMA cms.cms_owner TO ROLE cms_owner_role;
```

```
GRANT CREATE TABLE       ON SCHEMA cms.cms_owner TO ROLE cms_owner_role;
GRANT CREATE VIEW        ON SCHEMA cms.cms_owner TO ROLE cms_owner_role;
GRANT CREATE SEQUENCE    ON SCHEMA cms.cms_owner TO ROLE cms_owner_role;
GRANT CREATE FUNCTION    ON SCHEMA cms.cms_owner TO ROLE cms_owner_role;
GRANT CREATE PROCEDURE   ON SCHEMA cms.cms_owner TO ROLE cms_owner_role;
GRANT CREATE STREAM      ON SCHEMA cms.cms_owner TO ROLE cms_owner_role;
GRANT CREATE TAG         ON SCHEMA cms.cms_owner TO ROLE cms_owner_role;

GRANT USAGE   ON WAREHOUSE cms_wh TO ROLE cms_owner_role;
GRANT OPERATE ON WAREHOUSE cms_wh TO ROLE cms_owner_role;
```

Enable access to Account Usage Store:

```
GRANT IMPORTED PRIVILEGES ON DATABASE snowflake TO ROLE cms_owner_role;
```

Grant the role of cost owner to ACCOUNTADMIN:

```
GRANT ROLE cms_owner_role TO ROLE accountadmin;
```

Grant the role of cost owner to yourself:

```
GRANT ROLE cms_owner_role TO USER <Your User Here>;
```

CMS – Refreshing a Replicated Database

You first set your context to use your newly created database, cms:

```
USE ROLE      accountadmin;
USE WAREHOUSE cms_wh;
USE DATABASE  cms;
USE SCHEMA    cms.cms_owner;
```

Then you automate your imported database refresh at 08:00 GMT daily, allowing plenty of time for the spoke database to have loaded the previous day's costs:

```
CREATE TASK nuymclu_replication_tertiary_cost_reporting_task
WAREHOUSE            = compute_wh
SCHEDULE             = 'USING CRON 0 8 * * * GMT'
USER_TASK_TIMEOUT_MS = 14400000
AS
ALTER DATABASE nuymclu_replication_tertiary_cost_reporting REFRESH;
```

Enable the task:

```
ALTER TASK nuymclu_replication_tertiary_cost_reporting_task RESUME;
```

Prove the task is scheduled:

```
SELECT timestampdiff ( second, current_timestamp, scheduled_time ) as
next_run,
       scheduled_time,
       current_timestamp,
       name,
       state
FROM    TABLE ( information_schema.task_history())
ORDER BY completed_time DESC;
```

You should see the STATE attribute is set to SCHEDULED.

A more comprehensive, and readily extensible, scheduling reference implementation is provided in Chapter 12. We leave this for your further investigation.

You should also examine the status information available for the last database refresh:

```
SELECT phase_name,
       result,
       start_time,
       end_time,
       details
FROM TABLE ( information_schema.database_refresh_progress ( nuymclu_
replication_tertiary_cost_reporting ));
```

You should see all phase RESULTs have Succeeded.

CMS – Accessing an Ingested Database

After you ingest your spoke database, you prove the data matches the spoke by executing this query:

```
SELECT *
FROM    nuymclu_replication_tertiary_cost_reporting.cost_owner.stg_metering_
daily_history;
```

Imported databases are owned by ACCOUNTADMIN and no user-defined roles with entitlement are granted, so let's implement grant entitlement to use your new database, nuymclu_replication_tertiary_cost_reporting:

```
USE ROLE securityadmin;
```

```
GRANT USAGE ON DATABASE nuymclu_replication_tertiary_cost_
reporting               TO ROLE cms_owner_role;
GRANT USAGE ON SCHEMA   nuymclu_replication_tertiary_cost_reporting.cost_
owner TO ROLE cms_owner_role;
```

Repeat this grant for every object within your imported spoke database:

```
GRANT SELECT ON nuymclu_replication_tertiary_cost_reporting.cost_owner.
stg_metering_daily_history
TO ROLE cms_owner_role;
```

Switch roles to cms_owner_role and prove you can access your staging table stg_metering_daily_history:

```
USE ROLE cms_owner_role;
```

```
SELECT *
FROM   nuymclu_replication_tertiary_cost_reporting.cost_owner.stg_metering_
daily_history;
```

Your results should match those seen above for the same query.

CMS – Creating Consumption Metrics Objects

Figure 4-11 shows the objects you have created and the new objects to be created within this section where you focus your attention on the CMS.

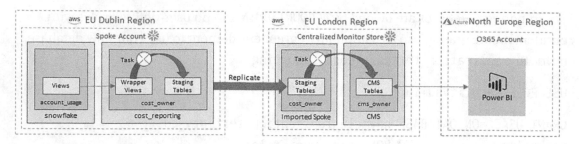

Figure 4-11. *Imported replication database*

You now create the CMS objects to store all data delivered by each spoke. Start by creating a table to hold all `metering_daily_history` data. To differentiate table names, prefix all tables with `cms_`.

```
CREATE OR REPLACE TABLE cms_metering_daily_history
(
spoke_name           VARCHAR ( 255 ) NOT NULL,
current_region       VARCHAR ( 30  ) NOT NULL,
current_account      VARCHAR ( 30  ) NOT NULL,
service_type         VARCHAR ( 30  ) NOT NULL,
usage_date           DATE,
credits_billed       NUMBER ( 38, 10 ),
extract_timestamp    TIMESTAMP_LTZ  NOT NULL,
ingest_timestamp     TIMESTAMP_LTZ  DEFAULT current_timestamp() NOT NULL
);
```

In order to seamlessly move data into the CMS table, create a stream and a task on the spoke table:

```
CREATE STREAM strm_nuymclu_replication_tertiary_cost_reporting_metering_
daily_history
ON TABLE nuymclu_replication_tertiary_cost_reporting.cost_owner.stg_
metering_daily_history;
```

If change tracking has not been enabled for a replicated object, then do so in the spoke database and refresh the replicated database within CMS, before attempting to create the stream.

With your stream created you now create a task that is triggered by the presence of data within the source table. Note the use of the `ACCOUNTADMIN` role.

```
USE ROLE        accountadmin;
USE WAREHOUSE cms_wh;
USE DATABASE    cms;
USE SCHEMA      cms.cms_owner;
```

Set the SCHEDULE according to your preference:

```
CREATE OR REPLACE TASK task_load_nuymclu_replication_tertiary_cost_
reporting_metering_daily_history
WAREHOUSE = cms_wh
SCHEDULE  = '5 minute'
WHEN system$stream_has_data ( 'strm_nuymclu_replication_tertiary_cost_
reporting_metering_daily_history' )
AS
INSERT INTO cms_metering_daily_history
SELECT 'nuymclu_replication_tertiary_cost_reporting_metering_daily_
history',
        current_region,
        current_account,
        service_type,
        usage_date,
        credits_billed,
        extract_timestamp,
        current_timestamp()
FROM    strm_nuymclu_replication_tertiary_cost_reporting_metering_daily_
history;
```

Enable the task:

```
ALTER TASK task_load_nuymclu_replication_tertiary_cost_reporting_metering_
daily_history RESUME;
```

Prove the task is scheduled:

```
SELECT timestampdiff ( second, current_timestamp, scheduled_time ) as
next_run,
        scheduled_time,
        current_timestamp,
```

```
        name,
        state
FROM    TABLE ( information_schema.task_history())
ORDER BY completed_time DESC;
```

A returned STATE of SKIPPED indicates either a previous run has not completed before the next valid run time starts or there is no data within the stream to process.

You must test your automated ingestion. The steps required are the following:

- Insert data into the spoke table stg_metering_daily_history, either manually or as created by the corresponding task.

- Refresh the spoke database imported into CMS, either manually or as refreshed by the corresponding task.

- Confirm the automated ingestion stream and task load data into table cms_metering_daily_history.

CMS – Summary of Consumption Metrics

You can prove data has been transferred from spoke to CMS by running

```
USE ROLE cms_owner_role;
```

```
SELECT * FROM cms_metering_daily_history;
```

You should see a response similar Figure 4-12 (noting timestamps have been omitted).

SPOKE_NAME	CURRENT_REGION	CURRENT_ACC	SERVICE_TYPE	USAGE_DATE	CREDITS_BILLED
nuymclu_replication	AWS_EU_WEST_1	PC52900	WAREHOUSE_METERIN	2023-02-17	0.48111111
nuymclu_replication	AWS_EU_WEST_1	PC52900	SERVERLESS_TASK	2023-02-17	0.003749273
nuymclu_replication	AWS_EU_WEST_1	PC52900	WAREHOUSE_METERIN	2023-02-16	0.405
nuymclu_replication	AWS_EU_WEST_1	PC52900	SERVERLESS_TASK	2023-02-16	0.000042778

Figure 4-12. *Sample cms_metering_daily_history content*

The baseline data in table cms_metering_daily_history can be abstracted to deliver monthly summary consumption metrics:

```
CREATE OR REPLACE VIEW v_cms_monthly_metering_daily_history
AS
SELECT spoke_name,
        current_region()  AS current_region,
        current_account() AS current_account,
        service_type,
        DECODE ( EXTRACT ( 'month', usage_date ),
                        1,  'January',
                        2,  'February',
                        3,  'March',
                        4,  'April',
                        5,  'May',
                        6,  'June',
                        7,  'July',
                        8,  'August',
                        9,  'September',
                        10, 'October',
                        11, 'November',
                        12, 'December')  AS month_of_year,
        TO_CHAR ( DATE_PART ( 'year', usage_date ))   AS billing_year,
        SUM   ( credits_billed ) AS sum_credits_billed,
        current_timestamp()        AS extract_timestamp
FROM    cms_metering_daily_history
GROUP BY spoke_name,
         service_type,
         DATE_PART ( 'Month', usage_date ),
         DATE_PART ( 'year',  usage_date );
```

Check if you see the expected results, which should summarize those from Figure 4-12:

```
SELECT * FROM v_cms_monthly_metering_daily_history;
```

Figure 4-13 shows the monthly summarized sample output.

SPOKE_NAME	CURRENT_REGION	CURRENT_/	SERVICE_TYPE	MONTH_OF_YE	BILLING_YE/	SUM_CREDITS_BILLED
nuymclu_replicatic	AWS_EU_WEST_2	XL29287	WAREHOUSE_METERIN	February	2023	0.88611111
nuymclu_replicatic	AWS_EU_WEST_2	XL29287	SERVERLESS_TASK	February	2023	0.003792051

Figure 4-13. *Summarized sample v_cms_metering_daily_history content*

Reporting Consumption Metrics

Exposing cost consumption metrics to your reporting tools is now a simple exercise. Using PowerBI as the example reporting tool, the steps are the following:

- Create a user and role for the PowerBI gateway connection.

- Grant SELECT on the required CMS reporting objects to the PowerBI role.

- Assign a warehouse to the PowerBI role.

- Provision a PowerBI workspace and associate the gateway connection with the workspace.

- Develop PowerBI reports and configure PowerBI users to access the reports.

External Cost Monitoring

Delivering a centralized cost monitoring presents the opportunity to capture costs incurred outside of Snowflake. By extending the centralized cost monitoring capability, it is possible to capture CSP costs, ingest into Snowflake, collate across data sets, then report centrally.

Figure 4-14 depicts costs incurred by using AWS capabilities. In this example, using S3 buckets and lambdas incurs costs, which along with other AWS tooling, results in consolidated billing within the AWS Billing Conductor.

You may wish to consolidate all AWS billing from the AWS Billing Conductor into each hub Snowflake and then reuse the existing data sharing mechanism to bring all AWS and Snowflake billing into your CMS.

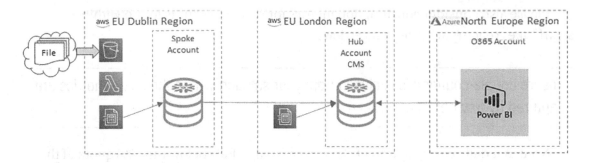

Figure 4-14. *External cost capture*

In addition to AWS, the same pattern can be implemented for both Azure and GCP. You may extend data capture to other data platforms as your need evolves.

Maintenance

Snowflake is constantly evolving with new attributes being added to views and new views appearing often either unnoticed or unannounced. We recommend a monthly review of all Snowflake Account Usage Store views where cost reporting is available.

At the time of writing, this is the list of Account Usage Store views provision cost reporting metrics:

- `automatic_clustering_history`
- `database_storage_usage_history`
- `data_transfer_history`
- `materialized_view_refresh_history`
- `metering_daily_history`
- `pipe_usage_history`
- `replication_usage_history`
- `search_optimization_history`
- `stage_storage_usage_history`
- `warehouse_metering_history`

Please also consider consumption metrics for reader accounts found within
`snowflake.reader_account_usage` views.

Ensure double counting is excluded from your solution where the same metrics are
captured by tow objects.

Note that daily consumption of cloud services that falls below the 10% quota of the
daily usage of the compute resources accumulates no cloud services charges. Further
details can be found here: `https://docs.snowflake.com/en/user-guide/cost-`
`understanding-compute#label-cloud-services-credit-usage`.

Summary

The scope of this chapter was to identify how to monitor, control, and manage your
Snowflake costs. You saw where consumed costs are made available by Snowflake.

You then worked through an in-depth, hands-on practical implementation of cost
reporting from the generation of consumption metrics and then consolidation into a
single repository.

Later, you learned how to report against your cost monitoring service using available
out-of-the-box components. You briefly explored how to extend your solution to
consume externally generated metrics for consolidated reporting.

Lastly, you learned about maintenance of your cost reporting estate, which should
form part of your business-as-usual activities.

You now move onto a subject of interest to your Marketing and Sales colleagues:
share utilization.

CHAPTER 5

Share Utilization

As Snowflake data sharing becomes more commonplace, your business colleagues will define increasingly more complex reporting requirements to identify not only the data sets consumed, but also how the data sets are consumed.

Snowflake documentation found at `https://docs.snowflake.com/en/user-guide/data-sharing-intro` shows these object types are sharable:

- Tables

- External tables

- Secure views

- Secure materialized views

- Secure UDFs

In this chapter, we discuss the "art of the possible" by identifying and accessing the Snowflake-supplied share monitoring data sets, along with other relevant and available information. Snowflake is continually enhancing the available data sets and we illustrate one such example later where Snowflake monitoring capability has been affected.

Let's now consider the purpose of Secure Direct Data Share, Data Exchange, and Data Marketplace. Figure 5-1 showcases each option compared to each other providing context for this chapter.

© Andrew Carruthers and Sahir Ahmed 2023
A. Carruthers and S. Ahmed, *Maturing the Snowflake Data Cloud*, https://doi.org/10.1007/978-1-4842-9340-9_5

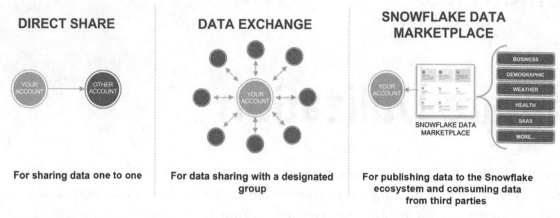

Figure 5-1. *Snowflake data interchange options*

From a data consumption perspective, the authors acknowledge Snowflake has a balancing act to perform and a fine line to tread: Consumers do not wish to expose their bespoke algorithms consuming providers data, whereas for providers, having access to consuming algorithms provides business intelligence and value.

Putting aside security concerns as they are both expected in organizations large and small and therefore should be a given, a sensible compromise is in place, offering both producers and consumers some of the information they want without exposing the "secret sauce" or intellectual property of the consuming parties. And this is what we discuss here: the challenge of obtaining as much information as possible from several perspectives, along with new features presenting opportunity to monetize your data. For those of us who have been around Snowflake for a while, we realize the pace of change is accelerating and data sharing is no exception.

We also discuss data clean rooms (DCRs) but due to space constraints do not deep-dive into their implementation. Instead, we offer insights, advice, and reference materials for DCRs, leaving this for your further investigation.

Data Sharing Options

Before considering what you will monitor, you must consider where your monitoring will occur. A brief recap of the Snowflake options for sharing data follows.

All forms of data sharing utilize the same underlying mechanics explained here.

Data sharing is a huge topic and Snowflake is constantly enhancing its approach, so the information presented here is correct at the time of writing but will almost certainly have changed by the time this book is published. Further information can be found here: `https://docs.snowflake.com/en/user-guide/data-sharing-product-offerings.html`.

Same Region and Same CSP

When Snowflake accounts are colocated within the same region and CSP, Secure Direct Data Share (SDDS) may be used to deliver both data and functionality, as shown conceptually in Figure 5-2.

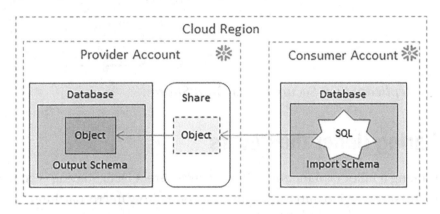

Figure 5-2. *Secure Direct Data Share conceptual model*

Different Region or Different CSP

When Snowflake accounts are located within the different regions or CSPs, you must first replicate before Secure Direct Data Share may be used to deliver data and functionality, as shown conceptually in Figure 5-3.

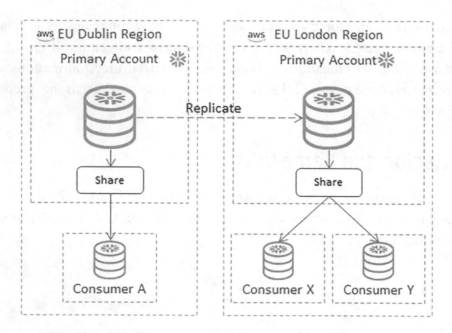

Figure 5-3. *Replicate then Secure Direct Data Share*

Data Sharing Under the Covers

For those seeking a fuller explanation of how Secure Direct Data Sharing is achieved from within the same region and CSP, Figure 5-4 shows how micro-partitions are shared from the Provider account to the Consuming account, noting only the current micro-partitions are available to the consuming share. The net effect of data sharing is that all changes to data made by the provider are immediately available to the consumer, yes, in real time!

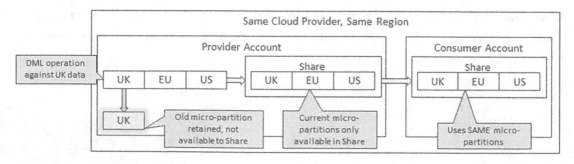

Figure 5-4. *Secure Direct Data Share micro-partition sharing*

While the underlying mechanics of data sharing apply to the differing mechanisms identified within this chapter, the metrics capture options differ across each mechanism.

Let's discuss how differing approaches affect the monitoring of shared data.

Reader Accounts

Reader accounts access the Provider account; therefore consumption and SQL statements lie within the Provider account. With this in mind, the Account Usage Store `query_history` view will contain all reader account SQL statements from which all inbound share utilization monitoring can be derived.

Note that there is a hard limit of 20 managed accounts. Please see the documentation found here: `https://docs.snowflake.com/en/sql-reference/sql/create-managed-account.html#usage-notes`.

Approach

Our approach within this chapter is to identify reporting capabilities for share utilization available within Snowflake. We assume everyone is familiar with the Snowflake-supplied user interface in both Classic UI and SnowSight form. Code within this chapter is compatible with SnowSQL. Please refer to Chapter 1 for installation instructions and to get started.

As previously discussed, Snowflake is moving the user experience to SnowSight, from which you may view Share utilization metrics. Snowflake is developing a web interface for shared data which at the time of writing is in public preview; details can be found here: `https://docs.snowflake.com/en/user-guide/data-sharing-web-interface.html#web-interface-for-shared-data`.

But SnowSight may not be your preferred method of producing and consuming share utilization metrics. Furthermore, your organization's infrastructure may not immediately allow access to SnowSight as it is accessible via a different URL.

You may also wish to collate metrics from a variety of Snowflake accounts and consolidate within a single reporting environment. Also, you may prefer to expose not only share utilization metrics but all available metrics discussed within this book using your preferred tooling. For this reason, we assume each organization requires all metrics to be exposed via a suite of components readily accessible for consumption. Adopting a more generic approach provides flexibility and can be readily tailored to meet your specific organization's needs.

We focus on providing a suite of SQL queries to expose all available consumption metrics, noting Snowflake is continually evolving their data sets and recommend frequent checks of the Data Sharing Usage Store to ensure all changes are captured. As you will read later within this chapter, not all information has been published by Snowflake. We have made every effort to include what is available, noting some information will have changed by the time you read this chapter.

As an example of how Snowflake monitoring has changed over time, please see Figure 5-5, which shows the available Data Sharing Usage views at the time of writing. We highlight LISTING_ACCESS_HISTORY, which was added more recently, and separately highlight four views only visible for Snowflake accounts that have been accepted into the Snowflake Marketplace.

Figure 5-5. *Data Sharing Usage evolution*

As an aside, we also point to the BCR_ROLLOUT schema and associated view relating to behavior change management for which documentation can be found here: https://docs.snowflake.com/en/user-guide/managing-behavior-change-releases.html#bcr-2022-07-ddm-rollout-view-reference. We discuss BCR_ROLLOUT in a later chapter.

This chapter reuses the Snowflake monitoring setup developed in Chapter 3. Please ensure these components have been deployed to your Snowflake account before attempting the code delivered within this chapter. A quick check is to run these commands and resolve any errors:

```
USE ROLE      monitor_owner_role;
USE WAREHOUSE monitor_wh;
USE DATABASE  monitor;
USE SCHEMA    monitor.monitor_owner;
```

Data Sharing Monitoring

As you have just seen, Snowflake data sharing implements peer-to-peer data delivery in real time but only for the current micro-partitions. To historize shared data, the consumer should implement their own mechanism to preserve data.

All data providers want to gather metrics on how their data is consumed, and as you progress through this chapter, you'll see template SQL to both identify and present consumption metrics.

Secure Direct Data Share

Figure 5-6 depicts Secure Direct Data Share showing the one-way approach from a producer sharing data to a single consumer.

DIRECT SHARE

For sharing data one to one

Figure 5-6. *Secure Direct Data Share*

Let's identify all active data shares within your account. For this, use the ACCOUNTADMIN role:

```
USE ROLE accountadmin;
```

Depending upon the role used to define shares, your results may differ. During testing of the code samples within this chapter, we found SYSADMIN did not identify shares created by the ACCOUNTADMIN role.

Accepting share creation is affected by the owning role. Attempt to identify all declared data shares:

```
SHOW shares;
```

Figure 5-7 illustrates partial results from a trial account where user-defined shares have not been configured and both INBOUND shares are provisioned at account creation time, noting the created_on timestamp. Your results may differ according to the shares active within your account.

created_on	kind	name	database_name	to	owner	comment	listing_global_name
2018-08-14 21:36:0...	INBOUND	SNOWFLAKE.ACCOUNT_USAGE	SNOWFLAKE				
2018-02-14 09:57:2...	INBOUND	SFSALESSHARED.SFC_SAMPLES_AWS...					

Figure 5-7. *Default inbound data sharing*

Your SHOW shares result set can be changed into a SQL-usable format:

```
SELECT "created_on", "kind", "name",
       "database_name", "to", "owner", "listing_global_name"
FROM TABLE ( RESULT_SCAN ( last_query_id()));
```

List the consumers of a named share:

```
SHOW GRANTS OF SHARE SNOWFLAKE.ACCOUNT_USAGE;
```

You may also use DESCRIBE instead of SHOW but the results are less useful.

List the objects enabled for sharing to a named share:

```
SHOW GRANTS TO SHARE SNOWFLAKE.ACCOUNT_USAGE;
```

As before, you can convert the output using RESULT_SCAN:

```
SELECT OBJECT_CONSTRUCT ( * )
FROM    TABLE ( RESULT_SCAN ( last_query_id()));
```

Snowflake does not list any built-in table functions exposing share consumption metrics. Documentation can be found here: https://docs.snowflake.com/en/sql-reference/functions-table.html#table-functions.

Without resorting to a user-defined function using RESULT_SCAN and then storing results into a table for onward consumption, you are very limited in scope for share utilization reporting. You cannot see any consumption metrics as there are none exposed.

Let's define a JavaScript stored procedure to return information on objects enabled to a named share:

```
CREATE OR REPLACE PROCEDURE sp_get_share_information ( P_PARAMETER STRING )
RETURNS VARIANT
LANGUAGE javascript
EXECUTE AS CALLER
AS
$$
   var sql_stmt    = "SHOW GRANTS TO SHARE " + P_PARAMETER;
   var stmt        = snowflake.createStatement ({ sqlText:sql_stmt });

   var sql_stmt_1 = `SELECT OBJECT_CONSTRUCT ( * ) FROM   TABLE ( RESULT_
   SCAN ( last_query_id()))`
   var stmt_1      = snowflake.createStatement ({ sqlText:sql_stmt_1 });

   var row_as_json   = {};
   var array_of_rows = [];
   var table_as_json = {};

   recset = stmt.execute();  // SHOW GRANTS...

   try
   {
      recset = stmt_1.execute();
      while(recset.next())
      {
         row_as_json  = recset.getColumnValue(1)
         array_of_rows.push(row_as_json);
      }
   }
   catch ( err )
   {
       result  = sql_stmt;
```

```
        result += "\nCode: "          + err.code;
        result += "\nState: "         + err.state;
        result += "\nMessage: "       + err.message;
        result += "\nStack Trace:\n" + err.stackTraceTxt;
    }

    table_as_json[P_PARAMETER] = array_of_rows;

    return table_as_json;
$$;
```

Then call your new stored procedure:

```
CALL sp_get_share_information ( 'SNOWFLAKE.ACCOUNT_USAGE' );
```

Figure 5-8 illustrates the first record within the returned data set.

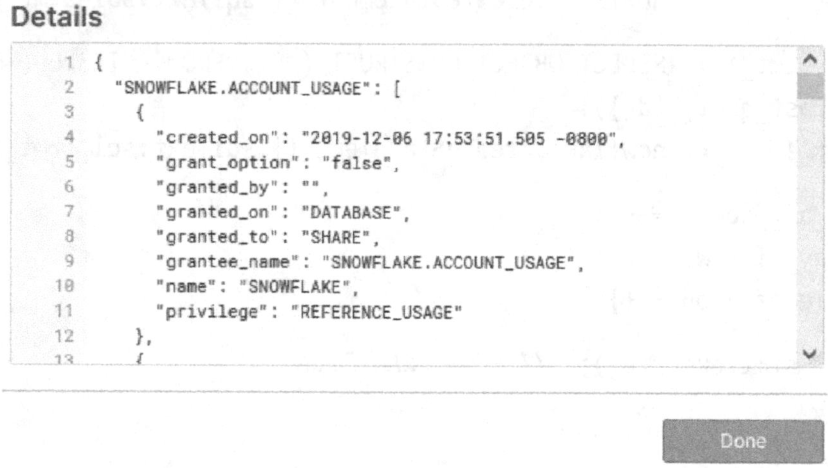

Details

```
 1  {
 2    "SNOWFLAKE.ACCOUNT_USAGE": [
 3      {
 4        "created_on": "2019-12-06 17:53:51.505 -0800",
 5        "grant_option": "false",
 6        "granted_by": "",
 7        "granted_on": "DATABASE",
 8        "granted_to": "SHARE",
 9        "grantee_name": "SNOWFLAKE.ACCOUNT_USAGE",
10        "name": "SNOWFLAKE",
11        "privilege": "REFERENCE_USAGE"
12      },
13      {
```

Done

Figure 5-8. *Grants to a Snowflake account usage share*

We leave the consumption of generated data sets for your further consideration.

Snowflake is not actively working on exposing usage metrics for Secure Direct Data Share. In general, Snowflake customers are maturing their data sharing offerings away from Secure Direct Data Share and instead prefer Private Listings and Snowflake Marketplace, both of which offer a better customer experience.

To remove your new stored procedure:

```
DROP PROCEDURE sp_get_share_information ( STRING );
```

Private Listings

Previously known as Data Exchange and now known as Private Listings, Figure 5-9 illustrates the hub-and-spoke model implemented. In contrast with Secure Direct Data Share where only one-way data transit is supported, when using Private Listings all subscribers to the exchange can be both publishers and consumers. The distinction is important: We implement data clean rooms (DCRs) using bidirectional data sharing and discuss DCRs later within this chapter.

DATA EXCHANGE

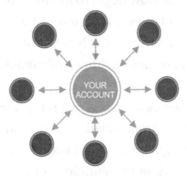

For data sharing with a designated group

Figure 5-9. *Private Listings (previously Data Exchange)*

Private Listings relies upon the same technology as previously explained for Secure Direct Data Share and for this reason is equally as secure.

For progressive organizations looking to address the absence of consumption metrics with Secure Direct Data Share, we suggest moving to establishing Private Listings where data usage monitoring options are readily available.

Private Listings is not available by default to all accounts. Snowflake may need to enable this feature. See the Appendix for details.

At the time of writing, this Snowflake Knowledge Base article was released showing changes to data sharing usage views: https://community.snowflake.com/s/article/Update-usage-views-to-return-snowflake-region-instead-of-cloud-region.

While provisioning Private Listings remains within the constraints of SnowSight, we illustrate how to programmatically access monitoring data. We do not cover the creation of Private Listings using SnowSight but the Appendix offers a starting point.

Building upon previous code, change context before executing later SQL commands:

```
USE ROLE       monitor_owner_role;
USE WAREHOUSE monitor_wh;
USE DATABASE   monitor;
USE SCHEMA     monitor.monitor_owner;
```

With your context set, you can now begin to identify consumption metrics for Private Listings. Start by asking a series of questions of interest to your business colleagues and find the answers within the available metrics. Note the list of questions we pose within this section is not exhaustive, and your business colleagues will certainly find more questions to ask.

Asking your business colleagues what they want is preferrable to technical staff serving up metrics. The former approach leads to successful business outcomes whereas the latter approach has merit but often does not address the real business need.

You may prefer to create views for the sample queries offered below instead of embedding business logic within your reporting tools. We prefer to reuse components and database views offer an ideal, well understood, and simple way to move business logic into Snowflake.

While we do not configure Private Exchange within this section, we do rely upon the presence of established private listings for the following queries to work.

Data sharing usage view latency information can be found here: `https://docs. snowflake.com/en/sql-reference/data-sharing-usage.html#data-sharing- usage-views`.

Accessed Objects

In this section you address how frequently data sets were accessed and by whom. You seek to identify those users and customers accessing specific datasets in order to identify upselling opportunities and justification for retaining the data set listing. By examining data access metrics, you may also propose data sets to populate DCRs, which we discuss later in this chapter.

You also must consider the regions and CSPs where data is made available as this knowledge will inform your data replication strategy. From a technical perspective, another successful business outcome is to optimize your egress costs based upon consumer data consumption profile.

The below query utilizes LISTING_ACCESS_HISTORY, which at the time of writing is in public preview and identifies most of your requirements. According to Snowflake documentation, the latency period for LISTING_ACCESS_HISTORY is up to 2 days and data is retained for 365 days. You may wish to retain data sets for longer than 365 days; refer to Chapter 3 where we explain how to persist LOGIN_HISTORY data along with a few gotchas to consider.

You do not identify specific users but instead identify where your data is consumed and by which customers. Note the query limits returned data to the past 7 days. Modify it to suit your requirements.

```
SELECT DATE_TRUNC ( 'DAY', query_date )        AS query_day,
       SUM ( 1 )                               AS num_requests,
       listing_objects_accessed,
       exchange_name,
       cloud_region,
       listing_global_name,
       provider_account_locator,
       provider_account_name,
       share_name,
       consumer_account_locator,
       consumer_account_name,
       consumer_account_organization
FROM   snowflake.data_sharing_usage.listing_access_history
WHERE  DATE_TRUNC ( 'DAY', query_date ) >=
       DATEADD ( day, -7, DATE_TRUNC ( 'DAY', current_date()))
GROUP BY DATE_TRUNC ( 'DAY', query_date ),
         listing_objects_accessed,
         exchange_name,
         cloud_region,
         listing_global_name,
         provider_account_locator,
         provider_account_name,
```

```
        share_name,
        consumer_account_locator,
        consumer_account_name,
        consumer_account_organization
ORDER BY DATE_TRUNC ( 'DAY', query_date ),
        listing_objects_accessed ASC;
```

Attribute LISTING_OBJECTS_ACCESSED is an ARRAY so it requires further processing to normalize the data. The documentation is very helpful and offers examples for your further consideration and can be found here: https://docs.snowflake.com/en/sql-reference/data-sharing-usage/listing-access-history.html#listing-access-history-view.

Usage and Billing

Adopting a different approach to LISTING_ACCESS_HISTORY, the LISTING_CONSUMPTION_DAILY view offers visibility of the number of times jobs were run against a listing. There is no provision to identify the actual objects accessed, therefore the LISTING_CONSUMPTION_DAILY view provides abstract or summary information on data access.

According to Snowflake documentation, the latency period for LISTING_ACCESS_HISTORY is up to 2 days and data is retained for 365 days.

The below query is largely the same as the query for LISTING_ACCESS_HISTORY, noting the absence of the LISTING_OBJECTS_ACCESSED ARRAY and some attribute names differ. You do not identify specific users but instead identify where your data is consumed and by which customers. Note the query limits returned data to the past 7 days. Modify it to suit your requirements.

```
SELECT DATE_TRUNC ( 'DAY', event_date )      AS event_date,
       SUM ( 1 )                             AS num_requests,
       exchange_name,
       snowflake_region,
       listing_global_name,
       provider_account_locator,
       provider_account_name,
       share_name,
       consumer_account_locator,
       consumer_account_name,
```

```
       consumer_organization
FROM   snowflake.data_sharing_usage.listing_consumption_daily
WHERE  DATE_TRUNC ( 'DAY', event_date ) >=
       DATEADD ( day, -7, DATE_TRUNC ( 'DAY', current_date()))
GROUP BY DATE_TRUNC ( 'DAY', event_date ),
       exchange_name,
       snowflake_region,
       listing_global_name,
       provider_account_locator,
       provider_account_name,
       share_name,
       consumer_account_locator,
       consumer_account_name,
       consumer_organization
ORDER BY DATE_TRUNC ( 'DAY', event_date ) ASC;
```

The documentation is very helpful and offers examples for your further consideration and can be found here: https://docs.snowflake.com/en/sql-reference/data-sharing-usage/listing-consumption-daily.html#listing-consumption-daily-view.

Consumption of Shares

Returning a single row per day for each consumer account accessing a named share allows you to graph data access by consumer.

According to Snowflake documentation, the latency period for LISTING_EVENTS_ DAILY is up to 2 days and data is retained for 365 days.

You do not identify specific users but instead identify where your data is consumed and by which customers. Note the query limits returned data to the past 7 days. Modify it to suit your requirements.

```
SELECT DATE_TRUNC ( 'DAY', event_date )      AS event_date,
       SUM ( 1 )                             AS num_requests,
       exchange_name,
       snowflake_region,
       listing_name,
       listing_display_name,
```

```
        listing_global_name,
        consumer_account_locator,
        consumer_account_name,
        consumer_organization,
        consumer_email
FROM    snowflake.data_sharing_usage.listing_events_daily
WHERE   DATE_TRUNC ( 'DAY', event_date ) >=
        DATEADD ( day, -7, DATE_TRUNC ( 'DAY', current_date()))
GROUP BY DATE_TRUNC ( 'DAY', event_date ),
        exchange_name,
        snowflake_region,
        listing_name,
        listing_display_name,
        listing_global_name,
        consumer_account_locator,
        consumer_account_name,
        consumer_organization,
        consumer_email
ORDER BY DATE_TRUNC ( 'DAY', event_date ) ASC;
```

Documentation can be found here: https://docs.snowflake.com/en/sql-reference/data-sharing-usage/listing-events-daily.html#listing-events-daily-view.

Daily and 28-Day Telemetry

Daily and 28-day telemetry data by data exchange and region can be obtained from the LISTING_TELEMETRY_DAILY view, which returns a row for each data exchange in your organization and each region where that data exchange is available.

You do not identify specific users but instead identify where your data is consumed and by which customers. Note the query limits returned data to the past 7 days. Modify it to suit your requirements.

According to Snowflake documentation, the latency period for LISTING_TELEMETRY_DAILY is up to 2 days and data is retained for 365 days. However, during real-world testing, history was observed for more than two years for a data exchange proof of concept conducted in August 2020.

```
SELECT DATE_TRUNC ( 'DAY', event_date )        AS event_date,
       exchange_name,
       snowflake_region,
       listing_name,
       listing_display_name,
       listing_global_name,
       event_type,
       action,
       consumer_accounts_daily,
       consumer_accounts_28D
FROM   snowflake.data_sharing_usage.listing_telemetry_daily
WHERE  DATE_TRUNC ( 'DAY', event_date ) >=
       DATEADD ( day, -7, DATE_TRUNC ( 'DAY', current_date()))
ORDER BY DATE_TRUNC ( 'DAY', event_date ) ASC;
```

Documentation can be found here: https://docs.snowflake.com/en/sql-reference/data-sharing-usage/listing-telemetry-daily.html#listing-telemetry-daily-view.

Snowflake Marketplace

With your investigation of Private Listings complete, let's discuss the Snowflake Marketplace, shown in Figure 5-10.

Figure 5-10. *Snowflake Marketplace*

Snowflake Marketplace relies upon the same technology as previously explained for Secure Direct Data Share and for this reason is equally as secure.

Snowflake Marketplace is not available by default to all accounts. Snowflake requires a separate legal agreement for this feature.

While provisioning Snowflake Marketplace remains within the constraints of SnowSight, we illustrate how to programmatically access monitoring data. We do not cover the creation of Snowflake Marketplace using SnowSight but the Appendix offers a starting point.

Building upon previous code, change context before executing later SQL commands:

```
USE ROLE       monitor_owner_role;
USE WAREHOUSE  monitor_wh;
USE DATABASE   monitor;
USE SCHEMA     monitor.monitor_owner;
```

With your context set, you can now begin to identify consumption metrics for Snowflake Marketplace.

Snowflake documentation URLs may change between the time of writing and the time you read this chapter.

Snowflake offers historical usage data for paid listings from two places within the imported Snowflake database: ORGANIZATION_USAGE and DATA_SHARING_USAGE. In this section we focus on DATA_SHARING_USAGE as this location offers a wider selection of views to consume from. Documentation can be found here: https://other-docs. snowflake.com/en/marketplace/monetization-provider-views.html#data-providers-monetization-usage-views.

Paid Listing Earnings

The view marketplace_disbursement_report is not available unless your organization has been accepted into the Snowflake Marketplace.

This query returns Snowflake Marketplace earnings from your organization's paid listing over time:

```
SELECT event_date,
       event_type,
       invoice_date,
       listing_name,
       listing_display_name,
       listing_global_name,
       charge_type,
       gross,
       fees,
       taxes,
       net_amount,
       currency
FROM   snowflake.data_sharing_usage.marketplace_disbursement_report;
```

Latency is stated to be up to 48 hours.

Further information can be found here: https://other-docs.snowflake.com/en/marketplace/monetization-provider-views.html#marketplace-disbursment-report-view.

Listing Subscriptions

The view marketplace_paid_usage_daily is not available unless your organization has been accepted into the Snowflake Marketplace.

This query returns Snowflake Marketplace earnings detail from your organization's paid listing over time:

```
SELECT report_date,
       usage_date,
       provider_name,
       provider_account_name,
       provider_account_locator,
       provider_organization_name,
       listing_display_name,
       listing_global_name,
       database_name,
       po_number,
       pricing_plan,
```

```
        charge_type,
        units,
        unit_price,
        charge,
        currency
FROM    snowflake.data_sharing_usage.marketplace_paid_usage_daily;
```

Further information can be found here: https://other-docs.snowflake.com/en/
marketplace/monetization-consumer-views.html#id4.

Figure 5-11 displays a sample subset of attributes returned for the above query
showing a free subscription to Snowflake Financial Statements.

PROVIDER_NAME	PROVIDER_ACCOUNT_NAME	PROVIDER_ACCOU	PROVIDER_ORGANI	LISTING_DISPLAY_NAME	LISTING_GLO
Snowflake Inc.	SNOWHOUSE_AWS_US_WEST_2	SNOWHOUSE	SFCOGSOPS	Snowflake Financial Statements	GZSNZ2TO5

Figure 5-11. *Snowflake Financial Statement subscription*

Daily Purchases

The view marketplace_purchase_events_daily is not available unless your
organization has been accepted into the Snowflake Marketplace.

While marketplace_purchase_events_daily appears within the snowflake.data_
sharing_usage schema, there is no supporting documentation so we can reasonably
assume the view definition is subject to change.

This query is correct at the time of writing and returns a Snowflake Marketplace
pricing plan and user metadata from your organization's paid listing over time. Please
check Snowflake documentation before use.

```
SELECT event_date,
       event_type,
       consumer_account_locator,
       consumer_account_name,
       listing_name,
       listing_display_name,
       listing_global_name,
       pricing_plan,
       user_metadata
FROM   snowflake.data_sharing_usage.marketplace_purchase_events_daily;
```

Listings Published

The view `monetized_usage_daily` is not available unless your organization has been accepted into the Snowflake Marketplace.

Given the

```
SELECT  report_date,
        usage_date,
        listing_name,
        listing_display_name,
        listing_global_name,
        consumer_account_locator,
        consumer_account_name,
        consumer_organization_name,
        consumer_snowflake_region,
        pricing_plan,
        charge_type,
        units,
        unit_price,
        gross_charge,
        currency
FROM    snowflake.data_sharing_usage.monetized_usage_daily;
```

Latency is stated to be up to 24 hours.

Figure 5-12 illustrates a partial response from the above query and is included to show the difference between the output in Figure 5-11.

REPORT_DATE	USAGE_DATE	LISTING_NAME	LISTING_DISPLAY_	...	CONSUMER_SNOV	PRICING_PLAN	CHARGE_TYPE
2022-04-03	2022-04-03	TEST_LISTING	Test Listing		AWS_US_EAST_1	NULL	SAMPLE
2022-07-19	2022-04-03	TEST_LISTING	Test Listing		AWS_US_EAST_1	NULL	SAMPLE

Figure 5-12. *Snowflake test listing*

Further information can be found here: `https://other-docs.snowflake.com/en/marketplace/monetization-provider-views.html#id11`.

Data Clean Rooms

Disclaimer: We do not provide a deep dive into DCRs. Instead, we explain why DCRs are a good idea along with providing some resources for further investigation.

What Are Data Clean Rooms?

DCRs enable organizations to interact within strictly controlled and totally isolated environments. With a full understanding of the data sharing options explained earlier in this chapter, you can quickly grasp the concept of a bidirectional data share. DCRs implement both producer and consumer functionality within two Snowflake accounts, one for your organization (which we deem to be the primary) and one for your customer (which we deem to be the secondary). In real-world use, there may be more than one secondary account participating in a DCR, but to keep the explanation simple, we only consider two participants for now.

Conceptually, DCRs are implemented as shown in Figure 5-13, operating as the intersection between two (or more) Snowflake accounts. Code isolation and data security are guaranteed as each participant specifies their boundaries and interfaces. Participants agree upon the interface parameters and returned result sets. In this manner, it can be seen that both participants retain full control over their own environments.

To further ensure data security is maintained, double-blind matching guarantees both source and destination data points remain completely anonymous.

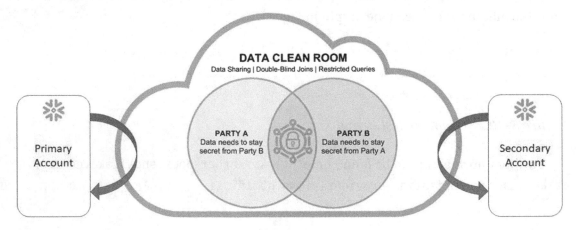

Figure 5-13. *DCR conceptual representation*

All DCR interactions occur within the context of the data shares. All SQL operations are tightly controlled via the objects shared with their counterpart.

Each participant has access to the data sharing views within their own Snowflake account, rendering the same metric data and information as articulated previously within this chapter. Neither party has visibility to any information belonging to the other party.

Why Data Clean Rooms?

DCRs offer several advantages over data sharing.

As you have seen, provisioning Secure Direct Data Sharing does not allow metrics to be captured using the Snowflake data sharing usage views. SDDS is unidirectional, thus preventing bidirectional interaction. As a consequence, many organizations are moving away from using SDDS to publish data directly to a consumer and instead are seeking a more collaborative approach to sharing data.

Try before you buy. A great sales and marketing tool for every organization is to showcase its data offerings using a subset of their data. DCRs offer the opportunity to interact more closely with clients, enable a more rapid turn-around of ideas, allow innovation to flourish, and encourage a data ecosystem to develop.

Organizations can use DCRs to bring data together in an agreed, defined, and controlled environment for joint analysis. The data does not move, and data security remains with the source. Data can be obfuscated, masked, or restricted at the source with the provider in full control of both data and intellectual property. All data privacy practices such as GDPR, PII, HIPAA, and others are honored as enforced by each data provider.

Provisioning DCRs can be scripted, leading to semiautomated deployment. Access control to DCRs is simple and is readily delegated to Operations staff. Assuming a degree of automation, DCRs do not require highly technical or skilled personnel to implement.

DCRs offer a route to more rapidly monetizing of data. The quicker you can reach commercial agreements, the more you realize the value of your data. With DCRs you also facilitate the breaking down of data silos in a safe and controlled manner. The cost of acquiring and using data is also significantly reduced, lowering the barriers to entry.

Establishing a Data Clean Room

Establishing a DCR is relatively easy and readily implemented by following these steps:

- Identify the member Snowflake accounts. See the Appendix for supporting information.

- Identify each member data sets for inclusion into the DCR.

- Create a Private Listing (formerly Data Exchange) for the DCR. See the Appendix for supporting information.

- Create bespoke secure functions and secure joins to guarantee data isolation.

- Conduct DCR activities.

- Remove DCR constructs and accounts ready to implement the next use case.

In principle, the establishment of DCRs is simple, requiring two Snowflake accounts, both of which can be trial accounts. Snowflake support is required to set up the Private Listing. See the Appendix for information.

Resources

Space does not permit a full walk-through of the process to create DCR functionality. We therefore offer these resources to assist your further research:

This post by Data Superhero Frank Bell is very informative containing several linked resources: `https://snowflakesolutions.net/snowflake-data-clean-rooms/`.

Snowflake offer this article suggesting business growth via DCRs: `www.snowflake.com/trending/data-clean-room-for-business-growth`.

A more detailed implementation focused article from Justin Langseth can be found here: `www.snowflake.com/blog/the-power-of-secure-user-defined-functions-for-protecting-shared-data/`.

Data Governance Perspective

We include this section as a reminder of your responsibility to secure your data. This chapter does not explain entitlement to view objects because this is covered elsewhere, but it is important to note all imported shares require a robust role-based access control (RBAC) policy along with any other necessary controls for your organization.

Many Information Technologists have the same view of data: Within the boundaries of appropriate and permissible usage, data should be freely available. Note the caveats "appropriate" and "permissible." We are not advocating that all data should be available to everyone on demand. Instead, data should be made available in a controlled manner for approved consumers, which leads us to our first question of "who can see what?" closely followed by our second question, "what did the viewer do with the data they saw?" By implementing robust RBAC policies, this chapter addresses the first question. For the second question, our industry is trying to figure out answers .

Summary

You began this chapter with a recap of the data sharing options before defining an approach focused on programmatic exposure of metrics to facilitate consumption by any tool rather than being forced into adopting SnowSight.

With this approach defined, you then investigated the data sharing usage schema views, noting a key difference in the provision of metrics between those accounts enabled for Snowflake Marketplace and those where Snowflake Marketplace has not been enabled.

Noting that shares may be provisioned by differing Snowflake administrative roles, you attempted with little success to use in-built Snowflake capability to develop interfaces allowing easy reporting of usage metrics. Furthermore, Snowflake is not enhancing Secure Direct Data Share telemetry but instead is promoting Private Listings and Snowflake Marketplace as its preferred data distribution capabilities. To assist with configuration, the Appendix provides more information.

Working from assumed business requirements, for both Private Listings and Snowflake Marketplace you identified how to provide metrics to your business colleagues with the expectation Snowflake will enhance the available views over time.

You explored an overview of data clean rooms along with summary benefits of implementing DCRs. While space does not permit a deep dive into the low-level implementation of a DCR, we do offer links to resources to investigate DCRs further.

You now move onto the next subject of interest: Snowflake metrics.

Appendix

We recognize not every organization has established Private Listings (previously Data Exchange). For those learning Snowflake, this section articulates the steps required to establish a simple Private Listing between two Snowflake accounts and later for demonstrating DCRs.

In this Appendix, we focus on creating a simple Private Listing by selectively displaying screens used throughout the setup process, relying upon you to investigate further.

Private Listings and Snowflake Marketplace are undergoing change. Details within this chapter correct at the time of writing.

Naturally, you will need a second Snowflake account. Refer to the initial Snowflake account as the primary and the newly created Snowflake account as the secondary. DCRs require bidirectional data sharing, so each Snowflake account will perform both Publisher and Consumer roles.

To create your Secondary Snowflake account, navigate to `https://signup.snowflake.com/` and populate the forms, not forgetting to select **Business Critical edition** and **AWS** as shown in Figure 5-14.

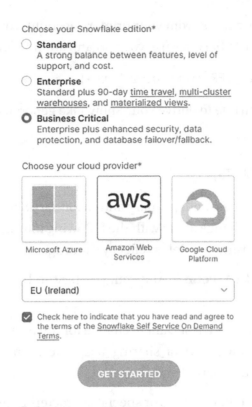

Figure 5-14. *New Snowflake account options*

After activating your secondary Snowflake account, log in and retain SnowSight as your user interface.

See the Snowflake documentation for further information on

- Private Listings: `https://docs.snowflake.com/en/user-guide/ui-snowsight-private-sharing.html#snowsight-data-private-sharing`

- Snowflake Marketplace Monetization: `https://other-docs.snowflake.com/en/marketplace/monetization.html`

The primary and secondary accounts must be enabled for Data Exchange, so you will need to raise a support case as shown in Figure 5-15. Navigate to Help & Support ➤ Support ➤ + Support Case.

Account identifiers can be found for each account by executing:

```
SELECT current_account();
```

For those using trial accounts, from your primary account, you will need to grant the orgadmin role to your user. Otherwise, use your organization's identifier.

```
GRANT ROLE orgadmin TO USER <Your User Here>;
```

Switch to the orgadmin role to derive your Snowflake generated organization ID:

```
USE ROLE orgadmin;
```

```
SHOW ORGANIZATION ACCOUNTS;
```

In the support case, request both primary and secondary Snowflake accounts be added to a named Data Exchange along with the following information. Note these are examples I used, please ensure text fields reflect your usage.

- **List of Snowflake Accounts**: Add your primary and secondary accounts.

- **Description**: "This is a test Data Exchange for the purposes of documenting snowflake.data_sharing_usage views and will be removed when trial period ends."

- **Name**: Cannot contain spaces or special characters except underscores, as in Maturing_the_Snowflake_Data_Cloud_ Apress_2023

- **Display Name**: Cannot contain spaces or special characters except underscores, as in Maturing_the_Snowflake_Data_Cloud_ Apress_2023

- **Account URL**: Use your primary account URL in the format `https:// rawdpwn-cb64333.snowflakecomputing.com/console/login`.

- **Organization ID**: FGMQDEY

Figure 5-15 illustrates how to raise a Snowflake support ticket.

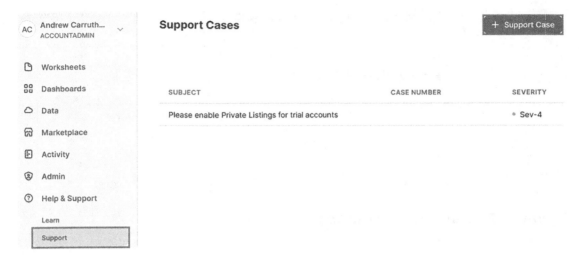

Figure 5-15. *Raising a support ticket*

The Snowflake support team will respond in due course to the support ticket raised, hopefully with confirmation that your Data Exchange has been created. If you experience any issues accepting the invitation for your new Data Exchange, run the following command in a Snowflake Worksheet to accept the Data Exchange invitation:

```
USE ROLE accountadmin;

SELECT SYSTEM$ACCEPT_ADMIN_INVITATION ( 'Maturing_the_Snowflake_Data_Cloud_
Apress_2023' );
```

Figure 5-16 shows the expected response to accepting the Data Exchange invitation:

Figure 5-16. *Enabling Data Exchange*

Once you perform the above, you may need to log out of Snowflake and the Data Marketplace, close your web browser, open a new one, and log back in for the above change to take effect.

Figure 5-17 shows a new Data Exchange available with Admin rights.

Figure 5-17. *Available Data Exchange(s)*

CHAPTER 6

Usage Reporting

In highly secure environments such as you will encounter within your organizations, at some point in the lifecycle of your data warehouse you will be asked to provision self-service of data sets. Your approach must be to provision access to only those data sets to which an end user is properly approved and entitled to view, and nothing more.

You may think the problem space of identifying "who can see what" is easily solved because you have a plethora of tooling available including role-based access control (RBAC), data masking policies, row-level security (RLS), object tagging, and (soon) tag-based masking available in your technical toolkit to protect your organization's data. The intersection of these tools provides everything you need to control access to your data. However, the issue we address in this chapter is not the actual provisioning of data but the proof of how data is accessed.

Let's consider how users interact with Snowflake and dig a little deeper into their needs while retaining a security mindset. After all, the success or failure of your applications is entirely dependent upon how you deliver (or not) successful business outcomes, and security is the biggest concern.

We are happy to create the perception of having lots of data domains within our data warehouses, but from an end user perspective, and to paraphrase Samuel Taylor Coleridge, "data, data everywhere and not a drop to use." How do we deliver data in a manner acceptable to all while proving our data remains secure?

Our response to the question of data security, as a minimum, must address these questions:

- **Who can see what?** Roles, objects, and attributes accessible by a user

- **Who can do what?** Roles and objects a user can manipulate

- **Who has seen what?** The historical evidence of data access

- **Who has done what?** The historical evidence of object manipulation

© Andrew Carruthers and Sahir Ahmed 2023
A. Carruthers and S. Ahmed, *Maturing the Snowflake Data Cloud*, https://doi.org/10.1007/978-1-4842-9340-9_6

Each question may have multiple parts. In this chapter, you will progressively build them up into a single unified SQL statement from which you can determine the proof your security requirements are met.

However, there are limitations with both point-in-time and historical SQL statement capture, perhaps the Achilles' heel of all monitoring: SQL statement capture does not guarantee the same data will be returned when replayed later as the underlying data will change over time. In order to ensure data consistency at the point of replay, you must ensure that your underlying data is not subject to update within its temporal boundary. That is, the data can't change without the temporal boundary changing too, and for this reason we prefer Slowly Changing Dimension 2 (SCD2) data sets, as explained here: `https://en.wikipedia.org/wiki/Slowly_changing_dimension#Type_2:_add_new_row`.

SCD2 is no protection if you allow your data warehouse contents to be changed in an ad-hoc manner by users or support staff. We recommend a thorough review of all data manipulation paths for your environment before relying upon the patterns expressed thin this chapter. At the initial point of creation of your data warehouse, it is far easier to establish the principle of no ad-hoc data manipulation than to attempt enforcement at some later date, by which time the integrity of your data warehouse will have been compromised. You must have fidelity back to the original received data.

Furthermore, entitlements to row-level security roles may change between the time you capture the SQL statement and the time at which the SQL statement was executed. You can't rely upon being able to exactly replicate the original scenario unless you conduct a thorough investigation and ensure no changes have been made that may affect results.

Nobody said this was going to be easy...

The answer to self-service must be to identify and report all data access including any changes to the role used while taking into consideration the temporal nature of your data. You can use the Snowflake Account Usage Store to source all required information. While some answers are seemingly better answered from each database `information_schema`, remember that the `information_schema` scope is limited to an individual database and is time-limited to 14 days. We suggest the use of `information_schema` is a poor choice for whole account data capture unless you either cycle through each database `information_schema` in turn or `UNION` all required database `information_schemas` into a single accessible view.

Account Usage Store view latency varies between 45 minutes and up to 3 hours. For data sharing, latency can be up to 2 days. Your testing may therefore be impacted.

Throughout this chapter you will establish your test environment and then build out test cases to validate your approach. You will use your Snowflake trial account and are not concerned with the AWS account within this chapter. All code has been tested with Firefox and Chrome, although other browsers are expected to work equally well; if not, raise a ticket with Snowflake support stating your specific testing environment.

All of the code referenced in this chapter is available in the accompanying SQL script of the same name.

In order to provide empirical answers, start with regulations and your organization's approach to governing data.

Operating Environment

Within this section we describe some common considerations affecting all data access. Yours may differ.

Data Usage Constraints

Your organization is certain to have policies determining how you secure data across a variety of dimensions. In this section, we list several policies and best practices. You may add to the list or remove according to your circumstances, and we do not claim the list is exhaustive.

- The best practice for storing data within Snowflake is to segregate by domain or application source and only make visible the minimum data set to satisfy business consumers or downstream applications.

- You will also find your privacy colleagues mandate all personally identifiable information (PII, any data that could potentially identify a specific individual) either be partially or fully obfuscated to prevent inadvertent disclosure or remain totally inaccessible.

- Each organization will also specify its own internal security classifiers according to how they perceive each attribute's sensitivity and will require appropriate masking, encryption, or another form of protection.

- PII data, along with many other data classifications, may also be tagged, which provides an alternative means of both identifying and extracting data.

- Your data governance colleagues will have several policies impacting your ability to freely access data, such as restrictions on data sharing which may not be programmatically enforced. Data catalogs may expose metadata and provide information on the canonical notation for object location, but simply knowing where an object resides does not imply entitlement to access the object.

- Organizations may also be subject to one or more regulatory regimes dependent upon the jurisdictions in which they conduct business, and furthermore, regulations may change according to the type of markets they operate in.

- Among other responsibilities, your Information Security Officers are mandated to ensure conformance to your organization's policies, assessing and managing information security risk for the business.

- Data Custodians, the nominated individuals who are responsible for data, must give their approval as a mandatory requirement before access to data sets is provisioned, representing a fundamental bedrock of your security profile: ensuring you have permission. This may require two approvers. While some organizations adopt a policy where membership of a given domain automatically entitles access to all data within that domain, the model is not common.

- Aside from all the above, Snowflake as a product is constantly evolving. We know security remains a constant. New features add more layers of complexity, and it is our job to maintain consistency.

Toxic Entitlement Combinations

Under certain circumstances, toxic entitlement combinations can be inadvertently created. It is typical for a single role to be either object owning or data manipulating (DML), but not to have entitlement for both. In this chapter, you will expose the underlying data to enable detection of toxic entitlement combinations.

You might also see toxic role combinations where the law of unintended consequences has resulted in both owner and DML entitlement granted to different roles granted to the same user. In this scenario, manually switching between roles can create toxic role combinations, albeit with a degree of separation that may be acceptable for certain support staff.

A less obvious scenario, or perhaps more explicit scenario by design, is where secondary roles are enabled, meaning all roles that have been granted to the user in addition to the current active primary role are active. Under this scenario, there is one caveat: Entitlement to create objects is provided by the primary role and not by any secondary role.

Secondary roles are documented here: `https://docs.snowflake.com/en/sql-reference/sql/use-secondary-roles.html#use-secondary-roles`. You can enable secondary roles thus:

```
USE SECONDARY ROLES ALL;
```

Likewise, you can disable secondary roles:

```
USE SECONDARY ROLES NONE;
```

We discourage the use of secondary roles as a potential source of toxic combinations but instead prefer a more considered approach of defining roles as discrete containers of entitlement explicitly granted to roles.

Approach

This chapter is written thematically where specific perspectives and objectives are outlined and all converge at the end. To aid understanding, Figure 6-1 provides an overview of the themes discussed within this chapter with a high-level breakdown of each theme. We regard this diagram as an aid to understanding the component under construction as you progress through this chapter.

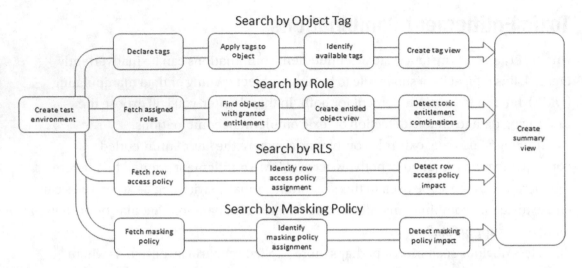

Figure 6-1. *Chapter overview*

We assume everyone is familiar with the Snowflake-supplied UI in both Classic UI and SnowSight form. Code within this chapter is compatible with SnowSQL. Please refer to Chapter 1 for installation instructions and to get started.

This chapter reuses the Snowflake monitoring setup developed in Chapter 3. Please ensure these components have been deployed to your Snowflake account before attempting the code delivered within this chapter. A quick check is to run these commands and resolve any errors:

```
USE ROLE       monitor_owner_role;
USE WAREHOUSE monitor_wh;
USE DATABASE   monitor;
USE SCHEMA     monitor.monitor_owner;
```

Use your newly created database and schema along with supplied warehouse compute_wh when setting your context.

Create an example table named employee:

```
CREATE OR REPLACE TABLE employee
(
preferred_name          VARCHAR(255),
surname_preferred       VARCHAR(255),
forename_preferred      VARCHAR(255),
gender                  VARCHAR(255)
);
```

Create a view named v_employee which you will use later to demonstrate show tags are propagated to dependent objects:

```
CREATE OR REPLACE VIEW v_employee
AS
SELECT *
FROM    employee;
```

And now some tags noting the requirement to maintain tag uniqueness:

```
CREATE OR REPLACE TAG PII
COMMENT = 'Personally Identifiable Information';

CREATE OR REPLACE TAG PII_S_Name
COMMENT = 'Personally Identifiable Information -> Sensitive -> Name';

CREATE OR REPLACE TAG PII_N_Gender
COMMENT = 'Personally Identifiable Information -> Non-Sensitive -> Gender';
```

Then apply tags to your example table employee:

```
ALTER TABLE monitor.monitor_owner.employee
SET TAG PII = 'Personally Identifiable Information';

ALTER TABLE monitor.monitor_owner.employee
MODIFY COLUMN preferred_name
SET TAG PII_S_Name = 'Personally Identifiable Information -> Sensitive
-> Name';

ALTER TABLE monitor.monitor_owner.employee
MODIFY COLUMN surname_preferred
SET TAG PII_S_Name = 'Personally Identifiable Information -> Sensitive
-> Name';

ALTER TABLE monitor.monitor_owner.employee
MODIFY COLUMN forename_preferred
SET TAG PII_S_Name = 'Personally Identifiable Information -> Sensitive
-> Name';
```

```
ALTER TABLE monitor.monitor_owner.employee
MODIFY COLUMN gender
SET TAG PII_N_Gender = 'Personally Identifiable Information -> Non-
Sensitive -> Gender';
```

Who Can See What

This section addresses how to identify canonical objects and data visible to an end user.

We assume these objectives are desired:

- Metadata search by object tag

- Self-discovery of canonical objects

- Data set retrieval from identified canonical objects

Let's investigate the data sources from which you can identify objects exposing entitlement information. For the investigation, you will focus on objects that return information on data access, and not object definitions.

At all times you must remember Snowflake security will only allow visibility to objects, attributes, and data for the in-scope role.

Searching by Object Tag

Your first objective is to identify the object tags of interest. Here are two ways to do this:

```
SHOW tags IN SCHEMA monitor.monitor_owner;

SELECT *
FROM    snowflake.account_usage.tags
WHERE   deleted IS NULL
ORDER BY tag_name;
```

Figure 6-2 illustrates sample results, noting there will be more attributes in your actual results returned.

TAG_NAME	TAG_SCHEMA	TAG_DATAE	TAG_OWNER	TAG_COMMENT
PII	MONITOR_OWNER	MONITOR	MONITOR_OWNER_ROLE	"Personally Identifiable Information"
PII_N_GENDER	MONITOR_OWNER	MONITOR	MONITOR_OWNER_ROLE	"Personally Identifiable Information → Non-Sensitive
PII_S_NAME	MONITOR_OWNER	MONITOR	MONITOR_OWNER_ROLE	"Personally Identifiable Information → Sensitive → Nɑ

Figure 6-2. *Custom tags*

Account Usage Store latency for `tag_references` and tags views is 2 hours.

Please also see the documentation found here: `https://docs.snowflake.com/en/`
`sql-reference/account-usage.html#account-usage-views`.

You can now investigate which objects and attributes have been tagged:

```
SELECT *
FROM    snowflake.account_usage.tag_references
WHERE   object_deleted IS NULL
ORDER BY tag_name;
```

Figure 6-3 illustrates sample results showing the tagged objects and attributes,
noting there will be more attributes in actual results returned.

TAG_DATA	TAG_SCHEMA	TAG_NAME	TAG_VALUE	OBJECT_SCHEMA	COLUMN_NAME
MONITOR	MONITOR_OWNER	PII	Personally Identifi	MONITOR_OWNER	null
MONITOR	MONITOR_OWNER	PII_N_GENDER	Personally Identifi	MONITOR_OWNER	GENDER
MONITOR	MONITOR_OWNER	PII_S_NAME	Personally Identifi	MONITOR_OWNER	FORENAME_PREFE
MONITOR	MONITOR_OWNER	PII_S_NAME	Personally Identifi	MONITOR_OWNER	SURNAME_PREFER
MONITOR	MONITOR_OWNER	PII_S_NAME	Personally Identifi	MONITOR_OWNER	PREFERRED_NAME

Figure 6-3. *Custom tagged objects and attributes*

View `v_employee` does not have tags applied.

Object Tag Propagation

Tags applied to the employee table are not inherited by the view v_employee by default. This behavior is explicitly called out within the Snowflake documentation found here: https://docs.snowflake.com/en/user-guide/object-tagging.html#tag-lineage. The implications of tags not propagating are quite profound. You might reasonably expect tag inheritance to be automatically supported; after all, when views are declared, each attribute is validated against each underlying object and attribute, but the absence of tag inheritance appears to be by design.

You must therefore implement manual deployment of tags on view v_employee to ensure consistency of tags to objects:

```
ALTER TABLE monitor.monitor_owner.v_employee
SET TAG PII = 'Personally Identifiable Information';

ALTER TABLE monitor.monitor_owner.v_employee
MODIFY COLUMN preferred_name
SET TAG PII_S_Name = 'Personally Identifiable Information -> Sensitive
-> Name';

ALTER TABLE monitor.monitor_owner.v_employee
MODIFY COLUMN surname_preferred
SET TAG PII_S_Name = 'Personally Identifiable Information -> Sensitive
-> Name';

ALTER TABLE monitor.monitor_owner.v_employee
MODIFY COLUMN forename_preferred
SET TAG PII_S_Name = 'Personally Identifiable Information -> Sensitive
-> Name';

ALTER TABLE monitor.monitor_owner.v_employee
MODIFY COLUMN gender
SET TAG PII_N_Gender = 'Personally Identifiable Information -> Non-
Sensitive -> Gender';
```

Tags are not automatically propagated from tables to views

Creating View v_tag_references

For later use. create a view named **v_tag_references**, which brings together attributes into a more consistent and usable format for later reuse:

```
CREATE OR REPLACE VIEW v_tag_references COPY GRANTS
AS
SELECT tag_name,
       object_database||'.'||
       object_schema  ||'.'||
       object_name              AS target_object,
       tag_database||'.'||
       tag_schema   ||'.'||
       tag_name                 AS tag_source,
       CASE
          WHEN column_name IS NULL THEN object_database||'.'||object_
          schema||'.'||object_name
          ELSE object_database||'.'||object_schema||'.'||object_
          name||'.'||column_name
       END                      AS target,
       domain
FROM    snowflake.account_usage.tag_references
WHERE   object_deleted IS NULL
ORDER BY tag_name, target_object, target;
```

Please pay particular attention to the attribute named `target_object` as you will use this attribute later to join with role information.

While writing this chapter my testing was impacted due to Account Usage Store latency for `tag_references` and `tags` views of up to 2 hours.

Prove your new view correctly returns expected content:

```
SELECT *
FROM   v_tag_references;
```

Figure 6-4 illustrates the expected output.

TAG_NAME	TARGET_OBJECT	TAG_SOURCE	TARGET	...	DOMAIN
PII	MONITOR.MONITOR_OWNER.EMPLC	MONITOR.MONITOR_OWNER.PII	MONITOR.MONITOR_OWNER.EMPLOYEE		TABLE
PII	MONITOR.MONITOR_OWNER.V_EMP	MONITOR.MONITOR_OWNER.PII	MONITOR.MONITOR_OWNER.V_EMPLOYEE		TABLE
PII_N_GENDER	MONITOR.MONITOR_OWNER.EMPLC	MONITOR.MONITOR_OWNER.PII_N_	MONITOR.MONITOR_OWNER.EMPLOYEE.GENDER		COLUMN
PII_N_GENDER	MONITOR.MONITOR_OWNER.V_EMP	MONITOR.MONITOR_OWNER.PII_N_	MONITOR.MONITOR_OWNER.V_EMPLOYEE.GENDER		COLUMN
PII_S_NAME	MONITOR.MONITOR_OWNER.EMPLC	MONITOR.MONITOR_OWNER.PII_S_	MONITOR.MONITOR_OWNER.EMPLOYEE.FORENAME_PREFERRED		COLUMN
PII_S_NAME	MONITOR.MONITOR_OWNER.EMPLC	MONITOR.MONITOR_OWNER.PII_S_	MONITOR.MONITOR_OWNER.EMPLOYEE.PREFERRED_NAME		COLUMN
PII_S_NAME	MONITOR.MONITOR_OWNER.EMPLC	MONITOR.MONITOR_OWNER.PII_S_	MONITOR.MONITOR_OWNER.EMPLOYEE.SURNAME_PREFERRED		COLUMN
PII_S_NAME	MONITOR.MONITOR_OWNER.V_EMP	MONITOR.MONITOR_OWNER.PII_S_	MONITOR.MONITOR_OWNER.V_EMPLOYEE.FORENAME_PREFERRED		COLUMN
PII_S_NAME	MONITOR.MONITOR_OWNER.V_EMP	MONITOR.MONITOR_OWNER.PII_S_	MONITOR.MONITOR_OWNER.V_EMPLOYEE.PREFERRED_NAME		COLUMN
PII_S_NAME	MONITOR.MONITOR_OWNER.V_EMP	MONITOR.MONITOR_OWNER.PII_S_	MONITOR.MONITOR_OWNER.V_EMPLOYEE.SURNAME_PREFERRED		COLUMN

Figure 6-4. *Object tags deployed to target objects*

You now have a single point of reference to determine object tags deployed to target objects for late use.

Tag Visibility by Role

The previous section declared tags within `monitor.monitor_owner` owned by role `self_service_owner_role`, but for best practice you should create a reader role named `self_service_reader_role` with entitlement to access `monitor.monitor_owner` only with entitlement to SELECT from summary VIEWS of which you created the first above, `v_tag_references`:

```
USE ROLE securityadmin;

CREATE OR REPLACE ROLE self_service_reader_role;

GRANT USAGE    ON DATABASE  MONITOR                   TO ROLE self_service_
reader_role;
GRANT USAGE    ON WAREHOUSE monitor_wh                TO ROLE self_service_
reader_role;
GRANT OPERATE ON WAREHOUSE monitor_wh                 TO ROLE self_service_
reader_role;
GRANT USAGE    ON SCHEMA     monitor.monitor_owner TO ROLE self_service_
reader_role;
```

Grant select on all current and future views in `monitor.monitor_owner` to role `self_service_reader_role`:

```
GRANT SELECT ON ALL    VIEWS IN SCHEMA monitor.monitor_owner TO ROLE self_
service_reader_role;
GRANT SELECT ON FUTURE VIEWS IN SCHEMA monitor.monitor_owner TO ROLE self_
service_reader_role;
```

Grant role self_service_reader_role to sysadmin (in line with Snowflake best practice recommendations) and to your user:

```
GRANT ROLE self_service_reader_role TO ROLE sysadmin;
GRANT ROLE self_service_reader_role TO USER <Your User Here>;
```

Set context:

```
USE ROLE      self_service_reader_role;
USE WAREHOUSE monitor_wh;
USE DATABASE  monitor;
USE SCHEMA    monitor.monitor_owner;
```

Attempt to view tags in schema:

```
SHOW tags IN SCHEMA monitor.monitor_owner;
```

Figure 6-5, repeated from above, illustrates sample results, noting there will be more attributes in actual results returned.

name	database_n	schema_name	owner	comment
PII	MONITOR	MONITOR_OWNER	MONITOR_OWNER_ROLE	Personally l(
PII_N_GENDER	MONITOR	MONITOR_OWNER	MONITOR_OWNER_ROLE	Personally l(
PII_S_NAME	MONITOR	MONITOR_OWNER	MONITOR_OWNER_ROLE	Personally l(

Figure 6-5. *Custom tags*

Let's test which Snowflake roles can see your custom tags. We leave you to prove the assertion for each role, but here are our results:

- ACCOUNTADMIN: SHOW tags IN SCHEMA monitor.monitor_owner works fine.

- SECURITYADMIN: SHOW tags IN SCHEMA monitor.monitor_owner works fine.

- SYSADMIN: `SHOW tags IN SCHEMA monitor.monitor_owner`
  works fine.

- USERADMIN: `SHOW tags IN SCHEMA monitor.monitor_owner` fails.

- PUBLIC: `SHOW tags IN SCHEMA monitor.monitor_owner` fails.

As seen above, your custom role of `self_service_reader_role` works fine too.

Entitlement to use a database and schema are prerequisites to SHOW custom tags.

However, using your custom role of `self_service_reader_role`, you will not be able to view the tags using Snowflake Account Usage Store unless you explicitly grant entitlement, as this fragment illustrates:

```
USE ROLE self_service_reader_role;

SELECT *
FROM   v_tag_references;
```

Did you notice anything unusual?

The role `self_service_reader_role` does not have entitlement to view Snowflake Account Usage Store objects but the view `v_tag_references` is based upon `snowflake.account_usage.tag_references` and inherits the defining role (`self_service_owner_role`) entitlement.

The comparable query throws an error: `SQL compilation error: Database 'SNOWFLAKE' does not exist or not authorized`.

```
SELECT *
FROM   snowflake.account_usage.tag_references
ORDER BY tag_name;
```

Having delivered the capability to search for objects by tag, let's move on to discuss self-discovery of accessible objects.

Searching by RBAC

In this section, you deep dive into searching for accessible objects and attributes, representing an assumed starting point for self-service of searchable content from entitled objects and attributes using RBAC roles as a starting point.

As an aside, let's look at how to identify the roles available to your user. Your results may differ according to how your user has been provisioned.

```
SELECT current_available_roles();
```

In this example, the user is the default for a trial account, therefore all Snowflake-supplied roles plus those created for this chapter are available, as shown in Figure 6-6.

CURRENT_AVAILABLE_ROLES()
["ACCOUNTADMIN", "ORGADMIN", "PUBLIC", "SECURITYADMIN", "SELF_SERVICE_OWNER_ROLE", "SELF_SERVICE_READER_ROLE", "SYSADMIN", "USERADMIN"]

Figure 6-6. *Available user roles*

The next step is to find all objects with both DML and Data Definition Language (DDL) entitlement granted.

You must revert your role to `monitor_owner_role`, which has entitlement to access the Snowflake Account Usage Store:

```
USE ROLE monitor_owner_role;
```

To determine objects accessible by role `self_service_reader_role`, you must first grant entitlement:

```
GRANT SELECT, REFERENCES ON v_tag_references TO ROLE self_service_
reader_role;
```

Wait for latency of up to 2 hours to be resolved as defined here: `https://docs.snowflake.com/en/sql-reference/account-usage/grants_to_roles.html#grants-to-roles-view`.

Find objects with entitlement granted to the `self_service_reader_role` role:

```
SELECT *
FROM   snowflake.account_usage.grants_to_roles
WHERE  deleted_on IS NULL
AND    table_schema = 'MONITOR_OWNER'
```

```
AND     grantee_name = 'SELF_SERVICE_READER_ROLE'
AND     granted_to   = 'ROLE'
ORDER BY name;
```

A subset of the attributes returned from the above query is shown in Figure 6-7.

PRIVILEGE	GRANTEE	NAME	TABLE_CAT	TABLE_SCHEMA···	GRANTEE_NAME	GRANTED_BY
USAGE	SCHEMA	MONITOR_OWNER	MONITOR	MONITOR_OWNER	SELF_SERVICE_READER_ROLE	SYSADMIN
SELECT	VIEW	V_EMPLOYEE	MONITOR	MONITOR_OWNER	SELF_SERVICE_READER_ROLE	MONITOR_OWNER_ROLE
REFERENCES	VIEW	V_TAG_REFERENCES	MONITOR	MONITOR_OWNER	SELF_SERVICE_READER_ROLE	MONITOR_OWNER_ROLE
SELECT	VIEW	V_TAG_REFERENCES	MONITOR	MONITOR_OWNER	SELF_SERVICE_READER_ROLE	MONITOR_OWNER_ROLE

Figure 6-7. *Objects accessible by role self_service_reader_role*

Creating View v_role_objects

For ease of later use, create a view named v_role_objects:

```
CREATE OR REPLACE VIEW v_role_objects COPY GRANTS
AS
SELECT grantee_name,
       granted_by,
       granted_to,
       table_catalog||'.'||
       table_schema ||'.'||
       name                    target_object,
       granted_on,
       privilege,
       grant_option
FROM    snowflake.account_usage.grants_to_roles
WHERE   deleted_on IS NULL
ORDER BY name;
```

And now prove the results obtained from view v_role_objects are as expected:

```
SELECT *
FROM   v_role_objects;
```

Figure 6-8 illustrates a subset of the results obtained.

GRANTEE_NAME	GRANTED_BY	TARGET_OBJECT	GRANTED_ON	PRIVILEGE	GRANT_OPTION
ACCOUNTADMIN	ACCOUNTADMIN	null	USER	OWNERSHIP	TRUE
MONITOR_OWNER_ROLE	MONITOR_OWNER_ROLE	MONITOR.MONITOR_OWNER.EMPLOYEE	TABLE	OWNERSHIP	TRUE
SECURITYADMIN	null	null	ACCOUNT	CREATE POLIC	TRUE
ACCOUNTADMIN	null	null	ACCOUNT	EXECUTE TAS	TRUE
SECURITYADMIN	null	null	ACCOUNT	ATTACH POLIC	TRUE
ORGADMIN	null	null	ACCOUNT	MANAGE ORG	TRUE

Figure 6-8. *Role object view results*

From the results shown in Figure 6-8 you can derive the following information:

- All Snowflake-supplied roles are represented with their entitlement grants.

- All user-defined roles are represented with both their role and object entitlement grants.

- From the entitlement granted to each role, using the grant_option attribute, you can determine whether entitlement can be granted to other roles and users.

You may be tempted to apply filters into the view v_role_objects declaration to remove known "good" entitlement combination subsets. We caution against filtering as an open un-filtered approach offers the greatest utility for your view. Instead, we prefer to leave any filtering to consumption from v_role_objects, as the next section demonstrates.

You should maintain a known "good" list of all entitlements granted by default when an account is created and use it as a master reference in the event additional entitlement is granted to a Snowflake-supplied role. One real-world scenario where we have encountered additional entitlement being granted to Snowflake-supplied roles was when investigating the creation of Private Listings. The impact of the additional entitlement grant was the code would not deploy to other environments without identical entitlement being granted. For this reason, you should avoid granting additional entitlement to Snowflake-supplied roles.

Identifying Toxic Entitlement Combinations

Using view v_role_objects you can begin to address toxic entitlement combinations, which for the purposes of your investigation you define as those roles or combination of roles that entitle a user to conduct both DML and DDL operations against the same objects.

You first fetch the roles assigned to each user, excluding deleted roles and those roles granted on startup:

```
SELECT role,
       grantee_name
FROM   snowflake.account_usage.grants_to_users
WHERE  deleted_on IS NULL
AND    granted_by IS NOT NULL;
```

Using the returned roles for each user, you check that the entitlements do not clash by joining to your newly defined view v_role_objects. Note the SQL statement supplied next is indicative only and the results require interpretation. Furthermore, the results have been filtered by explicitly excluding entitlement granted by the SYSADMIN role. You may wish to remove the SYSADMIN filter and experiment with the view declaration.

```
SELECT ro.grantee_name,
       ro.target_object,
       ro.granted_on,
       ro.privilege,
       ro.grant_option,
       ro.granted_by,
       rg.granted_to,
       rg.grantee_name
FROM   v_role_objects                        ro,
       snowflake.account_usage.grants_to_users rg
WHERE  ro.grantee_name = rg.role
AND    rg.deleted_on IS NULL
AND    rg.granted_by IS NOT NULL
AND    ro.granted_by != 'SYSADMIN'
ORDER BY ro.target_object, ro.privilege ASC;
```

Figure 6-9 shows the expected resultset.

GRANTEE_NAME	TARGET_OBJECT	GRANTE	PRIVILEGE	GRANT_(	GRANTED_BY	GRANTED_	GRANTEE
MONITOR_OWNER_ROLE	MONITOR.MONITOR_OWNER.EMPLOYEE	TABLE	OWNERSHIP	TRUE	MONITOR_OWNE	USER	ANDYC
MONITOR_OWNER_ROLE	MONITOR.MONITOR_OWNER.V_EMPLOYEE	VIEW	OWNERSHIP	TRUE	MONITOR_OWNE	USER	ANDYC
SELF_SERVICE_READER_R	MONITOR.MONITOR_OWNER.V_EMPLOYEE	VIEW	SELECT	FALSE	MONITOR_OWNE	USER	ANDYC
MONITOR_OWNER_ROLE	MONITOR.MONITOR_OWNER.V_ROLE_OBJECTS	VIEW	OWNERSHIP	TRUE	MONITOR_OWNE	USER	ANDYC
SELF_SERVICE_READER_R	MONITOR.MONITOR_OWNER.V_ROLE_OBJECTS	VIEW	SELECT	FALSE	MONITOR_OWNE	USER	ANDYC
MONITOR_OWNER_ROLE	MONITOR.MONITOR_OWNER.V_TAG_REFERENCES	VIEW	OWNERSHIP	TRUE	MONITOR_OWNE	USER	ANDYC
SELF_SERVICE_READER_R	MONITOR.MONITOR_OWNER.V_TAG_REFERENCES	VIEW	REFERENCES	FALSE	MONITOR_OWNE	USER	ANDYC
SELF_SERVICE_READER_R	MONITOR.MONITOR_OWNER.V_TAG_REFERENCES	VIEW	SELECT	FALSE	MONITOR_OWNE	USER	ANDYC

Figure 6-9. *Entitlements by role and user*

Having obtained the results, what does Figure 6-9 tell you? The first important point is to note the canonical notation for `target_object`; this is where objects reside within your account. Then you must check the `grantee_name`, which is the role to which entitlement is granted and the privilege (entitlement) granted.

With these three pieces of information, you can compare information across rows and determine whether any toxic combinations exist. In this example, all is well; there are no toxic combinations.

To illustrate a toxic combination, let's create one:

```
USE ROLE securityadmin;
```

```
GRANT INSERT ON monitor.monitor_owner.employee TO ROLE self_service_
reader_role;
```

Reset your role:

```
USE ROLE monitor_owner_role;
```

After waiting for Account Usage Store latency, rerun the query from above. Figure 6-10 shows the expected resultset.

GRANTEE_NAME	TARGET_OBJECT	GRANTED	PRIVILEGE	GRANT_C	GRANTED_BY	GRANTED_	GRANTEE
SELF_SERVICE_READER_R	MONITOR.MONITOR_OWNER.EMPLOYEE	TABLE	INSERT	FALSE	MONITOR_OWNER_ROLE	USER	ANDYC
MONITOR_OWNER_ROLE	MONITOR.MONITOR_OWNER.EMPLOYEE	TABLE	OWNERSHIP	TRUE	MONITOR_OWNER_ROLE	USER	ANDYC
MONITOR_OWNER_ROLE	MONITOR.MONITOR_OWNER.V_EMPLOYEE	VIEW	OWNERSHIP	TRUE	MONITOR_OWNER_ROLE	USER	ANDYC
SELF_SERVICE_READER_R	MONITOR.MONITOR_OWNER.V_EMPLOYEE	VIEW	SELECT	FALSE	MONITOR_OWNER_ROLE	USER	ANDYC
MONITOR_OWNER_ROLE	MONITOR.MONITOR_OWNER.V_ROLE_OBJECTS	VIEW	OWNERSHIP	TRUE	MONITOR_OWNER_ROLE	USER	ANDYC
SELF_SERVICE_READER_R	MONITOR.MONITOR_OWNER.V_ROLE_OBJECTS	VIEW	SELECT	FALSE	MONITOR_OWNER_ROLE	USER	ANDYC
MONITOR_OWNER_ROLE	MONITOR.MONITOR_OWNER.V_TAG_REFERENCES	VIEW	OWNERSHIP	TRUE	MONITOR_OWNER_ROLE	USER	ANDYC
SELF_SERVICE_READER_R	MONITOR.MONITOR_OWNER.V_TAG_REFERENCES	VIEW	REFERENCES	FALSE	MONITOR_OWNER_ROLE	USER	ANDYC
SELF_SERVICE_READER_R	MONITOR.MONITOR_OWNER.V_TAG_REFERENCES	VIEW	SELECT	FALSE	MONITOR_OWNER_ROLE	USER	ANDYC

Figure 6-10. *Toxic entitlement combination identified*

In order to explain Figure 6-10, you should understand that the role of self_
service_reader_role is intended to enable SELECT access to specific objects only.
The role must not be able to manipulate data within the database; therefore, having
the INSERT privilege breaches the role intent. Strict enforcement of entitlement to roles
according to their intended function is a core tenet of your security model.

Note this is a point-in-time check. Entitlements may change over time. We
recommend that role entitlements be captured along with timestamp and therefore
propose a view named v_role_object_entitlement to encapsulate a check query:

```
CREATE OR REPLACE VIEW v_role_object_entitlement COPY GRANTS
AS
SELECT ro.grantee_name,
       ro.target_object,
       ro.granted_on,
       ro.privilege,
       ro.grant_option,
       ro.granted_by,
       rg.granted_to,
       rg.grantee_name      AS grant_to_role_or_user,
       current_timestamp()  AS extract_timestamp
FROM   v_role_objects                          ro,
       snowflake.account_usage.grants_to_users rg
WHERE  ro.grantee_name = rg.role
AND    rg.deleted_on IS NULL
AND    rg.granted_by IS NOT NULL
AND    ro.granted_by != 'SYSADMIN'
ORDER BY ro.target_object, ro.privilege ASC;
```

For the sake of completeness, enable role `self_service_reader_role` to access view `v_role_object_entitlement`:

```
GRANT SELECT, REFERENCES ON monitor.monitor_owner.v_role_object_entitlement
TO ROLE self_service_reader_role;
```

And prove you can access view `v_role_object_entitlement` from role `self_service_reader_role`:

```
USE ROLE self_service_reader_role;
```

```
SELECT *
FROM    v_role_object_entitlement;
```

Note that you should not expect to see view `v_role_object_entitlement` in the resultset until latency has expired.

Extending Monitoring Capability

Identifying toxic role combinations is only one use case for the data set returned by view `v_role_object_entitlement`. You can also use the returned data set for

- Identifying object entitlement to roles

- Identifying roles granted to other roles or users

- Identifying object types for declared objects

- Identifying whether entitlement can be propagated or devolved to other roles and users

We leave you to determine the appropriate controls as required by your organization and amend or extend the template code supplied within this section.

Integrating Object Tags and Object Access

You might have entitlement to view tags, or entitlement to view objects, but not see both tags AND entitlement within a single role. What you need is a single view that combines both tag information and role information.

In the next query you perform an OUTER JOIN using the (+) syntax, which may be unfamiliar. This is a legacy construct from Oracle SQL, we believe, as a shortcut, and it supports porting of Oracle code to Snowflake without refactoring to use ANSI standard syntax. You may wish to refactor the code sample.

The next SQL query identifies all objects regardless of whether they have been tagged or not and is readily extended to meet your purposes by adding any or all attributes from source views.

```
SELECT  tr.tag_name,
        roe.target_object,
        tr.tag_source,
        tr.target,
        tr.domain,
        roe.granted_by,
        roe.granted_to,
        roe.privilege
FROM    v_role_object_entitlement roe,
        v_tag_references           tr
WHERE   roe.target_object = tr.target_object (+)
AND     roe.privilege      = 'SELECT';
```

To persist the above SQL statement as a view, you first change role to the monitor_ owner_role and then create view v_tag_role_object_entitlement:

```
USE ROLE monitor_owner_role;
```

This example excludes OWNERSHIP and REFERENCES entitlements granted to roles:

```
CREATE OR REPLACE VIEW v_tag_role_object_entitlement COPY GRANTS
AS
SELECT  tr.tag_name,
        roe.target_object,
        tr.tag_source,
        tr.target,
        tr.domain,
        roe.granted_by,
        roe.granted_to,
        roe.privilege
```

```
FROM    v_role_object_entitlement roe,
        v_tag_references            tr
WHERE   roe.target_object           = tr.target_object (+)
AND     roe.privilege       NOT IN ( 'OWNERSHIP', 'REFERENCES' );
```

Then grant SELECT to role self_service_reader_role:

```
GRANT SELECT ON monitor.monitor_owner.v_tag_role_object_entitlement TO ROLE
self_service_reader_role;
```

And prove that the role self_service_reader_role can access view v_tag_role_object_entitlement:

```
USE ROLE self_service_reader_role;
```

```
SELECT *
FROM    v_tag_role_object_entitlement
ORDER BY target_object ASC;
```

Figure 6-11 illustrates sample results showing the tags, objects, and entitlement, noting there will be more attributes in your actual results returned.

TAG_NAME	TARGET_OBJECT	TAG_SOURCE	DOMAIN	GRANT	PRIVILE(
PII	MONITOR.MONITOR_OWNER.EMPLOYEE	MONITOR.MONITOR_OWNER.PII	TABLE	USER	INSERT
PII_S_NAME	MONITOR.MONITOR_OWNER.EMPLOYEE	MONITOR.MONITOR_OWNER.PII_S_NAME	COLUMN	USER	INSERT
PII_S_NAME	MONITOR.MONITOR_OWNER.EMPLOYEE	MONITOR.MONITOR_OWNER.PII_S_NAME	COLUMN	USER	INSERT
PII_S_NAME	MONITOR.MONITOR_OWNER.EMPLOYEE	MONITOR.MONITOR_OWNER.PII_S_NAME	COLUMN	USER	INSERT
PII_N_GENDER	MONITOR.MONITOR_OWNER.EMPLOYEE	MONITOR.MONITOR_OWNER.PII_N_GENC	COLUMN	USER	INSERT
PII	MONITOR.MONITOR_OWNER.V_EMPLOYEE	MONITOR.MONITOR_OWNER.PII	TABLE	USER	SELECT
PII_N_GENDER	MONITOR.MONITOR_OWNER.V_EMPLOYEE	MONITOR.MONITOR_OWNER.PII_N_GENC	COLUMN	USER	SELECT
PII_S_NAME	MONITOR.MONITOR_OWNER.V_EMPLOYEE	MONITOR.MONITOR_OWNER.PII_S_NAME	COLUMN	USER	SELECT
PII_S_NAME	MONITOR.MONITOR_OWNER.V_EMPLOYEE	MONITOR.MONITOR_OWNER.PII_S_NAME	COLUMN	USER	SELECT
PII_S_NAME	MONITOR.MONITOR_OWNER.V_EMPLOYEE	MONITOR.MONITOR_OWNER.PII_S_NAME	COLUMN	USER	SELECT
null	MONITOR.MONITOR_OWNER.V_ROLE_OBJECTS	null	null	USER	SELECT
null	MONITOR.MONITOR_OWNER.V_ROLE_OBJECT_EN'	null	null	USER	SELECT
null	MONITOR.MONITOR_OWNER.V_TAG_REFERENCES	null	null	USER	SELECT

Figure 6-11. *Objects with tags and entitlements*

With the unified view `v_tag_role_object_entitlement` you are able to see both tagged and untagged objects. When the view attribute list is extended, you can also determine toxic entitlement combinations along with any extended monitoring capability as required by your organization.

Searching by Row-Level Security (RLS)

RLS is implemented using row access policies. Essentially, RLS allows data subsets to be accessed according to specified criteria. Historically, you might have deployed views to explicitly filter each subset of data and serve content to disparate audiences, but now Snowflake has provisioned capability to prefilter datasets using RLS.

In this section, you focus on a simple RLS implementation to satisfy usage reporting only. For a detailed explanation of row access policies, please refer to the documentation found here: `https://docs.snowflake.com/en/user-guide/security-row-intro.html#understanding-row-access-policies`.

The aim is to create a row access policy that will contain very limited functionality. It is the row access policy as a container you are interested in, and not the inherent capability.

For row access policies, there are two Account Usage Store views from which you source information: `row_access_policies` and `policy_references`.

Account Usage Store latency for `row_access_policies` and `policy_references` views is 2 hours. Your testing may therefore be impacted.

Setup

Before attempting to identify row access policies, you must set up your test environment.

You must entitle your role `monitor_owner_role` to create row access policies:

```
USE ROLE securityadmin;

GRANT CREATE ROW ACCESS POLICY ON SCHEMA monitor.monitor_owner TO ROLE
monitor_owner_role;
```

Now apply row access policies, noting you do this at the ACCOUNT level. For other options, refer to the documentation found here: https://docs.snowflake.com/en/ user-guide/security-row-intro.html#summary-of-ddl-commands-operations-and-privileges.

```
USE ROLE accountadmin;

GRANT APPLY ROW ACCESS POLICY ON ACCOUNT TO ROLE monitor_owner_role;
```

With the administration complete, revert to the monitor role:

```
USE ROLE      monitor_owner_role;
USE WAREHOUSE monitor_wh;
USE DATABASE  monitor;
USE SCHEMA    monitor.monitor_owner;
```

We assume there are no rows in the employee table. If you have entered data, TRUNCATE the employee table first:

```
INSERT INTO employee
VALUES
('Jonathan Doe', 'Doe', 'John', 'Male'   ),
('Jane Doe',     'Doe', 'Jane', 'Female' );
```

Then check that two rows are returned:

```
SELECT * FROM employee;
```

Before you define your row access policy, please be aware that you can't reference an object within the code to which you will apply the policy. Attempting to do so results in this error: Policy body contains a UDF or Select statement that refers to a Table attached to another Policy.

Now define a row access policy, noting the overly simple limitation that data is returned only if the consuming role is self_service_reader_role:

```
CREATE OR REPLACE ROW ACCESS POLICY monitor.monitor_owner.emp_policy AS
(gender VARCHAR)
RETURNS BOOLEAN ->
   'SELF_SERVICE_READER_ROLE' = current_role();
```

Apply your new policy emp_policy to table employee and attribute gender:

```
ALTER TABLE monitor.monitor_owner.employee ADD ROW ACCESS POLICY monitor_
owner.emp_policy ON (gender);
```

Now grant the SELECT privilege on the employee table to role self_service_
reader_role:

```
GRANT SELECT ON TABLE monitor.monitor_owner.employee TO ROLE self_service_
reader_role;
```

Confirm your current role is monitor_owner_role:

```
SELECT current_role();
```

Your row access policy should prevent data from being returned when selecting from the employee table:

```
SELECT *
FROM    monitor.monitor_owner.employee;
```

Change your context to use role self_service_reader_role:

```
USE ROLE self_service_reader_role;
```

You should now see two rows, proving your row access policy works as expected:

```
SELECT *
FROM    monitor.monitor_owner.employee;
```

Accessing Row Access Policies

With your setup complete, and proven to work as expected, you move on to using the Snowflake Account Usage Store to extend your usage reporting capability.

First, set your context to use role monitor_owner_role:

```
USE ROLE       monitor_owner_role;
```

Now create a view named v_row_access_policy to select all active row access policies. You may wish to modify the SQL to select all row access policies declared over time.

```
CREATE OR REPLACE VIEW v_row_access_policy COPY GRANTS
AS
SELECT policy_catalog    ||'.'||
       policy_schema     ||'.'||
       policy_name        AS row_access_policy_object,
       policy_owner,
       policy_signature,
       policy_return_type,
       policy_body,
       policy_id
FROM   snowflake.account_usage.row_access_policies
WHERE  deleted IS NULL;
```

Now create a view called v_policy_reference to expose where row access policies are applied:

```
CREATE OR REPLACE VIEW v_policy_reference COPY GRANTS
AS
SELECT policy_db         ||'.'||
       policy_schema     ||'.'||
       policy_name        AS row_access_policy_object,
       ref_database_name||'.'||
       ref_schema_name   ||'.'||
       ref_entity_name    AS object_applied_to,
       ref_entity_domain,
       ref_column_name,
       ref_arg_column_names,
       policy_id
FROM   snowflake.account_usage.policy_references;
```

Join the above views together into view v_row_access_policy_reference to provide a single object where both row access policy declaration and usage are available:

```
CREATE OR REPLACE VIEW v_row_access_policy_reference COPY GRANTS
AS
SELECT rap.row_access_policy_object,
       rap.policy_owner,
       rap.policy_signature,
```

```
        rap.policy_return_type,
        rap.policy_body,
        pr.object_applied_to,
        pr.ref_entity_domain,
        pr.ref_column_name,
        pr.ref_arg_column_names,
        pr.policy_id
FROM    v_row_access_policy rap,
        v_policy_reference  pr
WHERE   rap.policy_id        = pr.policy_id;
```

Prove your results are as expected. You should see a single row as declared within your setup:

```
SELECT *
FROM    v_row_access_policy_reference
ORDER BY row_access_policy_object ASC;
```

Row Access Policy Impact

Determining the impact of a row access policy programmatically is not readily achievable. One exception is where limited, and well understood, use cases are encapsulated within the row access policies with bespoke code used to programmatically interpret the ref_arg_column_names and policy_body attributes. Your mileage will vary because developers don't always implement code according to the local development approach in force, and any change to the local development approach may affect all row access policy implementations.

Noting the point-in-time nature of the sample code and the possibility of change over time, we suggest row access policy interpretation is left to your subject matter experts with reference to documentation.

Entitlement, Tags, and Row Access Policies

Extending your usage reporting capability is relatively simple. The following view called v_tag_role_object_entitlement_policy replaces the v_tag_role_object_ entitlement declared earlier to display row access policies applied to objects:

232

```
CREATE OR REPLACE VIEW v_tag_role_object_entitlement_policy COPY GRANTS
AS
SELECT tr.tag_name,
       roe.target_object,
       tr.tag_source,
       tr.target,
       tr.domain,
       roe.granted_by,
       roe.granted_to,
       roe.privilege,
       rapr.row_access_policy_object,
       rapr.ref_arg_column_names,
       rapr.policy_body,
       rapr.policy_id
FROM   v_role_object_entitlement      roe,
       v_tag_references               tr,
       v_row_access_policy_reference rapr
WHERE  roe.target_object             = tr.target_object      (+)
AND    roe.target_object             = rapr.object_applied_to (+)
AND    roe.privilege       NOT IN ( 'OWNERSHIP', 'REFERENCES' );
```

Grant SELECT to role self_service_reader_role:

```
GRANT SELECT ON monitor.monitor_owner. v_tag_role_object_entitlement_policy
TO ROLE self_service_reader_role;
```

And prove role self_service_reader_role can access view v_tag_role_object_entitlement_policy:

```
USE ROLE self_service_reader_role;
```

```
SELECT *
FROM   v_tag_role_object_entitlement_policy
ORDER BY target_object ASC;
```

Cleanup

To remove your row access policy, you must first change the role to `monitor_owner_role`:

```
USE ROLE        monitor_owner_role;
```

Then disassociate the row access policy from the `employee` table:

```
ALTER TABLE monitor.monitor_owner.employee DROP ROW ACCESS POLICY monitor_owner.emp_policy;
```

Finally, drop the row access policy:

```
DROP ROW ACCESS POLICY monitor.monitor_owner.emp_policy (VARCHAR);
```

Searching by Masking Policy

Dynamic data masking (DDM) is implemented using masking policies. Essentially, DDM allows individual attribute values to be either fully or partially masked according to the role accessing the data. Historically, you may have deployed views to explicitly mask specific attributes and serve content to disparate audiences, but this approach is manual, error prone, cumbersome, and hard to maintain in large estates. With DDM, Snowflake has provisioned in-built capability making the management, control, and deployment much easier.

In this section, you focus on a simple DDM implementation to satisfy usage reporting only. For a detailed explanation of DDM, please refer to the documentation found here: `https://docs.snowflake.com/en/user-guide/security-column-ddm-intro.html#understanding-dynamic-data-masking`.

The aim is to create a data masking policy that will contain very limited functionality. It is the data masking policy as a container you are interested in, and not the inherent capability.

For masking policies, there are two Account Usage Store views from which you source information: `masking_policies` and `policy_references`.

Setup

Before attempting to identify data masking policies, you must set up your test environment.

You must entitle your role `monitor_owner_role` to create row access policies:

```
USE ROLE securityadmin;
```

```
GRANT CREATE MASKING POLICY ON SCHEMA monitor.monitor_owner to ROLE
monitor_owner_role;
```

Now apply masking policies, noting you do this at the ACCOUNT level. For other options, refer to the documentation found here: `https://docs.snowflake.com/en/user-guide/security-column-ddm-use.html#id1`.

```
USE ROLE accountadmin;
```

```
GRANT APPLY  MASKING POLICY ON ACCOUNT TO ROLE monitor_owner_role;
```
With the administration complete, revert to the monitor role:

```
USE ROLE      monitor_owner_role;
USE WAREHOUSE monitor_wh;
USE DATABASE  monitor;
USE SCHEMA    monitor.monitor_owner;
```

Define a masking policy, noting the overly simple limitation that clear text data is returned if the consuming role is `monitor_owner_role`; otherwise data is masked and returned as ********.

```
CREATE OR REPLACE MASKING POLICY monitor.monitor_owner.masking_policy AS
(p_param VARCHAR)
RETURNS STRING ->
CASE
    WHEN current_role() IN ( 'MONITOR_OWNER_ROLE' ) THEN p_param
    ELSE '********'
END;
```

Attempt to apply your masking policy to `employee.gender`:

```
ALTER TABLE monitor.monitor_owner.employee
MODIFY COLUMN gender
SET MASKING POLICY masking_policy;
```

Assuming you have been working through this chapter sequentially, you should see this error: `Column 'GENDER' cannot be masked because it is used as argument by one or more masking/row-access policies.`

Referring back to the section on row access policies, you can see gender is already referenced so it causes the error message above.

Instead of referencing gender, let's use another attribute called `preferred_name`:

```
ALTER TABLE monitor.monitor_owner.employee
MODIFY COLUMN preferred_name
SET MASKING POLICY masking_policy;
```

Confirm your current role is `monitor_owner_role`:

```
SELECT current_role();
```

While you are testing your masking policy, the Our Row Access Policy will prevent data from being returned when selecting from the `employee` table. The absence of a result set is a consequence of the interaction between your row access policy and masking policy:

```
SELECT *
FROM    monitor.monitor_owner.employee;
```

Change your context to use role `self_service_reader_role`:

```
USE ROLE self_service_reader_role;
```

You should now see two rows with `preferred_name` set to ********, proving your masking policy works as expected:

```
SELECT *
FROM    monitor.monitor_owner.employee;
```

Figure 6-12 illustrates the effect of implementing DDM on the `employee.preferred_name` attribute.

PREFERRED_NAME	SURNAME_PREFERRED	FORENAME_PREFERRED	GENDER
********	Doe	John	Male
********	Doe	Jane	Female

Figure 6-12. *DDM applied to an attribute*

Accessing Data Masking Policies

With your setup complete, and proven to work as expected, you move on to using the Snowflake Account Usage Store to extend your usage reporting capability.

First, set your context to use role monitor_owner_role:

```
USE ROLE      monitor_owner_role;
```

Now create a view called v_masking_policy to select all active row access policies. You may wish to modify the SQL to select all row access policies declared over time.

```
CREATE OR REPLACE VIEW v_masking_policy COPY GRANTS
AS
SELECT policy_catalog    ||'.'||
       policy_schema     ||'.'||
       policy_name       AS masking_policy_object,
       policy_owner,
       policy_signature,
       policy_return_type,
       policy_body,
       policy_id
FROM   snowflake.account_usage.masking_policies
WHERE  deleted IS NULL;
```

View v_policy_reference will now contain two rows, one for your row access policy and one for your masking policy:

```
SELECT *
FROM   v_policy_reference;
```

Join the above views together into view v_masking_policy_reference to provide a single object where the masking policy declaration and usage are available:

```
CREATE OR REPLACE VIEW v_masking_policy_reference COPY GRANTS
AS
SELECT mp.masking_policy_object,
       mp.policy_owner,
       mp.policy_signature,
       mp.policy_return_type,
       mp.policy_body,
       pr.object_applied_to,
       pr.ref_entity_domain,
       pr.ref_column_name,
       pr.ref_arg_column_names,
       pr.policy_id
FROM   v_masking_policy   mp,
       v_policy_reference pr
WHERE  mp.policy_id        = pr.policy_id;
```

Prove your results are as expected. You should see a single row as declared within your setup:

```
SELECT *
FROM   v_masking_policy_reference
ORDER BY masking_policy_object ASC;
```

Masking Policy Impact

Determining the impact of a masking policy programmatically is not readily achievable. One exception is where limited, and well understood, use cases are encapsulated within the masking policies with bespoke code used to programmatically interpret the ref_arg_column_names and policy_body attributes.

Entitlement, Tags, Row Access Policies, and Masking

Extending your usage reporting capability is relatively simple. The following view of v_tag_role_object_entitlement_policy_masking replaces the v_tag_role_object_entitlement_policy declared earlier to display row access policies applied to objects:

```
CREATE OR REPLACE VIEW v_tag_role_object_entitlement_policy_masking
COPY GRANTS
AS
SELECT  tr.tag_name,
        roe.target_object,
        tr.tag_source,
        tr.target,
        tr.domain,
        roe.granted_by,
        roe.granted_to,
        roe.privilege,
        rapr.row_access_policy_object,
        rapr.ref_arg_column_names       AS row_access_ref_arg_column_names,
        rapr.policy_body                AS row_access_policy_body,
        rapr.policy_id                  AS row_access_policy_id,
        mpr.masking_policy_object,
        mpr.ref_arg_column_names        AS masking_ref_arg_column_names,
        mpr.policy_body                 AS masking_policy_body,
        mpr.policy_id                   AS masking_policy_id
FROM    v_role_object_entitlement       roe,
        v_tag_references                tr,
        v_row_access_policy_reference    rapr,
        v_masking_policy_reference       mpr
WHERE   roe.target_object               = tr.target_object      (+)
AND     roe.target_object               = rapr.object_applied_to (+)
AND     roe.target_object               = mpr.object_applied_to  (+)
AND     roe.privilege       NOT IN ( 'OWNERSHIP', 'REFERENCES' );
```

Now grant SELECT to role self_service_reader_role:

```
GRANT SELECT ON monitor.monitor_owner. v_tag_role_object_entitlement_
policy_masking TO ROLE self_service_reader_role;
```

Prove role self_service_reader_role can access view v_tag_role_object_
entitlement_policy_masking:

```
USE ROLE self_service_reader_role;
```

```
SELECT *
FROM    v_tag_role_object_entitlement_policy_masking
ORDER BY target_object ASC;
```

Cleanup

To remove your masking policy, you must first change the role to monitor_owner_role:

```
USE ROLE       monitor_owner_role;
```

Then disassociate the masking policy from the employee table:

```
ALTER TABLE monitor.monitor_owner.employee
MODIFY COLUMN preferred_name
UNSET MASKING POLICY;
```

Finally, drop the row access policy:

```
DROP MASKING POLICY monitor.monitor_owner.masking_policy (VARCHAR);
```

Searching By Tag-Based Masking

An extension to Snowflake tags and masking policies is the ability to apply a masking policy via a tag. Space does not permit a full investigation of this capability but having articulated test cases for both searching by tag and searching by masking policy we believe you have the tools to further extend usage monitoring and leave this section for your further investigation. Documentation can be found here: https://docs. snowflake.com/en/user-guide/tag-based-masking-policies.html#tag-based-masking-policies.

Who Has Seen What

At the outset of this chapter, we identified several challenges with replicating the exact dataset that would be returned when later replaying captured SQL statements. We now explain how to mitigate risks and offer some information for your further investigation.

Historic Query Capture

Your starting point when determining "who has seen what" must start with capturing queries issued at some point in the past.

You know Snowflake has a 1-year immutable record of all activity conducted within an account. However, queries age out of the history, so to retain full history you must implement periodic archiving, which we address in Chapter 3 for login_history where the same techniques can be reused to archive snowflake.account_usage.query_history.

Use snowflake.account_usage.query_history for the purposes of your investigation and set your context:

```
USE ROLE       monitor_owner_role;
USE WAREHOUSE monitor_wh;
USE DATABASE   monitor;
USE SCHEMA     monitor.monitor_owner;
```

The following code is not exhaustive in scope; it's merely a starting point. The curious amongst us are advised to investigate snowflake.account_usage.query_history as attributes not exposed by the following query will reveal a lot of information for your further investigation.

Any investigation must have a focus, and for this one let's assume you are interested in the most recent 10 SELECT queries issued by a named user ANDYC against database MONITOR and schema MONITOR_OWNER:

```
SELECT query_text,
       role_name,
       execution_status
FROM   snowflake.account_usage.query_history
WHERE  database_name = 'MONITOR'
AND    schema_name   = 'MONITOR_OWNER'
AND    user_name     = 'ANDYC'
AND    query_type    = 'SELECT'
ORDER BY start_time DESC
LIMIT 10;
```

Your query result set will differ from mine according to the work conducted on your account.

With sample queries in hand, we suggest you

- Create some sample tables and populate with data sets.

- Apply a row access policy implementing a filter to subset your data sets.

- Capture sample queries and returned data sets.

- Vary the row access policy.

- Re-execute your query and see how the result sets change.

Slowly Changing Dimension 2 (SCD2)

Implementing SCD2 relies upon the ability to detect a change between an original record and a changed record. There are two approaches to solving this problem, but both rely upon the presence of either a primary key or unique key, both of which may be either a single attribute or composite of two or more attributes within the data set. A further simplification is if the source system identifies the record status (New, Updated, Deleted). We leave this for your further investigation.

A full implementation of SCD2 can be found within my previous book, *Building the Snowflake Data Cloud.*

Naturally, all queries reliant upon consistent unmodified data with SCD2 implemented must use the time band attributes to ensure the returned result sets are consistent.

Some implementations of SCD2 implement `current_flag` to indicate the most recent records. While convenient for any point-in-time queries returning the latest active data, the use of `current_flag` when attempting to retrieve historical data is almost certain to provide incorrect results. Please ensure queries are refactored to use time band attributes only for historical queries.

Row Access Policy Changes

As previously indicated, changes to row access policies between the time of the original query invocation and a later invocation may lead to data differences within the returned result set. You must therefore ensure the row access policy definition remains consistent across the lifetime of the dataset to ensure consistency of returned result set.

Whole Record Hashing

Another technique that may prove useful is to implement a hashing algorithm across all business attributes to ensure record uniqueness. With minimal re-engineering of the concepts presented in my previous book, *Building the Snowflake Data Cloud*, a whole record hashing algorithm may offer benefits.

But two notes of caution:

- The value of a hash is entirely dependent upon the underlying attribute values remaining consistent. Any change to attribute values renders the hash invalid.

- MD5 is quick to calculate but suffers from vulnerabilities including "hash clash." More information can be found here: `https:// en.wikipedia.org/wiki/MD5`.

Time Travel/Fail Safe

Assuming you have enabled Time Travel, you may be tempted to use this capability for reliable historical record retrieval. While attractive, the maximum Time Travel setting is 90 days, therefore all query reruns will suffer the inherent rolling time window limitation of 90 days.

We do not recommend the use of Time Travel to reliably return historical data sets, particularly where end user data access is required (who may be uninformed of the 90-day historical limitation).

Since Fail Safe is a Snowflake feature inaccessible without the involvement of Snowflake support, we exclude the use of Fail Safe as a viable capability for use.

What Did They Do With What They Saw?

After investigating how to identify both how and which data is accessed, you must consider the purpose to which the data was put. In other words, what happened to the data after you accessed it? This is a difficult topic to cover and answering the question will certainly involve other parts of your organization in developing a comprehensive suite of measures to protect your data.

The vast majority of data access within organizations occurs within the bounds of acceptable use and in accordance with the laws and constraints placed upon us. However, there are scenarios where either inadvertent or deliberate acts may cause data usage to exceed acceptable use boundaries and leak outside of our environment.

What we know is that people are inventive, and for those seeking ways to egress data from our organizations, we must safeguard our data. A full description of ways in which we protect our data are beyond the immediate scope of this chapter and some aspects are covered elsewhere within this book.

It is worth knowing that this question is one that hyperscalers such as Amazon, Microsoft, and Google are struggling to answer.

Summary

You began this chapter by exploring the concept of self-service and defining scenarios for both current and historical reporting. You now know there are some limitations and considerations when accessing historical datasets, noting a reliance upon not allowing ad-hoc updates to records within your data warehouse.

You saw a number of data usage constraints and information on how to identify toxic entitlement combinations. Moving onto the core theme of identifying "who can see what," you worked through the means by which data is made accessible to users.

Having identified how data is made accessible, you then investigated how to retrospectively identify SQL queries from the Account Usage Store along with some caveats that may affect your ability to accurately reconstruct accessible data sets back to any given point in time for which your system holds data.

Finally, you explored the difficulty of identifying how viewed data is used, a thorny topic that our industry is attempting to address and for which there are no easy answers.

Now turn your focus to your data catalog, the subject of the next chapter.

CHAPTER 7

Data Catalog

From your investigations so far you have discovered how to identify and interrogate the Snowflake Account Usage Store to derive information useful for a variety of purposes. Now you'll turn your attention to investigating a different aspect of information available from the Account Usage Store: exposing the metadata relating to the data content held within Snowflake. In this chapter, you're not interested in the attribute values themselves; you're much more interested in how to define the containers which hold the data.

For every object and attribute of business significance defined within Snowflake, your data governance colleagues want to know what each object and attribute means in both business and technical terms. To this end, you must implement the access paths to metadata and provision the means for tooling to extract identified metadata, after which, in other tooling, both business and technical information can be added.

Fortunately, accessing Snowflake metadata is quite simple and leverages your existing knowledge from working through this book systematically, and we now show you how.

Note that you may encounter latency when accessing the Account Usage Store views. Always check documentation, which can be found here: `https://docs.snowflake.com/en/sql-reference/account-usage.html#account-usage-views`.

Overview

In general terms, Figure 7-1 illustrates how to access object and attribute metadata. Access is via SQL which may be generated from a variety of tools with resultant data sets either consumed directly or saved to an S3 bucket (not shown) for onward consumption.

© Andrew Carruthers and Sahir Ahmed 2023
A. Carruthers and S. Ahmed, *Maturing the Snowflake Data Cloud*, https://doi.org/10.1007/978-1-4842-9340-9_7

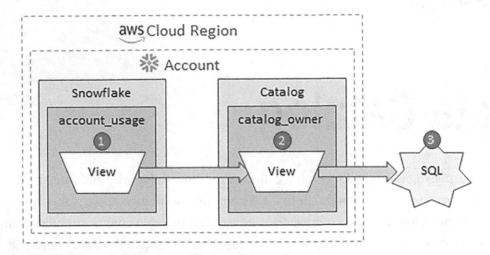

Figure 7-1. *Generic metadata consumption pattern*

Figure 7-1 requires some explanation, and each callout is referenced below.

1. Many views exist within the Account Usage Store. You're only interested in a few for consuming metadata.

2. Provisioning a bespoke data catalog database allows you to overlay specific Account Usage Store views and enrich them with user-defined attributes to meet specific consumption requirements and aid reporting.

3. We do not specify tooling except to state all data catalog database access will be via a role and use SQL. More on this below.

With your metadata access path defined, you can move on to the technical implementation.

Accessing Metadata

This section articulates the steps required to securely expose Snowflake metadata for consumption by internal tooling or vendor-supplied (third-party) tooling.

You begin by creating a Catalog database and associated objects to create a separate container for metadata access and then move on to create bespoke views for metadata consumption.

Catalog Database

Isolating metadata access can be done in several ways. Within this chapter, we prefer to create a Catalog database and offer the below scripts as a starting point. Your implementation may differ. This is a guideline only.

During testing with Snowsight we found deployment was made easier by running sections of the code sequentially rather than attempting to execute all at once.

Start with declarations, which enable later automation (if you wish):

```
SET catalog_database        = 'CATALOG';
SET catalog_owner_schema    = 'CATALOG.catalog_owner';
SET catalog_warehouse       = 'catalog_wh';
SET catalog_owner_role      = 'catalog_owner_role';
SET catalog_reader_role     = 'catalog_reader_role';
```

Change the role to sysadmin and create a database, warehouse, and schema:

```
USE ROLE sysadmin;

CREATE OR REPLACE DATABASE IDENTIFIER ( $catalog_database ) DATA_RETENTION_
TIME_IN_DAYS = 90;

CREATE OR REPLACE WAREHOUSE IDENTIFIER ( $catalog_warehouse ) WITH
WAREHOUSE_SIZE      = 'X-SMALL'
AUTO_SUSPEND        = 60
AUTO_RESUME         = TRUE
MIN_CLUSTER_COUNT   = 1
MAX_CLUSTER_COUNT   = 4
SCALING_POLICY      = 'STANDARD'
INITIALLY_SUSPENDED = TRUE;

CREATE OR REPLACE SCHEMA IDENTIFIER ( $catalog_owner_schema );
```

Change the role to securityadmin, create two new roles named catalog_owner_role and catalog_reader_role, and then grant entitlement:

```
USE ROLE securityadmin;

CREATE OR REPLACE ROLE IDENTIFIER ( $catalog_owner_role  ) COMMENT =
'CATALOG.catalog_owner Role';
CREATE OR REPLACE ROLE IDENTIFIER ( $catalog_reader_role ) COMMENT =
'CATALOG.catalog_reader Role';

GRANT ROLE IDENTIFIER ( $catalog_owner_role  ) TO ROLE securityadmin;
GRANT ROLE IDENTIFIER ( $catalog_reader_role ) TO ROLE securityadmin;
```

Grant entitlement to your newly created roles:

```
GRANT USAGE   ON DATABASE  IDENTIFIER ( $catalog_database     ) TO ROLE
IDENTIFIER ( $catalog_owner_role );
GRANT USAGE   ON WAREHOUSE IDENTIFIER ( $catalog_warehouse    ) TO ROLE
IDENTIFIER ( $catalog_owner_role );
GRANT OPERATE ON WAREHOUSE IDENTIFIER ( $catalog_warehouse    ) TO ROLE
IDENTIFIER ( $catalog_owner_role );
GRANT USAGE   ON SCHEMA    IDENTIFIER ( $catalog_owner_schema ) TO ROLE
IDENTIFIER ( $catalog_owner_role );

GRANT USAGE   ON DATABASE  IDENTIFIER ( $catalog_database     ) TO ROLE
IDENTIFIER ( $catalog_reader_role );
GRANT USAGE   ON WAREHOUSE IDENTIFIER ( $catalog_warehouse    ) TO ROLE
IDENTIFIER ( $catalog_reader_role );
GRANT OPERATE ON WAREHOUSE IDENTIFIER ( $catalog_warehouse    ) TO ROLE
IDENTIFIER ( $catalog_reader_role );
GRANT USAGE   ON SCHEMA    IDENTIFIER ( $catalog_owner_schema ) TO ROLE
IDENTIFIER ( $catalog_reader_role );

GRANT CREATE TABLE          ON SCHEMA IDENTIFIER ( $catalog_owner_
schema    ) TO ROLE IDENTIFIER ( $catalog_owner_role );
GRANT CREATE VIEW           ON SCHEMA IDENTIFIER ( $catalog_owner_
schema    ) TO ROLE IDENTIFIER ( $catalog_owner_role );
```

```
GRANT CREATE SEQUENCE          ON SCHEMA IDENTIFIER ( $catalog_owner_
schema   ) TO ROLE IDENTIFIER ( $catalog_owner_role );
GRANT CREATE FUNCTION          ON SCHEMA IDENTIFIER ( $catalog_owner_
schema   ) TO ROLE IDENTIFIER ( $catalog_owner_role );
GRANT CREATE PROCEDURE         ON SCHEMA IDENTIFIER ( $catalog_owner_
schema   ) TO ROLE IDENTIFIER ( $catalog_owner_role );
GRANT CREATE STREAM            ON SCHEMA IDENTIFIER ( $catalog_owner_
schema   ) TO ROLE IDENTIFIER ( $catalog_owner_role );
GRANT CREATE MATERIALIZED VIEW ON SCHEMA IDENTIFIER ( $catalog_owner_
schema   ) TO ROLE IDENTIFIER ( $catalog_owner_role );
GRANT CREATE FILE FORMAT       ON SCHEMA IDENTIFIER ( $catalog_owner_
schema   ) TO ROLE IDENTIFIER ( $catalog_owner_role );
```

Now grant roles to yourself:

```
GRANT ROLE IDENTIFIER ( $catalog_owner_role  ) TO USER <Your User Here>;
GRANT ROLE IDENTIFIER ( $catalog_reader_role ) TO USER <Your User Here>;
```

Enable access to the Account Usage Store to the role catalog_owner_role:

```
GRANT IMPORTED PRIVILEGES ON DATABASE snowflake TO ROLE IDENTIFIER (
$catalog_owner_role );
```

Creating Catalog Schema Views

Figure 7-2 shows the previously defined roles overlaid on the Account Usage Store and Catalog database, illustrating the scope for which each role interacts with the components.

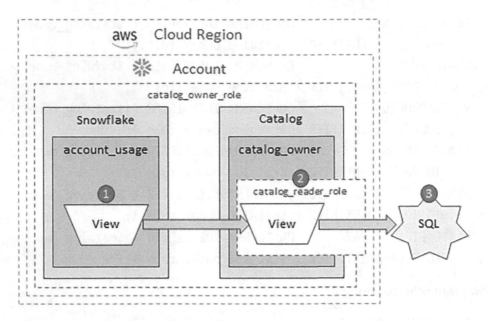

Figure 7-2. *Roles overlaid on catalog schema*

Figure 7-2 requires some explanation, and each callout is referenced below.

1. For a given view within the Account Usage Store, we identify the minimal subset of attributes required to satisfy the data catalog requirements.

2. We provision bespoke views, enriched with user-defined attributes, owned by `catalog_owner_role` and meeting the specific consumption requirements for reporting.

3. At this point, we do not specify tooling except to state all `Catalog` database access will be via the `catalog_reader_role` granted to specific users enabling reporting.

Implementing two distinct roles segregates object ownership from object usage, thereby preventing toxic entitlement combinations and mitigating the risk of inappropriate data access.

Having created the `Catalog` database, schema, and roles, you are now able to deploy bespoke views that implement your desired features. You adopt this approach for two reasons:

- To protect your environment by removing attributes from Account
  Usage Store source views that are not necessary to fulfil metadata
  reporting requirements

- To add user-defined attributes to enhance Account Usage
  Store source views, which add value to the metadata reporting
  requirements

You must also consider the possibility that third-party tooling may expect all view attributes to be present and will ignore any bespoke attributes you create. Your mileage will vary according to tooling chosen.

Metadata reporting consumes Snowflake credits. To segregate workload and later identify costs associated with metadata extract, you can create a separate warehouse.

To create your bespoke views, you first change the role and set your context:

```
USE ROLE      IDENTIFIER ( $catalog_owner_role   );
USE DATABASE  IDENTIFIER ( $catalog_database     );
USE SCHEMA    IDENTIFIER ( $catalog_owner_schema );
USE WAREHOUSE IDENTIFIER ( $catalog_warehouse    );
```

Using `catalog_owner_role`, confirm that you can view the Account Usage Store:

```
SELECT *
WHERE  database_name = 'SNOWFLAKE'
AND    deleted IS NULL;
```

You should see a single row returned for the SNOWFLAKE database.

Earlier in this chapter we mentioned the creation of bespoke attributes. Now we'll demonstrate how they can be created plus their use and rationale for inclusion.

The starting point is to assume that all third-party tools require every attribute. This may not be a correct assumption for your use cases, but we cannot know your organization-specific tooling and thus leave the attribute selection to your discretion.

To cater for future Account Usage Store enhancements where attributes may be added to views, take the unusual step of using SELECT * as a precondition for all wrapper views. As a general principle, we do not recommend the use of SELECT * as this is bad practice; however, you must balance the overhead of continual maintenance of bespoke views in the event you decide to explicitly identify each attribute.

Furthermore, there one more condition to consider: The addition of bespoke attributes must not impact any code where positional notation is used within queries, therefore SELECT * must come before any bespoke attributes as this SECURE VIEW, which returns both object and attribute information demonstrates:

```
CREATE OR REPLACE SECURE VIEW v_secure_enriched_aus_columns
AS
SELECT c.*,
       current_region()||'.'||
       current_account()||'.'||
       c.table_catalog||'.'||
       c.table_schema||'.'||
       c.table_name                      AS path_to_object,
       current_region()||'.'||
       current_account()||'.'||
       'snowflake.account_usage.columns' AS data_source,
       'v_secure_enriched_aus_columns'   AS data_object
FROM   snowflake.account_usage.columns   c
WHERE  c.deleted IS NULL
ORDER BY c.table_catalog,
         c.table_schema,
         c.table_name,
         c.ordinal_position ASC;
```

You use a SECURE VIEW to protect the view declaration from being accessible. Documentation for SECURE VIEWs can be found here: https://docs.snowflake.com/en/user-guide/views-secure.html#working-with-secure-views.

With your view declared, prove the contents are as expected:

```
SELECT *
FROM   v_secure_enriched_aus_columns;
```

Provisioning Read Access

Having successfully created your metadata reporting view named v_secure_enriched_aus_columns, you must grant entitlement to your reader role named catalog_reader_role:

```
GRANT SELECT ON VIEW v_secure_enriched_aus_columns TO ROLE catalog_reader_role;
```

Now switch the role to `catalog_reader_role` and prove your reader role can access the view:

```
USE ROLE IDENTIFIER ( $catalog_reader_role );

SELECT *
FROM    v_secure_enriched_aus_columns;
```

The partial result set shown in Figure 7-3 illustrates information returned from objects declared from an earlier chapter in this book.

COLUMN_ID	COLUMN_NAME	TABLE_NAME	TABLE_SCHEMA	TABLE_CATALOG	ORDINAL_POSITION
12294	EMPLOYEE_ID	EMPLOYEE	MONITOR_OWNER	MONITOR	1
12295	PREFERRED_NAME	EMPLOYEE	MONITOR_OWNER	MONITOR	2
12296	SURNAME_PREFERRED	EMPLOYEE	MONITOR_OWNER	MONITOR	3
12297	FORENAME_PREFERRED	EMPLOYEE	MONITOR_OWNER	MONITOR	4
12298	GENDER	EMPLOYEE	MONITOR_OWNER	MONITOR	5

Figure 7-3. *Partial view result set*

Extending Capability

Your first bespoke view contains all attributes from its underlying Account Usage Store view in order to incorporate core attributes and ensure backwards capability. You may also be required to deliver more views to extend the metadata coverage range of your environments, though a note of caution applies: You should give careful consideration before exposing procedure and function code as this may result in leakage of intellectual property, leading to a cyber security breach.

Additional capabilities may include summarization, aggregation, and time-banded and pre-filtered data sets. Typically, these additional capabilities are local reporting tool requirements defined by your business users.

We trust the example code provided will enable you to extend capability as required.

External Tooling

A wide variety of third parties await your engagement, offering plug-and-play capability ready to monitor, alert, and otherwise consume your Account Usage Store metadata. Most vendors make a simple mistake in assuming their tooling is both essential to the efficient operation and running of your Snowflake account (it isn't) and consider their tooling to be central to their chosen operating domain. As such, you will find vendors who by default will insist upon using the `accountadmin` role to install their tooling.

We strongly reject the assertion of the `accountadmin` role being necessary to install any software and recommend a thorough investigation of all vendor software requirements before allowing the installer to run.

We are of the opinion that vendors who insist upon the `accountadmin` role to install their software do not understand Snowflake RBAC model sufficiently well and typically exhibit a lack of core Snowflake competence by defaulting to use the `accountadmin` role. We understand the dilemma for vendors; from a third-party perspective, it is very difficult to spend significant hours, often months of hard work, developing the deep knowledge of every target system into which their tooling will integrate.

Our preferred approach is to conduct a thorough analysis of the minimal subset of objects and attributes required to achieve the tooling install, while retaining control of our security perimeter. In some scenarios, we have found configuration files can be altered to access our bespoke views instead of the Account Usage Store, thus negating the requirement to use the `accountadmin` role.

In summary, why bother to secure your Snowflake account and then give unlimited access to a third party without conducting full technical due diligence? It is incumbent upon you to fully understand every interface into and out of your Snowflake account and third-party tooling is no exception.

Referential Integrity

As you know, Snowflake allows the declaration of constraints but only `NOT NULL` is enforced. The behavior changes for Unistore, for which further information can be found here: `www.snowflake.com/en/data-cloud/workloads/unistore`. As Unistore is not yet fully functional or released, we put aside Unistore as a future enhancement.

Currently within Snowflake primary keys, foreign keys and referential integrity constraints are supported. One use case is for relationship self-discovery within tooling. In support of self-discovery and wherever possible supporting best practice, you should declare constraints that allow programmatic discovery of relationships between tables.

A cursory examination of the Account Usage Store reveals the presence of a referential_constraints view, which we leave for your further investigation.

Data Quality

Many data catalogs offer the capability to profile data with the ability to determine data quality by sampling stored data. Profiling data goes beyond the immediate concerns of this chapter and is included for completeness. We provide a pointer to *Building the Snowflake Data Cloud* where we built an example data quality JavaScript stored procedure.

Do you need data quality metrics? Well, the answer depends upon how comprehensive your data warehouse is to be and is beyond the scope of this book to determine. What you must decide is how comprehensive your data catalog is to be.

Generating Metadata

Within this chapter we have addressed how to access metadata for objects declared within Snowflake. We must also consider the opposite perspective: how to generate objects, attributes, and constraints within tooling external to Snowflake and then deploy and create these objects within Snowflake.

But why would we want to define objects, attributes, and constraints outside of Snowflake?

Our organizations contain a complex array of tooling, products, and capabilities whose interactions, dependencies, and relationships are not easily understood at the macro level. That is, when trying to understand "how" data flows between components, and the downstream impact of system failure, we often find ourselves acting with limited or partial information.

For those organizations that have fully implemented operational resilience and mapped their important business services, there are levels of detail buried within their systems' technical implementations. How much easier would our work be if we adopted a pattern-based approach to system architecture and used tooling to define, map, and then deploy objects and components? For those of us who have been around for a while, CASE tools, data modelling, and a plethora of associated disciplines have been in use for many years and have never gone out of fashion. Perhaps it's time to get back to analyst/ programmers and end-to-end system design according to first principles, after which we will have all the documentation we will ever need along with the ability to understand our systems at a detailed level.

Our point is simply this: Snowflake is not intended to be a central metadata repository for your organization but is a significant component that exists within your organization's infrastructure.

As such, we suggest removing as much repeatable hands-on coding from your day jobs, which provides you time to think more deeply and deliver more simple solutions. But to achieve a pattern-based approach, you must invest time and effort in building tooling to deliver the desired patterns. In *Building the Snowflake Data Cloud*, we investigated and built a simple code generator in support of a pattern-based approach to developing Snowflake systems. We do not repeat the work carried out previously but point to the existence of a proven code generation template in current use with the strong recommendation to either build or implement third-party tooling to achieve similar outcomes.

Summary

You began this chapter with an introduction to the subject of accessing all metadata relating to data domains of interest to your business colleagues. You explored the role third-party tooling plays in defining both business and technical definitions to Snowflake metadata.

Using a worked example, you identified an extensible design pattern to expose Snowflake metadata, which also provides some enriched attributes. The objective is to provision objects of light maintenance and take into consideration future Snowflake product enhancements without impacting your configuration management.

The cost cataloging solution developed in Chapter 5 to include usage cataloging may also be extended to incorporate metadata extract for centralized consumption by third-party tooling.

You explored some thoughts beyond the immediate subject of exposing Snowflake metadata. Our aim was not to divert attention away from the core theme of this chapter but instead to inform and stimulate your further investigation.

The next chapter explores organizations.

CHAPTER 8

Organization

The term "organization" is confusing because it refers to the corporate environment many of us work in. For example, the corporate environment the authors work for is the London Stock Exchange Group (also known as LSEG, `www.londonstockexchange.com/`). In this book we use the word "organization" to mean any corporate environment as a place of work.

In contrast, a Snowflake organization is an optional high-level container to manage all Snowflake accounts within any corporate structure. Within this chapter, we use the term "Snowflake organization" to differentiate use. You can visualize a Snowflake organization as the outer layer of the Matryoshka dolls, commonly referred to Russian dolls. See here for more information: `https://en.wikipedia.org/wiki/Matryoshka_doll`.

Whilst the Matryoshka doll analogy breaks down rather quickly, Figure 8-1 is useful as a starting point in explaining Snowflake containers, with Snowflake organization being the outermost.

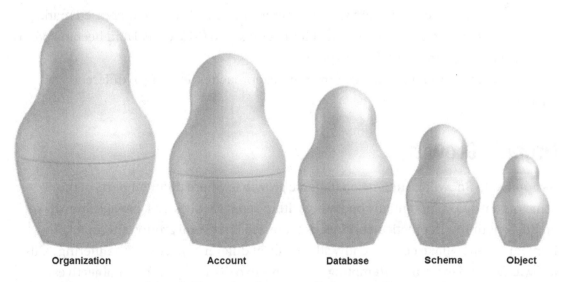

Organization   **Account**   **Database**   **Schema**   **Object**

Figure 8-1. *Matryoshka doll analog for Snowflake containers*

© Andrew Carruthers and Sahir Ahmed 2023
A. Carruthers and S. Ahmed, *Maturing the Snowflake Data Cloud*, https://doi.org/10.1007/978-1-4842-9340-9_8

As you have worked through the examples within this book, the most entitled role you have encountered is `accountadmin`, which you used in the context of administering your account. So where does the Snowflake organization feature come from, and how does this optional new feature work?

Throughout this chapter, we prefer to use SQL commands rather than Snowsight features. We adopt this approach for several reasons:

- Using the SQL commands often provides greater feature coverage than Snowsight "point and click" operations.

- Through experience gained over 30 years, a deeper understanding is gained through typing SQL commands.

- SQL commands can easily be scripted for later automation.

Each to their own, but for repeatable pattern-based deployment of your landing zone, it's hard to beat automation.

For information on how to use the Snowsight equivalents for SQL commands presented within this chapter, please refer to the documentation found here: `https://docs.snowflake.com/en/user-guide/organizations-manage-accounts.html#managing-accounts-in-your-organization`.

Impact of Organizational Growth

At some point, your organization will either grow organically due to increased market share or be acquired as part of a Merger and Acquisition (M&A) deal and become part of the inorganic growth of another business.

Almost all companies pursue both growth approaches, leading to different challenges.

Organic Growth

As applied to Snowflake utilization, organic growth is where different parts of the same organization, potentially operating within separate divisions, engage Snowflake at different times. In practice, this means there will be several groups attempting to implement Snowflake concurrently, all moving at different speeds, with differing skills, knowledge, and expertise attempting to achieve two distinctly different objectives:

1. The first objective is to embed the Snowflake platform into the organization infrastructure to upskill with Snowflake, comply with the security posture, educate others, and achieve many other tasks.

2. The second objective is to deliver a proof of concept using the Snowflake platform to demonstrate both Snowflake capability and business benefit.

From experience, we know just how hard it is to achieve these objectives. In large part, the rationale for writing this book is to mitigate the risks and issues others will encounter when embarking on the same (or similar) journey we did some three years ago.

In your hands is the distilled knowledge and experience with which your skills will enable you to successfully deliver Snowflake as a platform into your organization. But this is only part of the solution your organization needs.

Without a well-considered approach to provisioning Snowflake accounts and the continual maintenance of centralized components as discussed elsewhere within this book, your organization will struggle to control costs, maintain a consistent approach and standards across your Snowflake estate, and your approach will forever remain fragmented.

What you need is a single container under which some administrative functions and monitoring are centralized, thus enabling a single overview of your organization. Snowflake organizations can be complicated by the number, and variety, of Master Service Agreements (MSAs) each operating division has signed. We discuss MSAs later.

Inorganic Growth

The velocity of inorganic M&A growth can produce unexpected challenges in managing Snowflake environments, which we now discuss.

Due to M&A activity you may also inherit an MSA under which an acquired company operates that runs orthogonal to your organization's approach.

Furthermore, it is certain the security posture, security profile, and physical implementations will differ according to the governance approach and policies in force at the time an acquired Snowflake account was built.

The beauty of Snowflake is the open approach, which allows a multitude of "right and proper" implementations. The problem is, everyone has their own approach, each of which is "right and proper" at the time of implementation, and there is a distinct lack of commonality. We hope you are reading this book to establish a blueprint and mitigate future risks.

Master Service Agreements

An MSA is the over-arching agreement under which Snowflake provisions services to an organization. The MSA defines the service boundaries, remedies, permitted actions, and many more legal terms well beyond the scope of this chapter. We do not conduct a deep dive into MSAs; we simply state that the document exists and call out a few points of note below. The MSA is the legal framework for the relationship between your organization and Snowflake Inc.

Different forms of MSA arise according to how your organization engages with Snowflake Inc. The most common form of MSA is a Direct contract between your organization and Snowflake Inc. Another form of MSA is a Marketplace contract where a third party is involved, in this example, we assume a cloud service provider (CSP) Marketplace. There may be other forms of MSA unknown to the authors.

Direct MSAs are probably the easiest to negotiate as there are only two parties involved: your organization and Snowflake Inc. According to the proposed deal, Snowflake Inc. may offer professional services (PS) at a discounted rate or include a provision for staff over a period of time. Particularly for those organizations at the beginning of their Snowflake journey, we highly recommend the engagement of PS for a minimum of three months in order to accelerate your Snowflake experience.

Marketplace MSAs are driven by a third party acting as an agent for Snowflake Inc. and may involve CSP credits, which are a rebate or sum of Marketplace credits made available as a consequence of signing up to a Marketplace MSA.

Caveat Emptor: Marketplace agreements can be restrictive and may prevent the creation of Snowflake accounts on rival CSP platforms. Additionally, you may encounter an inability to transfer credits between Direct and Marketplace agreements. And you may face administrative issues when using a Snowflake organization to manage your accounts as a Snowflake organization is typically tied to a single MSA. Additional engineering effort may also be required to consolidate metrics to produce a single view of your Snowflake credit consumption. Finally, there may be a soft limit to the number of accounts allowed under the MSA.

Both Snowflake Inc. and organizations can spend considerable time consolidating MSAs post M&A activity. When merging MSAs you may find differing credit discount rates apply, a point for your negotiations. MSAs also make provision for divestments, where part of an organization is sold, to ensure continuity of service. Finally, all MSAs run for a specified period with exit clauses allowing egress from Snowflake.

According to projected consumption at the point of MSA creation, Snowflake will offer a discount for the prepurchase of credits. The alternative model is to pay as you go, for which there is no discount. Some basic information can be found here: `www.snowflake.com/pricing/`.

Quarterly Business Reviews

For organizations with MSAs in place, Snowflake Inc. usually conducts quarterly business reviews (QBRs). As the name suggests, QBRs are an opportunity for both your organization and Snowflake Inc. to reflect upon progress made within the past three-month period and course correct for the upcoming quarter.

Within this section we do not consider the full scope of QBRs, but instead call out a few items of interest, specifically

- New product features and capabilities are highlighted, some of which will be relevant to the Snowflake landing zone.

- Historic Snowflake credit consumption to date per line of business, operating division, or other pre-agreed metric is stated.

- Current capacity planning and projected credit consumption metrics are presented.

Aside from the in-house cost monitoring components built and deployed for your landing zone, via the QBR, you have opportunity to validate your consumption metrics against those produced by Snowflake Inc. You must take the opportunity to contrast, compare, and evaluate where you may experience a shortfall of credits against projected spend. In a dynamic operating environment, particularly where new demand for Snowflake accounts occurs unexpectedly, credit consumption will rise accordingly.

A further benefit derived from the QBR is a full list of Snowflake accounts associated with your organization. While you should be fully aware of all Snowflake accounts associated to your organization, the reality is often quite different. For a variety of reasons, particularly in an organization with federated Snowflake engagements, a single view of all accounts is not necessarily readily achievable.

You should also embrace the opportunity each QBR presents to ensure your landing zone accurately represents information presented.

Role Hierarchy

Snowflake organization membership is initiated by role orgadmin which, unlike accountadmin, is not granted to any user by default. Instead, orgadmin is an opt-in feature. That is, you must decide to join each account into a Snowflake organization and cannot assume your accounts belong to any organization other than their own, which is assigned on account creation.

Figure 8-2 illustrates where the orgadmin role fits into the Snowflake-supplied role hierarchy.

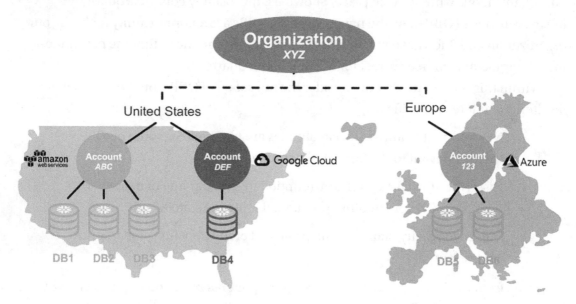

Figure 8-2. *Snowflake-supplied roles*

The following are descriptions of the Snowflake-supplied roles shown in Figure 8-2:

- ORGADMIN: Allows centralized account administration

- ACCOUNTADMIN: Account parameter declaration, guardrails, etc.

- SYSADMIN: Database, warehouse, and other object administration

- SECURITYADMIN: Role and entitlement administration

- USERADMIN: User administration

- PUBLIC: Pseudo-role that can own objects and is granted to all users

- Custom roles are those you create to manage applications. You have created custom roles in previous chapters of this book.

Snowflake Organization Features

Figure 8-3 illustrates how the Snowflake organization feature logically groups accounts provisioned across all supported CSPs geographically separated across two continents. In reality, all Snowflake points of presence across all locations are supported by Snowflake organizations.

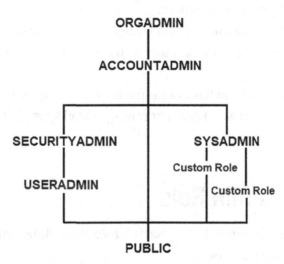

Figure 8-3. *Snowflake organization logical representation*

Having established where a Snowflake organization encapsulates all accounts for an organization, let's now examine the Snowflake organization features more closely with worked examples to illustrate usage.

At this point. you might be asking why we need the orgadmin role. In truth, orgadmin is not essential to the running of any single account, though the table shown in Table 8-1 illustrates purpose and benefits of using the orgadmin role when compared to using the accountadmin role.

Table 8-1. *Orgadmin and Accountadmin Role Comparison*

ORGADMIN	ACCOUNTADMIN
Manages operations **across** accounts	Manages operations **within** an account
Enables accounts **for** replication	Sets up replication **within** an account
Access usage **across** accounts	Access usage **within** an account
Creates, renames, and drops accounts	
Single view of **all** accounts	

Using the orgadmin role does not confer any other entitlement except as listed in Table 8-1. The use of orgadmin facilitates self-service along with features previously requiring Snowflake Support involvement.

We recommend the use of the orgadmin role as part of your landing zone initiative to both reduce your overall account administrative burden and provision centralized monitoring capabilities.

Further information on Snowflake organizations can be found here: `https://docs.snowflake.com/en/user-guide/organizations-gs.html#getting-started-with-organizations`.

Using the Orgadmin Role

In this section you deep dive into the orgadmin's role capabilities with hands-on walk through of how to use the features.

In general, you cannot administer an account while connected to the account. You must administer accounts from outside of the account being administered.

Identifying Your Default Organization

To take advantage of the orgadmin role's features, you must first ensure your user is entitled to use the role orgadmin:

```
USE ROLE orgadmin;
```

The error message `Requested role 'ORGADMIN' is not assigned to the executing user. Specify another role to activate.` means you are not entitled to use the role `orgadmin`.

Instead, you must grant the role `orgadmin` to either yourself or another role using the `accountadmin` role. In this example, you use your own user:

```
USE ROLE accountadmin;
```

Then grant the role `orgadmin` to your user:

```
GRANT ROLE orgadmin to USER <Your User Here>;
```

Before attempting to use your newly granted role of `orgadmin`:

```
USE ROLE orgadmin;
```

After which you can examine the accounts that constitute your Snowflake organization:

```
SHOW ORGANIZATION ACCOUNTS;
```

You should see a result set similar to that shown in Figure 8-4.

organization_n	account_n	snowflake_reg	edition	account_url	comment	account_loc
RAWDPWN	CB64333	AWS_EU_WEST	BUSINESS_CR.	https://rawdp	Created by Si	JH85671

Figure 8-4. *Sample Snowflake organization account information*

In our example, the `organization_name` is set to RAWDPWN. We cannot dynamically change the `organization_name` but instead must contact Snowflake Support: `https://community.snowflake.com/s/article/How-To-Submit-a-Support-Case-in-Snowflake-Lodge`.

Enabling Account Replication

The Snowflake replication story is still work in progress. Account replication and failover at the time of writing is in public preview. More details can be found here: `https://docs.snowflake.com/en/user-guide/account-replication-failover.html#account-replication-and-failover`.

We discuss available replication capabilities in a later chapter.

For now, you will examine how to enable replication, and for this you need to switch roles to accountadmin:

```
USE ROLE accountadmin;
```

```
SHOW REPLICATION ACCOUNTS;
```

Since you have not enabled your account for replication, you should not see any results.

Enable replication:

```
USE ROLE orgadmin;
```

```
SHOW ORGANIZATION ACCOUNTS;
```

Fetch the account locator as this is required to enable replication:

```
SELECT current_account();
```

In the next SQL statement, replace <your_account> with the response from the above SQL statement; ours is JH85671:

```
SELECT system$global_account_set_parameter ( '<your_account>', 'ENABLE_
ACCOUNT_DATABASE_REPLICATION', 'true' );
```

You should see a SUCCESS response similar to that shown in Figure 8-5.

Figure 8-5. *Enabling account replication*

Confirm replication is enabled:

```
USE ROLE accountadmin;
```

```
SHOW REPLICATION ACCOUNTS;
```

You should see a result set similar to the one shown in Figure 8-6.

snowflake_regi	account_name	account_locator	comment	organization_name	is_org_admin
AWS_EU_WEST	CB64333	JH85671	Created by Signup Service	RAWDPWN	true

Figure 8-6. *Account replication enabled*

Disable Account Replication

You may need to disable account replication, for which you use the following commands:

```
USE ROLE orgadmin;
```

```
SELECT system$global_account_set_parameter ( '<your_account>', 'ENABLE_
ACCOUNT_DATABASE_REPLICATION', 'false' );
```

You should see a SUCCESS response.
Confirm replication is disabled:

```
USE ROLE accountadmin;
```

```
SHOW REPLICATION ACCOUNTS;
```

Since replication has been disabled, you should not see any results.

Listing Organization Accounts

You encountered these commands above.
Set your role to orgadmin:

```
USE ROLE orgadmin;
```

After which you can examine the accounts that constitute your Snowflake organization:

```
SHOW ORGANIZATION ACCOUNTS;
```

Delivering an Account

Before you consider creating a new Snowflake account, there are several items you must consider. Let's go through them now.

Ownership

We do not advocate the creation of a centralized Snowflake account administration capability where all Snowflake account management and maintenance is carried out. Instead, we advocate a decentralized approach where each acquiring team remains responsible for their own Snowflake accounts. To this end, you must identify who is responsible for each Snowflake account and therefore accountable for maintenance, configuration, and, ultimately, credit consumption.

We propose that the named individual plus at least one other and a "break-glass" user are entitled with the accountadmin role at the point of account creation.

Naming Convention

Aside from the Snowflake default account naming convention, you must determine your local organization naming convention. To facilitate identification of each account, and allow consolidation with a wider infrastructure, we suggest the following abbreviations as a starting point:

- SNOW_: Indicates the type of account, in this case "Snowflake"

- LOB_: Line of business or division

- APP_: Application abbreviation

- ENV: Environment, which may be Production (PRD), Non-Production (NON-PRD), Development (DEV), Integration (INT), etc.

Feel free to amend the examples according to your organizational needs, which may include CSP and location information.

With the above abbreviations in mind, you can easily, and uniquely, identify provisioned accounts.

Credit Allocation

At the point of new Snowflake account creation, you must also consider how credits will be purchased and allocated for the new account. Through experience, this is typically an afterthought once the need for a Snowflake account has been identified.

But we must be clear when engaging with prospective consumers: Snowflake can be expensive if misconfigured, and setting reasonable expectations for budget allocation and consumption should be done up front.

CSP Provision

As previously discussed, your choice of CSP may be restricted according to the MSA in force at the time of account creation. Your colleagues in Procurement will have an interest in knowing which accounts have been created as they will manage the relationship with Snowflake.

As part of the cost monitoring discussed in a previous chapter, your Financial Operations (FinOps) colleagues will need to know account ownership details along with expected credit consumption in order to baseline their consumption expectation metrics and project future growth.

Upskilling

Our preferred route to success is to seed the account acquiring team with skill, knowledge, and experience. Through first-hand experience, the authors strongly recommend the early engagement of Snowflake PS resources to transition staff onto Snowflake. An easy mistake is to assume Snowflake works in the same way as other data warehouses. While superficially the same, you must tune your applications for the platform and not make assumptions on Snowflake behavior.

We make this point again, to remind you of the importance of provisioning resources to assist others in becoming successful.

Creating An Account

Set your role to `orgadmin` as this is the only role entitled to create a new Snowflake account:

```
USE ROLE orgadmin;
```

Now create a new account, recognizing this will take up to a minute to provision. Don't forget to replace the ADMIN_PASSWORD and EMAIL information.

In our example, and to comply with our naming convention, we use SNOW_SALES_ SWH_DEV representing:

- SNOW_: "Snowflake" is the type of account.

- SALES_: "Sales" is the Line of Business.

- SWH_: "Sales Warehouse" is the application abbreviation.

- DEV: "Development" environment

Ensure EDITION is set to BUSINESS_CRITICAL because you will use features later:

```
CREATE ACCOUNT SNOW_SALES_SWH_DEV
ADMIN_NAME     = ANDYC
ADMIN_PASSWORD = '<Your Password Here>'
FIRST_NAME     = Andrew
LAST_NAME      = Carruthers
EMAIL          = '<Your Email Here>'
EDITION        = BUSINESS_CRITICAL
REGION         = AWS_EU_WEST_1;
```

On successful account creation, you should see a response that looks like Figure 8-7.

status

{"accountLocator":"NW33776","accountLocatorUrl":"https://nw33776.eu-v

Figure 8-7. *Successful account creation*

Fully expanded, the returned JSON string looks like this:

```
{
  "accountLocator": "NW33776",
  "accountLocatorUrl": "https://nw33776.eu-west-1.snowflakecomputing.com",
  "accountName": "SNOW_SALES_SWH_DEV",
  "url": "https://rawdpwn- snow_sales_swh_dev.snowflakecomputing.com",
  "edition": "BUSINESS_CRITICAL",
  "regionGroup": "PUBLIC",
```

```
    "cloud": "AWS",
    "region": "AWS_EU_WEST_1"
}
```

Please check your email because Snowflake will have sent an introduction email to the email address supplied within the `CREATE ACCOUNT SQL` statement.

Using the URL provided (`https://rawdpwn-SNOW_SALES_SWH_DEV.snowflake computing.com`), you can log in using the admin credentials provided (ANDYC and your password) when creating the account. For this, we suggest using a different browser.

On successful login, the password will automatically expire, necessitating a password change, as shown in Figure 8-8.

Figure 8-8. *New account password reset*

For the sharp-eyed amongst us, the URL includes both the organization name (`rawdpwn`) and account name (`SNOW_SALES_SWH_DEV`).

From the account where you created the new account, confirm your new account, SNOW_SALES_SWH_DEV, is part of your organization:

```
SHOW ORGANIZATION ACCOUNTS;
```

The result set will now include your new account, SNOW_SALES_SWH_DEV, as shown in Figure 8-9.

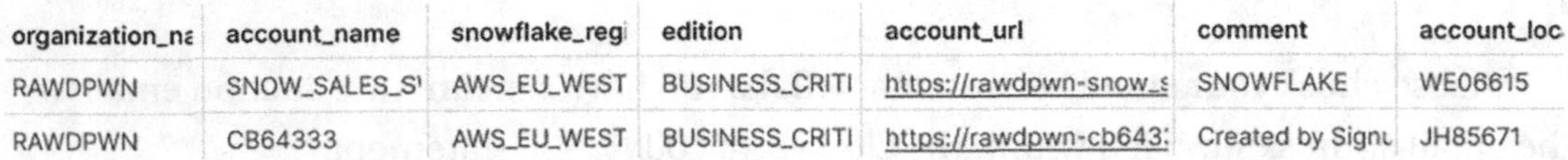

organization_na	account_name	snowflake_regi	edition	account_url	comment	account_loc
RAWDPWN	SNOW_SALES_S\	AWS_EU_WEST	BUSINESS_CRITI	https://rawdpwn-snow_s	SNOWFLAKE	WE06615
RAWDPWN	CB64333	AWS_EU_WEST	BUSINESS_CRITI	https://rawdpwn-cb643:	Created by Signu	JH85671

Figure 8-9. New organization account

Please also refer to documentation at `https://docs.snowflake.com/en/sql-reference/sql/create-account.html#create-account`.

Checklist

We recommend a checklist be populated as a precursor to later automation of account creation and offer Table 8-2 as a starting point.

Table 8-2. Account Creation Checklist

Item	Value
Account Name	SNOW_SALES_SWH_DEV
Purpose	Sales Data Warehouse
Admin User	AndyC
First Name	Andrew
Surname	Carruthers
Admin Email	<Your Email Here>
Edition	BUSINESS_CRITICAL
Region	AWS_EU_WEST_1
CSP	AWS
Credits Purchased	10000
Replication Enabled?	True

Your table may include bespoke information such as project code, expected growth per annum, and so on.

Renaming An Account

Let's move on to discussing how to rename a Snowflake account, and for this, you set your role to orgadmin:

```
USE ROLE orgadmin;
```

Create a new account to demonstrate account renaming:

```
CREATE ACCOUNT MTSDC
ADMIN_NAME      = ANDYC
ADMIN_PASSWORD = '<Your Password Here>'
FIRST_NAME      = Andrew
LAST_NAME       = Carruthers
EMAIL           = '<Your Email Here>'
EDITION         = BUSINESS_CRITICAL
REGION          = AWS_EU_WEST_1;
```

Check that your new account has been created, paying specific attention to the account_url value, which in this case is https://rawdpwn-mtsdc. snowflakecomputing.com:

```
SHOW ORGANIZATION ACCOUNTS;
```

You may choose to retain the original URL while renaming your account mtsdc to SNOW_SALES_SWH_PROD; in which case SAVE_OLD_URL is set to TRUE. A good reason for retaining the original URL is for backwards compatibility to ensure downstream dependencies upon the existing URL remain intact pending later cleanup.

```
ALTER ACCOUNT MTSDC RENAME TO SNOW_SALES_SWH_PROD
SAVE_OLD_URL = TRUE;
```

When you check the account attributes, you will find old_account_url has been set to the old value of https://rawdpwn-mtsdc.snowflakecomputing.com:

```
SHOW ORGANIZATION ACCOUNTS;
```

And later drop the old URL:

```
ALTER ACCOUNT SNOW_SALES_SWH_PROD DROP OLD URL;
```

Checking that the account attributes show that old_account_url has been set to NULL:

```
SHOW ORGANIZATION ACCOUNTS;
```

To continue your testing, reset the account name of SNOW_SALES_SWH_PROD to mtsdc:

```
ALTER ACCOUNT SNOW_SALES_SWH_PROD RENAME TO MTSDC
SAVE_OLD_URL = FALSE;
```

Confirm the account name has been reset:

```
SHOW ORGANIZATION ACCOUNTS;
```

Rename account mtsdc to SNOW_SALES_SWH_PROD, noting SAVE_OLD_URL is now set to FALSE:

```
ALTER ACCOUNT MTSDC RENAME TO SNOW_SALES_SWH_PROD
SAVE_OLD_URL = FALSE;
```

Check that account_name has been renamed, the account_url has been updated, and old_account_url is set to NULL:

```
SHOW ORGANIZATION ACCOUNTS;
```

You should see that account_url has changed to https://rawdpwn-snow_sales_swh_prod.snowflakecomputing.com, noting your account_url will differ.

After renaming an account, ensure all referenced documentation including the checklist is updated.

Please also refer to the documentation at https://docs.snowflake.com/en/sql-reference/sql/alter-account.html.

Dropping An Account

At the time of writing, dropping an account can only be performed by Snowflake Support. However, work is in progress to enable the orgadmin role to delete an account by dropping it.

There are several implications to consider before attempting to drop an account:

- Before sending a deletion request to Snowflake Support, or later when account deletion becomes an `orgadmin` role capability, all users should have been disabled for several weeks to ensure there is no unexpected account activity.

- All required data should be copied to another account.

- Secure Direct Data Shares. Private Listings, Snowflake Marketplace, and Replication should be removed before account is deleted.

- Account deletion should not happen without appropriate checks, balances, and approvals in order to prevent malicious actions.

- For deleted accounts removed by Snowflake Support, there is a 30-day grace period during which the account can be recovered. This period may change for self-service via the `orgadmin` role.

- Where created, cost monitoring metadata should be removed.

- After dropping an account, ensure all referenced documentation including checklists is updated.

Naturally, the above list is not exhaustive but serves as a starting point for your further consideration and enhancement.

Please also refer to the documentation at `https://docs.snowflake.com/en/user-guide/organizations-manage-accounts.html#deleting-an-account`.

Snowflake Organization Metrics

Having worked through features enabled by the `orgadmin` role, you might expect to have entitlement to view associated metrics using the `orgadmin` role but, as you will see, this is not the case.

All Snowflake organization metrics are available through the Snowflake imported database. Throughout this book we have made repeated reference to the Account Usage Store, and for Snowflake organization metrics, we continue to make use of `snowflake.organization_usage` views.

Orgadmin Account Usage Store Access

Set your role and attempt to access snowflake.organization_usage views:

USE ROLE orgadmin;

Then open your database browser. You may need to refresh the left-hand pane of your browser first. You may be surprised to see the omission of organization_usage from Figure 8-10.

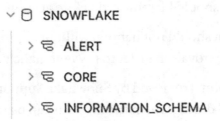

Figure 8-10. *Orgadmin Account Usage Store*

To understand why you can't access organization_usage, change the role to accountadmin and retry:

USE ROLE accountadmin;

You should now see organization_usage, as shown in Figure 8-11.

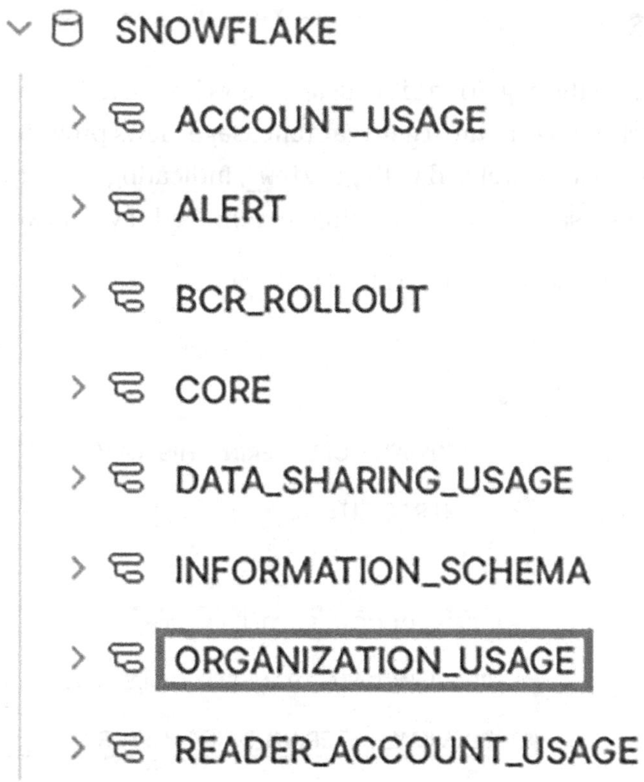

Figure 8-11. *Accountadmin Account Usage Store*

A quick and easy fix is to grant the `orgadmin` role entitlement to access the Account Usage Store, but this will expose additional schemas that are not relevant to Snowflake organizations and may introduce a security breach.

Most probably, and in the authors' opinions, the Account Usage Store entitlement for `orgadmin` is work in progress and may later change to include `organization_usage` schema access.

Organization Usage Views

At the time of writing, the `organization_usage` schema is undergoing change with more capability being incrementally released. You may wish to check with your Snowflake Sales Engineer before spending a great deal of time developing code that may either soon require refactoring or overlap with newly provisioned Snowflake monitoring capability.

Preview Views

In order to understand the `organization_usage` views better, let's consider their purpose and features. A quick glance at the `organization_usage` views provisioned to date shows there are a number of views prefixed with `preview_`, indicating they are works in progress. Figure 8-12 illustrates a sample subset (noting we removed several views from the image).

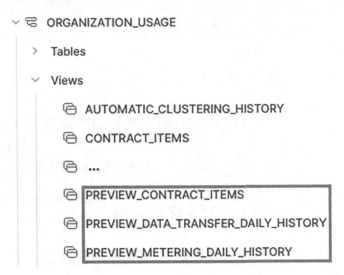

Figure 8-12.   Organization usage preview views

As the `preview_` prefix suggests, these views are not formally documented, though reference to some named views can be found within the June 2021 release notes, for which further information can be found here: `https://docs.snowflake.com/en/release-notes/2021-06.html#organization-usage-updates`.

We suggest `preview_` views are useful to indicate the future direction of product feature availability but should not be relied upon for final functionality or behavior. We therefore recommend you not rely on `preview_` views for production-grade code. They may be of use in prototyping.

Other Views

A reasonable working assumption for the `organization_usage` schema is that all views not prefixed with `preview_` are stable. However, a quick check of the documentation may identify views under development. It is true that the documentation can sometimes lag behind releases, but in the authors' experiences, only by a matter of days or few weeks at most.

As a matter of good practice, the authors refer to documentation before beginning work just to check that features have not changed and the declared objects match the documentation. We may also check with our Snowflake Sales Engineer for any notified changes, private previews, and sample code before proceeding.

View Comparison

So what are the differences between views presented under the organization_usage schema and any other schema? The more you understand how Snowflake provisions metadata and under what circumstances, the better you will be informed to make decisions.

Let's take one example view: AUTOMATIC_CLUSTERING_HISTORY. Start by setting your context:

```
USE ROLE       accountadmin;
```

Assume catalog_wh has not been dropped; otherwise, create an XSmall warehouse called catalog_wh:

```
USE WAREHOUSE catalog_wh;
```

Then create a SQL statement to find all occurrences of the AUTOMATIC_CLUSTERING_ HISTORY view within the Account Usage Store. Use the information_schema.views view as it contains all view object declarations within the Account Usage Store:

```
SELECT LOWER ( table_catalog||'.'||
               table_schema ||'.'||
               table_name ) AS view_location
FROM   snowflake.information_schema.views
WHERE  table_name = 'AUTOMATIC_CLUSTERING_HISTORY'
ORDER BY 1 ASC;
```

The SQL provides the canonical location for the returned data set. You should find two declarations:

- snowflake.account_usage.automatic_clustering_history

- snowflake.organization_usage.automatic_clustering_history

Navigate to each identified view in turn.

Within Snowsight there is a neat feature to add both object name and object attributes into the worksheet, as shown in Figure 8-13.

Figure 8-13. *Snowsight's Add Columns in Editor feature*

As a quick comparison check, for both identified views, add the columns into the editor. For illustration purposes, we do this in two parts. The first section is shown in Figure 8-14 where we use spaces to align the first common attribute, CREDITS_USED, at the word boundary.

```
snowflake.account_usage.automatic_clustering_history
                                               START_TIME, END_TIME, CREDITS_USED,

snowflake.organization_usage.automatic_clustering_history
ORGANIZATION_NAME, ACCOUNT_NAME, ACCOUNT_LOCATOR, REGION, USAGE_DATE, CREDITS_USED,
```

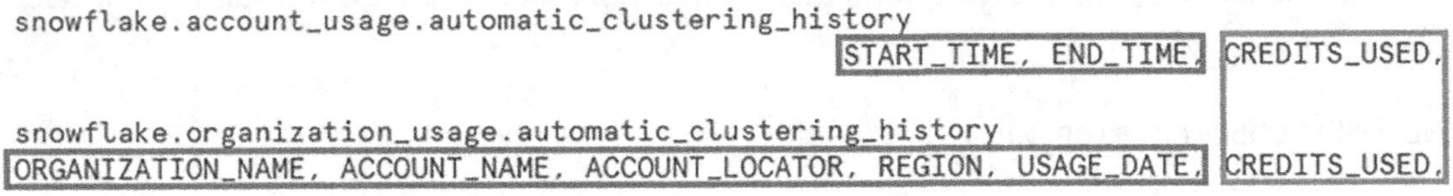

Figure 8-14. *First word alignment*

Recommencing on the first common attribute, CREDITS_USED, at the word boundary as shown in Figure 8-15 demonstrates that all remaining attributes match.

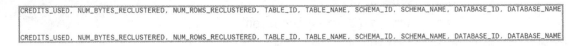

Figure 8-15. *Second word alignment*

We can infer that USAGE_DATE is a summary column. We speculate this attribute utilizes the date stamp component of START_DATE to populate.

We show Figure 8-14 and Figure 8-15 within the accompanying SQL script for this chapter.

Having identified two views with the same name, you can see the attribute sets differentiate usage. This leads us to an important point: Don't assume there is only one definition of an object. Always check. The context is important.

View Coverage

Our investigation confirms the correct view definition for deriving Snowflake organization information is found at `snowflake.organization_usage.automatic_clustering_history`.

If you compare the list of views available under `snowflake.organization_usage` with the list of views available under `snowflake.account_usage`, you will identify gaps in provision. At the time of writing, two notable exceptions are

- `QUERY_ACCELERATION_HISTORY`

- `SERVERLESS_TASK_HISTORY`

We expect these two views to be provisioned soon.

We suggest you review `snowflake.organization_usage` views every month for changes. A quick way to identify changes is to write a script to compare the current view definitions against a snapshot of the previous script run output. We leave this task for your further consideration.

With due consideration for work completed in earlier chapters, where you centralized various monitoring activities, you can readily see how monitoring capabilities provisioned under Snowflake organizations can overlap or make obsolete earlier work. You must ensure that your monitoring capabilities are planned to coincide with feature releases in order to prevent continual rework.

You must also consider the impact of multiple MSAs, which may prevent the consolidation of accounts into a single organization structure.

View Usage

Using `snowflake.organization_usage.automatic_clustering_history` as an example and extending the `catalog` database defined in the previous chapter, let's add new monitoring capability.

Start with declarations that enable later automation (if you wish):

```
SET catalog_database      = 'CATALOG';
SET catalog_owner_schema  = 'CATALOG.catalog_owner';
SET catalog_warehouse     = 'catalog_wh';
SET catalog_owner_role    = 'catalog_owner_role';
SET catalog_reader_role   = 'catalog_reader_role';
```

Then change the role and set your context:

```
USE ROLE      IDENTIFIER ( $catalog_owner_role   );
USE DATABASE  IDENTIFIER ( $catalog_database     );
USE SCHEMA    IDENTIFIER ( $catalog_owner_schema );
USE WAREHOUSE IDENTIFIER ( $catalog_warehouse    );
```

Create an enriched secure view named v_secure_enriched_ou_ach:

```
CREATE OR REPLACE SECURE VIEW v_secure_enriched_ou_ach
AS
SELECT c.*,
       c.organization_name||'.'||
       c.account_name||'.'||
       c.database_name||'.'||
       c.schema_name||'.'||
       c.table_name                    AS path_to_object,
       'v_secure_enriched_ou_ach'      AS data_object
FROM   snowflake.organization_usage.automatic_clustering_history c
ORDER BY c.organization_name,
         c.account_name,
         c.database_name,
         c.schema_name,
         c.table_name ASC;
```

Check that you can execute a query against your new view, v_secure_enriched_ou_ach:

```
SELECT * FROM v_secure_enriched_ou_ach;
```

You may not see any results due to inactivity. Declaring your view v_secure_enriched_ou_ach is to demonstrate how to build out capability.

Provision Read Access

Having successfully created your metadata reporting view v_secure_enriched_ou_ach, you must grant entitlement to your reader role catalog_reader_role:

```
GRANT SELECT ON VIEW v_secure_enriched_ou_ach TO ROLE catalog_reader_role;
```

Now switch roles to catalog_reader_role and prove your reader role can access the view:

```
USE ROLE IDENTIFIER ( $catalog_reader_role );

SELECT *
FROM   v_secure_enriched_ou_ach;
```

For consistency you should apply the same deployment pattern to all in-scope views found within snowflake.organization_usage.

Summary

You began this chapter by learning about Snowflake organization as a concept and then you explored the impact of organization growth and how MSAs can impact the use of the Snowflake organization feature and accounts in general. You then touched upon the QBR and learned of a few points of interest for your further consideration.

As the orgadmin role is a new feature, you expanded the Snowflake role hierarchy to see where and how the orgadmin role fits in along with exposing features supplied by the orgadmin role.

You then walked through each orgadmin role feature with hands-on explanations and accompanying feature-specific comments for your further consideration.

Later you briefly explored where to find the metadata for Snowflake organizations, noting this is an area of expansion for Snowflake where new capabilities are actively being delivered.

Reusing the catalog database built within an earlier chapter, you developed an enriched wrapper view for snowflake.organization_usage.automatic_clustering_history as the basis for your further enhancement.

Alerting is the subject of the next chapter.

CHAPTER 9

Alerting

Snowflake alerts is an emerging topic based upon new functionality currently available in public preview. In this chapter, first you'll learn how to use the in-built Snowflake email capability and create wrapper stored procedures in both JavaScript and Python for later use. Alerts provide in-built capability to create conditions, periodically monitor for when the condition is either met or exceeded, and implement an action. We walk through each step and demonstrate scenarios where alerts may prove useful.

For those familiar with event-based processing, Snowflake alerts are broadly comparable, noting the use of a monitoring schedule.

We suggest judicious use of Snowflake alerts. Naturally, Snowflake credit consumption will occur so consider the cost implementations as part of your analysis.

Within the context of centralized provisioning for a landing zone, we suggest the creation of email wrapper procedures and sample alert configuration code will be sufficient. We do not advocate a centralized alerting capability except for stored procedure maintenance. We also expect each account owner or tenant will develop alerting according to their own need.

Setup

For this chapter, you will reuse the CATALOG database and schema built in Chapter 7.

Creating Email Integration

With your environment set, now create an email notification. To do this, you utilize the built-in system$send_email() stored procedure, which is only available on AWS and regions

- us-west-2

- us-east-1

- eu-west-1

© Andrew Carruthers and Sahir Ahmed 2023
A. Carruthers and S. Ahmed, *Maturing the Snowflake Data Cloud*, https://doi.org/10.1007/978-1-4842-9340-9_9

Further information can be found here: https://docs.snowflake.com/en/user-guide/email-stored-procedures.

First, switch roles to ACCOUNTADMIN:

```
USE ROLE accountadmin;
```

Then create your NOTIFICATION INTEGRATION, noting the ALLOWED_RECIPIENTS is a comma-separated list and you should set your own email address:

```
CREATE OR REPLACE NOTIFICATION INTEGRATION email_test
TYPE                = EMAIL
ENABLED             = TRUE
ALLOWED_RECIPIENTS = ( '<Your email here>' );
```

Now check that your NOTIFICATION INTEGRATION exists:

```
SHOW NOTIFICATION INTEGRATIONS;
```

You should see your new integration, as shown in Figure 9-1.

name	type	category	enabled	created_on
EMAIL_TEST	EMAIL	NOTIFICATION	true	2023-01-16 12:29:40.336 -0800

Figure 9-1. *Email integration created*

You must allow your catalog_owner_role to use the email integration:

```
GRANT USAGE ON INTEGRATION email_test TO ROLE catalog_owner_role;
```

Environment

Switch the role to use the CATALOG database:

```
SET catalog_database        = 'CATALOG';
SET catalog_owner_schema    = 'CATALOG.catalog_owner';
SET catalog_warehouse       = 'catalog_wh';
SET catalog_owner_role      = 'catalog_owner_role';

USE ROLE      IDENTIFIER ( $catalog_owner_role   );
USE DATABASE  IDENTIFIER ( $catalog_database      );
```

```
USE SCHEMA    IDENTIFIER ( $catalog_owner_schema );
USE WAREHOUSE IDENTIFIER ( $catalog_warehouse    );
```

If any of the above SQL statements fail, please refer to Chapter 7 for setup configuration.

Test system$send_email

Having set your context, now attempt to send an email:

```
CALL system$send_email ( 'email_test',
                         '<Your email here>',
                         'Test email from system$send_email',
                         'Test payload' );
```

Assuming Snowsight is used to run the above code, you should see a TRUE response, as shown in Figure 9-2.

	SYSTEM$SEND_EMAIL
1	TRUE

Figure 9-2. *Email test response*

After a minute or so, check your email to ensure receipt of the test message.

Creating the JavaScript Email Procedure

Having proven you can successfully send email, you now create a stored procedure as a building block for later reuse.

One very good reason to create a wrapper stored procedure is to abstract the NOTIFICATION INTEGRATION name: You may change the NOTIFICATION INTEGRATION name and make one change to the stored procedure rather than searching and replacing many places within your source code.

```
CREATE OR REPLACE PROCEDURE sp_send_mail ( P_RECIPIENT STRING,
                                           P_SUBJECT   STRING,
                                           P_PAYLOAD   STRING )
                                           RETURNS STRING
```

```
LANGUAGE javascript
EXECUTE AS CALLER
AS
$$
   var sql_stmt   = "CALL system$send_email ( 'email_test'," +
                                       "'" + P_RECIPIENT + "'," +
                                       "'" + P_SUBJECT   + "'," +
                                       "'" + P_PAYLOAD   + "');"

   var stmt       = snowflake.createStatement ({ sqlText:sql_stmt });

   try
   {
      recset = stmt.execute();
      result = recset.next();
   }
   catch ( err )
   {
      result  = sql_stmt;
      result += "\nCode: "          + err.code;
      result += "\nState: "         + err.state;
      result += "\nMessage: "       + err.message;
      result += "\nStack Trace:\n" + err.stackTraceTxt;
   }

   return recset.getColumnValue(1);
$$;
```

Then call your new stored procedure, replacing the placeholder with your own email address:

```
CALL sp_send_mail ( '<Your email here>',
                    'Test email from sp_send_mail',
                    'Test payload' );
```

You should see a true response, as shown in Figure 9-3.

	SP_SEND_MAIL
1	true

Figure 9-3. *Email test response*

After a minute or so, check your email to ensure receipt of the test message, noting it will be from Snowflake Computing <no-reply@snowflake.net>.

Creating a Python Email Procedure

In order to introduce Python, and noting the current version is limited to 3.8, use this stored procedure:

```
CREATE OR REPLACE PROCEDURE py_send_mail ( p_recipient STRING,
                                           p_subject   STRING,
                                           p_payload   STRING )
RETURNS STRING
LANGUAGE PYTHON
RUNTIME_VERSION = 3.8
PACKAGES        = ( 'snowflake-snowpark-python', 'tabulate' )
HANDLER         = 'x'
EXECUTE AS CALLER
AS
$$
import snowflake

def x ( session, p_recipient, p_subject, p_payload ):
   session.call ( 'system$send_email',
               'email_test',
               p_recipient,
               p_subject,
               p_payload )
   return 'email sent:\n%s' % p_payload
$$;
```

When attempting to compile your Python stored procedure, you should see this error: `Anaconda terms must be accepted by ORGADMIN to use Anaconda 3rd party packages. Please follow the instructions at https://docs.snowflake.com/en/developer-guide/udf/python/udf-python-packages.html#using-third-party-packages-from-anaconda`.

You must enable Anaconda. Refer to the documentation at `https://docs.snowflake.com/en/developer-guide/udf/python/udf-python-packages.html#using-third-party-packages-from-anaconda`.

To enable Anaconda, and assuming Snowsight is used, click the home icon in the top left corner of the screen, as shown in Figure 9-4.

Figure 9-4. *Home icon*

Then click the down arrow next to your name, switch role, and select **ORGADMIN**, as shown in Figure 9-5.

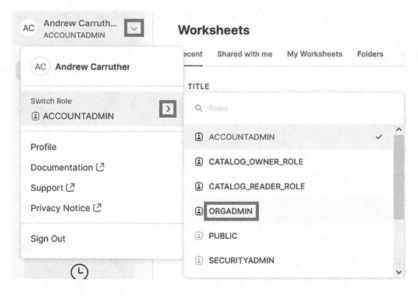

Figure 9-5. *Setting ORGADMIN role*

Accept the Anaconda billing and terms. For this, select **Admin ➤ Billing & Terms**, as shown in Figure 9-6.

Figure 9-6. *Selecting Billing & Terms*

Click the **Enable** button, as shown in Figure 9-7.

Figure 9-7. *Enable Billing & Terms*

A new dialog appears for you to accept, as shown in Figure 9-8. Click the **Acknowledge & Continue** button.

Figure 9-8. *Accepting Anaconda's additional terms*

After you have accepted the Anaconda additional terms, you should see the confirmation dialog shown in Figure 9-9.

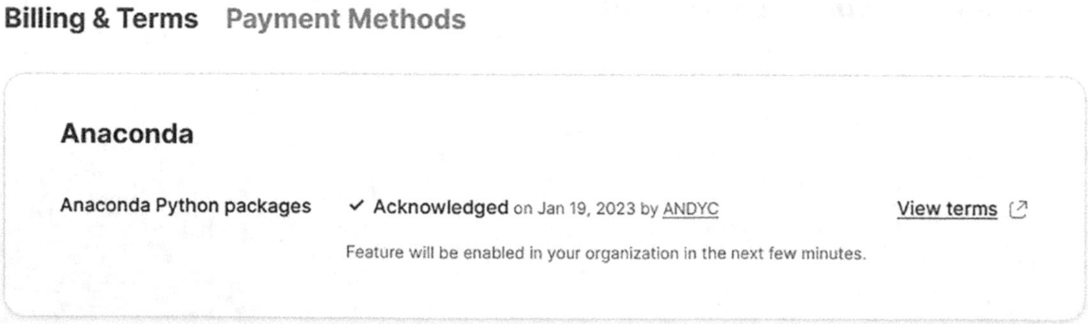

Billing & Terms   Payment Methods

Anaconda

Anaconda Python packages   ✓ Acknowledged on Jan 19, 2023 by ANDYC   View terms ↗

Feature will be enabled in your organization in the next few minutes.

Figure 9-9. *Anaconda additional terms confirmation*

Returning to your worksheet, you can now compile your Python stored procedure. Then call your new stored procedure, replacing the placeholder with your own email address:

```
CALL py_send_mail ( '<Your email here>',
                    'Test email from py_send_mail',
                    'Test payload' );
```

You should see the response shown in Figure 9-10.

	PY_SEND_MAIL
1	email sent: Test payload

Figure 9-10. *Email test response*

After a minute or so, check your email to ensure receipt of the test message, noting it will be from Snowflake Computing <no-reply@snowflake.net>.

Configuring Alerts

Creating alerts relies upon the active role having entitlement to execute alerts, an account level entitlement. To do this, extend your CATALOG_OWNER_ROLE.

Setting Up a Logging Table

To illustrate how an alert can log to a local table, you must create a local logging table. First, declare your variables:

```
SET catalog_database        = 'CATALOG';
SET catalog_owner_schema    = 'CATALOG.catalog_owner';
SET catalog_warehouse       = 'catalog_wh';
SET catalog_owner_role      = 'catalog_owner_role';
```

Now set your execution context:

```
USE ROLE      IDENTIFIER ( $catalog_owner_role   );
USE DATABASE  IDENTIFIER ( $catalog_database      );
USE SCHEMA    IDENTIFIER ( $catalog_owner_schema );
USE WAREHOUSE IDENTIFIER ( $catalog_warehouse     );
```

Create a logging table:

```
CREATE OR REPLACE TABLE logging
(
message            STRING,
insert_timestamp   TIMESTAMP DEFAULT current_timestamp()
);
```

Account Level Entitlement

Switch the role to ACCOUNTADMIN:

```
USE ROLE accountadmin;
```

Grant account-level entitlement to your role:

```
GRANT EXECUTE ALERT ON ACCOUNT TO ROLE IDENTIFIER ( $catalog_owner_role );
```

Example Account Alert Requirement

To create a meaningful alert, imagine you have a requirement to ensure your Snowflake account has a single application with a known number of objects. Many organizations have a requirement to ensure the configuration management of applications is managed by approved releases only, so any variance in the number of objects could represent an unauthorized change.

Using the above example, your alert will trigger when the number of objects detected on a sample schema differs from a known baseline figure. Using the CATALOG database and CATALOG_OWNER schema developed earlier in this book, you know there are two views:

- v_secure_enriched_ou_ach

- v_secure_enriched_aus_columns

To prove your assertion:

```
SELECT table_catalog||'.'||table_schema||'.'||table_name AS object_path,
       table_type
FROM   snowflake.account_usage.tables
WHERE  deleted IS NULL;
```

Figure 9-11 illustrates the expected response, noting you may experience up to 90 minutes latency when accessing the Account Usage Store views.

OBJECT_PATH	...	TABLE_TYPE
CATALOG.CATALOG_OWNER.V_SECURE_ENRICHED_AUS_COLUMNS		VIEW
CATALOG.CATALOG_OWNER.V_SECURE_ENRICHED_OU_ACH		VIEW
CATALOG.CATALOG_OWNER.LOGGING		BASE TABLE

Figure 9-11. *Account Usage Store objects*

The query returns three rows, which you now refactor into a SQL statement used within your alert. If your account differs, simply refactor to accommodate the number of rows to be compared against.

```
SELECT num_rows
FROM   (
       SELECT count(*) num_rows
       FROM   snowflake.account_usage.tables
       WHERE  deleted IS NULL
       )
WHERE  num_rows != 3;
```

The above query will return no results where the number of rows returned = 3, which is perfect for configuring your alert where any variance in the number of declared objects results in a returned row, which triggers the alert response.

A real-world alert would compare the num_rows to a reference table which is updated, post all change releases, and not use a hard-coded value as shown.

In order to bypass the 90-minute latency when accessing the Account Usage Store views, you may choose to use the CATALOG database and CATALOG_OWNER schema to trigger your alert. This query shows the equivalent query, noting you must exclude the INFORMATION_SCHEMA:

```
SELECT table_catalog||'.'||table_schema||'.'||table_name AS object_path,
       table_type
FROM   catalog.information_schema.tables
WHERE  table_schema != 'INFORMATION_SCHEMA';
```

The returned result set should match Figure 9-11.

The corresponding alert trigger query becomes

```
SELECT num_rows
FROM   (
       SELECT count(*) num_rows
       FROM   catalog.information_schema.tables
       WHERE  table_schema != 'INFORMATION_SCHEMA'
       )
WHERE  num_rows != 3;
```

For completeness, using a table function for Account Usage Store access does not bypass latency as the below SQL statement proves, noting the latency period may have expired, thus rendering this check invalid:

```
SELECT count(*) FROM TABLE('snowflake.account_usage.tables');
SELECT count(*) FROM TABLE('catalog.information_schema.tables');
```

The Account Usage Store approach is preferred because it covers all databases within an account subject to any locally applied filters whereas the information_schema approach only covers a single database.

Naturally, this test case is quite simple and used to illustrate capability. We leave it for you to further develop it into a robust alerting capability suitable for your organization. Later we suggest some tooling integrations for your consideration.

Configuring an Account Level Alert

Having identified the number of objects declared within your CATALOG database and CATALOG_OWNER schema, now create an alert. When declaring an alert, you can't use the IDENTIFIER notation to reference a warehouse, so you must use the actual name:

```
CREATE OR REPLACE ALERT test_alert
WAREHOUSE = catalog_wh
SCHEDULE  = '1 MINUTE'
IF (
   EXISTS
      (
      SELECT num_rows
```

```
    FROM    (
            SELECT  count(*) num_rows
            FROM    snowflake.account_usage.tables
            WHERE   deleted IS NULL
            )
    WHERE   num_rows != 3
    )
  )
THEN
    INSERT INTO logging VALUES ( 'ALERT: Object count mismatch!' );
```

Having created your alert, you should prove it exists:

```
SHOW ALERTS;
```

Figure 9-12 shows a partial returned record and indicates your alert has been created but is suspended.

created_on	name	schedule	state	condition	action
2023-01-23 12:11:...	TEST_ALERT	1 MINUTE	suspended	SELECT num_rows...	INSERT INTO loggi...

Figure 9-12. *Declared alert*

You may also use DESCRIBE ALERT test_alert; because it provides the same information.

To enable your alert:

```
ALTER ALERT test_alert RESUME;
```

Recheck the alert status where the state should now show as started:

```
SHOW ALERTS;
```

After 2 minutes, check the logging table for entries:

```
SELECT * FROM CATALOG.catalog_owner.logging;
```

For this next check, the Account Usage Store latency of up to 90 minutes may affect your results.

Depending upon whether the alert detects a mismatch in rows, you should see one or more records similar to the one shown in Figure 9-13. If no rows are recorded, create a table to force a mismatch against the expected number of objects (3) and recheck the logging table content.

MESSAGE	INSERT_TIMESTAMP
ALERT: Object count mismatch!	2023-01-23 12:41:04.504

Figure 9-13. *Logged alert data*

To suspend your alert:

```
ALTER ALERT test_alert SUSPEND;
```

Extending Alert Capability

Having proven your alert successfully inserts a record into a logging table, now refactor the alert to call a JavaScript stored procedure, noting you must change the recipient email address within the alert declaration shown next:

```
CREATE OR REPLACE ALERT test_alert
WAREHOUSE = catalog_wh
SCHEDULE  = '1 MINUTE'
IF (
   EXISTS
      (
      SELECT num_rows
      FROM   (
               SELECT count(*) num_rows
               FROM   snowflake.account_usage.tables
               WHERE  deleted IS NULL
             )
      WHERE  num_rows != 3;
      )
   )
```

```
THEN
   CALL sp_send_mail ( '<Your email here>',
                       'Test email from alert test_alert',
                       'ALERT: Object count mismatch!' );
```

Then enable your alert:

```
ALTER ALERT test_alert RESUME;
```

Recheck the alert status where the state should now show as `started`:

```
SHOW ALERTS;
```

After a minute or so, check your email to ensure receipt of the test message, noting it will be from `Snowflake Computing no-reply@snowflake.net`. Also, check the subject and body to ensure both correctly reflect the alert message.

Not forgetting to SUSPEND your alert to prevent excessive resource consumption:

```
ALTER ALERT test_alert SUSPEND;
```

You may wish to extend the JavaScript stored procedure to also log into the logging table, left for your further investigation.

Alert History

In support of monitoring alert execution, a new table function called `information_schema.ALERT_HISTORY` has been provisioned. This new capability provides the capability to examine alert executions within a specified time window (in our example, the past 24 hours):

```
SELECT *
FROM TABLE ( information_history.alert_history
           (
           scheduled_time_range_start =>
                  DATEADD ( 'hour', -24, current_timstamp())
           )
         )
ORDER BY scheduled_time DESC;
```

Figure 9-14 shows a subset of the data returned from the information_schema. ALERT_HISTORY table function.

NAME	CONDITION	ACTION	STATE	SCHEDULED_TIME
TEST_ALERT	SELECT num_rows FROM (...	INSERT INTO logging (mes...	SUCCEEDED	2023-01-23 12:40:42.326 +...

Figure 9-14. *Alert history*

Cleanup

To remove entitlement:

```
USE ROLE accountadmin;

DROP NOTIFICATION INTEGRATION email_test;
```

To revoke account-level entitlement:

```
REVOKE EXECUTE ALERT ON
ACCOUNT FROM ROLE IDENTIFIER ( $catalog_owner_role );
```

To remove an account-level alert:

```
DROP ALERT test_alert;
```

To remove the logging table and stored procedures:

```
USE ROLE      IDENTIFIER ( $catalog_owner_role   );
USE DATABASE  IDENTIFIER ( $catalog_database     );
USE SCHEMA    IDENTIFIER ( $catalog_owner_schema );

DROP TABLE logging;

DROP PROCEDURE sp_send_mail ( STRING,
                              STRING,
                              STRING );

DROP PROCEDURE py_send_mail ( STRING,
                              STRING,
                              STRING );
```

Summary

You started this chapter by implementing stored procedure wrappers for Snowflake supplying email functionality using both JavaScript and Python, noting the steps required to enable Anaconda.

You then identified how alerts can be enabled at the account level. Schema-level implementation is for your further investigation. As a test case, you developed a simple mechanism to detect when new objects are added to a schema with alerts logged into a table. You later extended the alert delivery to call your example JavaScript email stored procedure.

At the time of writing, email functionality is a preview feature; further information can be found here: `https://docs.snowflake.com/en/sql-reference/email-stored-procedures.html#sending-email-notifications`.

Alerts are also new functionality within Snowflake and, as you have seen, the email capability is reliant upon a corresponding `NOTIFICATION INTEGRATION`. Further information can be found here: `https://docs.snowflake.com/en/sql-reference/sql/create-notification-integration.html#create-notification-integration`. Note the scope of integrations is not limited to email, with a queue being the other available option left for your further investigation.

The focus now turns to Cloud Data Management Capabilities (CDMC), the subject of the next chapter.

CHAPTER 10

Cloud Data Management Capabilities (CDMC)

In order to correctly represent CDMC, the below selected quote and some content have been taken from `https://edmcouncil.org/` with the expectation of setting the correct context for all work conducted within this chapter.

> *The EDM Council is the Global Association created to elevate the practice of Data Management as a business and operational priority. The Council is the leading advocate for the development and implementation of Data Standards, Best Practices and comprehensive Training and Certification programs.*

We strongly recommend joining the EDM Council. We make reference to the CDMC Framework articulated within CDMC_Framework_V1.1.1.pdf available for download once signed in.

As the CDMC Framework is platform agnostic, there are no reference implementations for Snowflake to ensure conformance to the standards set out within the CDMC Framework. We seek to close this gap.

We do not attempt to provide a complete or exhaustive programmatic verification of proof for each CDMC control but instead pick selected controls for close examination.

To realize the greatest value from this chapter, we recommend working through the content systematically. We build CDMC monitoring capabilities progressively with the express aim of equipping you with both the rationale and tools to deliver your organization's bespoke implementation. We have chosen a few CDMC controls to

303

© Andrew Carruthers and Sahir Ahmed 2023
A. Carruthers and S. Ahmed, *Maturing the Snowflake Data Cloud*, https://doi.org/10.1007/978-1-4842-9340-9_10

implement, showcasing implementation patterns for your adaptation and extension. Each section is referenced back to CDMC_Framework_V1.1.1.pdf to both aid your understanding and to keep this chapter within an acceptable page count.

We also recognize not every control lends itself to programmatic verification of proof, and each organization will vary in terms of maturity, tooling, automation, and data structure. Consequently, we offer a sample repository where supporting data for controls can be loaded and from which programmatic verification of proof is possible. We also propose Snowflake internal structures to source data from and focus on establishing a limited baseline suite of Snowflake components readily adapted for your specific needs. Finally, we suggest how external documentation can be used to source data and defer to techniques found in *Building the Snowflake Data Cloud* to source such data.

CDMC is a wide and deep framework with a pathway showing the steps expected to achieve each level of compliance within each control. Where possible, we advise how controls can be implemented and scored according to the CDMC Scoring Guide found on page 14 of CDMC_Framework_V1.1.1.pdf, which we summarize here:

- **Not Initiated**: You aren't doing anything purposeful at all.
  Score = 1 point.

- **Conceptual**: You know that you need to do something and are just beginning to talk about the issues involved. Score = 2 points.

- **Developmental**: You are actively implementing controls to ensure compliance. Score = 3 points.

- **Defined**: Your controls are defined and approved and are ready to implement. Score = 4 points

- **Achieved**: Your controls are actively in place and being enforced.
  Score = 5 points.

- **Enhanced**: Employees see the controls as "just how things are done."
  Score = 6 points

We do not discuss CDMC certification, which requires an independent assessment of capabilities and controls. It is anticipated the information contained within this chapter will enable an organization to both assess and elevate its scoring. Our measure of success is whether we have been successful in educating and equipping our readers to ratchet the data management dial up in a measurable increment.

Naturally this chapter closely follows the CDMC_Framework_V1.1.1.pdf. For those seeking a more concise, high-level view of the CDMC controls, we recommend downloading CDMC_Key_Controls__V1.1.1.pdf, which focusses on the core 14 controls.

The key controls are summarized here:

- Governance of data within cloud environments including sourcing, management, ownership, and usage

- Creation and maintenance of glossaries and data catalogs along with data discovery capabilities

- Management and control of data access along with monitoring usage patterns

- Protection of sensitive data and prevention of unauthorized access and data leakage

- Ensuring data quality is both managed and maintained throughout its lifecycle

- Preserving the integrity of data movement and content within cloud environments

Overview

CDMC Framework is a suite of 14 capabilities and 37 sub-capabilities focused on identifying objectives, posing questions, and supporting artifacts. Our approach in this chapter focusses on understanding the advice given to both practitioners and providers, and then implementing a reference architecture for Snowflake illustrated by specific use cases.

The information provided within this chapter is not exhaustive but represents the authors' views to provide a starting point for both further investigation and extension. Put another way, the absence of a Snowflake reference implementation impedes every organization and denies it the opportunity to leverage, enhance, and develop its own proofs.

This chapter is intended to initiate conversation on CDMC capabilities and develop Snowflake-specific tooling to prove your applications are compliant. As we progress, we make constant reference to the corresponding CDMC controls. Our approach is to set the context for each control by adopting the same CDMC Capability numbering scheme from the CDMC_Key_Controls__V1.1.1.pdf. Note the full descriptions should be referenced using CDMC_Framework_V1.1.1.pdf.

All of the information offered in support of CDMC controls is our own original work and does not form part of the CDMC Framework. The work is our understanding and interpretation only.

In my experience of nearly 30 years working in IT primarily as an independent contractor, one factor occurs far more frequently than any other: The absence, or inadequacy, of system documentation for any platform or application. The CDMC Framework allows you to conduct self-discovery and scorecard your organization, leading to identification of gaps in controls along with opportunities to remediate absent system documentation.

We do not provide implementation detail for every control. Instead, we offer a pattern-based approach with worked examples facilitating expansion and delivery into your organization. Where possible for selected controls, we indicate which pattern may be most appropriate to meet the control requirements.

Data Catalog

Your organization must have a single central data catalog, an organized inventory of data assets in your organization, where your business definitions will also be recorded.

A data catalog is much more than just a repository for business definitions. With a little thought, you can (and should) extend your data catalog to include a wealth of information providing evidence of conformance to CDMC controls.

As a central repository, your data catalog allows you to answer fundamental questions such as

- **Ownership**: Stakeholders, custodians, repositories

- **Governance**: Entitlement groups, tagging, access policies

- **Management**: Collect, analyze, and maintain contextual information

- **Discovery**: Where data resides, how to access data sets

- **Audit**: Provide a full audit trail of changes

From the CDMC Framework perspective, a data catalog provides an audit trail of information providing evidence in support of control scoring.

For those organizations approaching the decision point to invest in a data catalog, within Appendix A we offer supporting documentation templates including the Business Case, Solution Design Document (SDD), and Feed Onboarding Contract (FOC) for your consideration to record information for later replay into a data catalog.

Data Ingestion Patterns

Depending upon where information exists within your organization, you have several options available to acquire evidence for the control points contained within the CDMC Framework. Figure 10-1 illustrates several options and suggests integrations to deliver proofs.

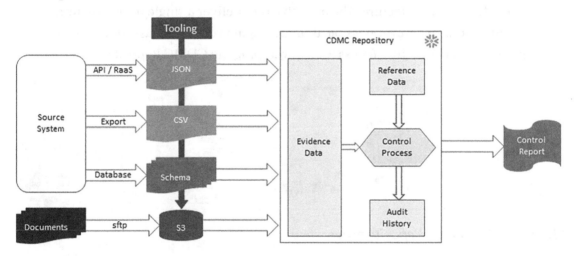

Figure 10-1. *Data integration options*

You may prefer different data integration patterns; those within Figure 10-1 are offered as a starting point. The timeliness of evidence data arrival into Snowflake is a matter for your discretion. We suggest a weekly refresh will meet most control reporting requirements. We leave implementation details for your further investigation and assume you have access to the evidence data from within Snowflake.

The sample code assumes evidence data is in an accessible table from which your control process joins local reference data and then generates an audit history. Run all control reports from data contained within the audit history.

All ingestion mechanisms referenced within Figure 10-1 are discussed within my previous book, *Building the Snowflake Data Cloud*, where full implementation details can be found.

Control processes are explained as you progress through this chapter and, as implied within Figure 10-1, a common audit history for all automated controls is provided, thus enabling a single common summary report. More on this later.

Reference Architecture

Throughout this chapter you use JSON to store data sets. You may find `https://jsonformatter.org/` useful to both examine and validate JSON structures when extending the examples found within this chapter.

Expanding upon your data ingestion patterns, let's focus on developing your Snowflake reference architecture. The objective is to deliver a single design pattern reusing components wherever possible thus leveraging your codebase; develop once, reuse many times. Figure 10-2 focusses on developing the CDMC Repository.

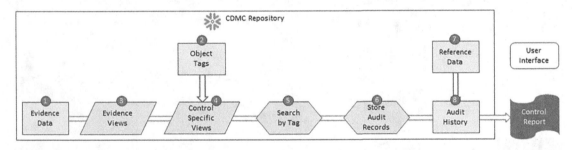

Figure 10-2. *CDMC Repository overview*

For the purpose of developing the reference architecture, we will shortly walk through all the components illustrated in Figure 10-2.

Item (1): We assume a suite of staging tables containing evidence data have been loaded from the Project Management Office (PMO) tooling and/or the business case template within Appendix A. Alternatively, we supply sample PMO data to showcase the "art of the possible" and later expand to use the Snowflake Account Usage Store to source evidence data.

Item (2): The creation of object tags allows you to define classifications for implementation across all your database objects. You later develop reporting views to deliver a consistent, repeatable experience. Note there is a hard limit of 20 unique tags per database object as specified here: `https://docs.snowflake.com/en/user-guide/object-tagging.html#tag-quotas-for-objects-columns`.

Item (3): The CDMC control implementations rely upon interpreting the available evidence and a degree of human interpretation. A suite of evidence views enables you to capture your human interpretation to conform the data into programmatically consumable artefacts. Item (3) is the place where bespoke code for each control is built and is intended to be replaced with your organization-specific implementation.

Item (4): Building upon your evidence views, you implement control-specific views that simplify reporting when generating JSON documents for later storage. Item (4) abstracts data surfaced by the bespoke views developed within Item (3) and is intended to be adapted for your organization-specific implementation.

Item (5): Utilizing a JavaScript stored procedure, you implement functionality to fetch tagged objects and attributes. While the code sample is restricted to a single tag, which returns a single object attribute set, our expectation is you will expand the code to suit your needs.

Item (6): You reuse the stored procedure from Item (6) to return a data set for storage in your audit trail. Wrapping your search stored procedure allows reuse for reporting tooling and provides a single implementation pattern for use across all controls.

Item (7): Local reference data may be used to enrich your reporting in Item (8) or provision filters for your reporting tooling.

Item (8): Every control process run will generate an audit trail. You use the audit trail to evidence your controls are both effective and to substantiate your organization is constantly improving conformance to the CDMC Framework.

We deliberately exclude specifications for the control report and leave these for you to determine.

Evidence Repository

Regardless of where source data exists, your Snowflake evidence data, tags, and controls should store results into Snowflake within a CDMC dedicated schema, which we now demonstrate.

Repository Overview

We assume familiarity with SQL and in the interests of brevity do not list all the commands but instead refer to the associated script called Chapter_10_CDMC.sql, which should be reviewed before use.

Please note the following points:

- We create a database, warehouse, and schema.

- Within the schema we create tables for reference data, sample evidence data, and results storage along with reference data for each control.

- The supplied script is intended to be adapted for your particular implementation requirements.

- A cleanup script is also provided.

We have not provided historization for reference data changes nor do we historize evidence data. We leave these for your future enhancement.

Repository Setup

Cut and paste the following commands to create the CDMC Snowflake repository:

```
SET cdmc_database        = 'CDMC';
SET cdmc_owner_schema    = 'CDMC.cdmc_owner';
SET cdmc_warehouse       = 'cdmc_wh';
SET cdmc_owner_role      = 'cdmc_owner_role';

USE ROLE sysadmin;

CREATE OR REPLACE DATABASE IDENTIFIER ( $cdmc_database )
DATA_RETENTION_TIME_IN_DAYS = 90;

CREATE OR REPLACE WAREHOUSE IDENTIFIER ( $cdmc_warehouse ) WITH
WAREHOUSE_SIZE       = 'X-SMALL'
AUTO_SUSPEND         = 60
AUTO_RESUME          = TRUE
MIN_CLUSTER_COUNT    = 1
MAX_CLUSTER_COUNT    = 4
```

```
SCALING_POLICY      = 'STANDARD'
INITIALLY_SUSPENDED = TRUE;

CREATE OR REPLACE SCHEMA IDENTIFIER ( $cdmc_owner_schema   );

USE ROLE securityadmin;

CREATE OR REPLACE ROLE IDENTIFIER ( $cdmc_owner_role  )
COMMENT = 'CDMC.cdmc_owner Role';

GRANT ROLE IDENTIFIER ( $cdmc_owner_role  ) TO ROLE securityadmin;

GRANT USAGE    ON DATABASE  IDENTIFIER ( $cdmc_database       ) TO ROLE
IDENTIFIER ( $cdmc_owner_role  );
GRANT USAGE    ON WAREHOUSE IDENTIFIER ( $cdmc_warehouse      ) TO ROLE
IDENTIFIER ( $cdmc_owner_role  );
GRANT OPERATE ON WAREHOUSE IDENTIFIER ( $cdmc_warehouse      ) TO ROLE
IDENTIFIER ( $cdmc_owner_role  );
GRANT USAGE    ON SCHEMA     IDENTIFIER ( $cdmc_owner_schema   ) TO ROLE
IDENTIFIER ( $cdmc_owner_role  );

GRANT USAGE                         ON SCHEMA IDENTIFIER
( $cdmc_owner_schema   ) TO ROLE IDENTIFIER ( $cdmc_owner_role );
GRANT MONITOR                       ON SCHEMA IDENTIFIER
( $cdmc_owner_schema   ) TO ROLE IDENTIFIER ( $cdmc_owner_role );
GRANT MODIFY                        ON SCHEMA IDENTIFIER
( $cdmc_owner_schema   ) TO ROLE IDENTIFIER ( $cdmc_owner_role );
GRANT CREATE TABLE                  ON SCHEMA IDENTIFIER
( $cdmc_owner_schema   ) TO ROLE IDENTIFIER ( $cdmc_owner_role );
GRANT CREATE VIEW                   ON SCHEMA IDENTIFIER
( $cdmc_owner_schema   ) TO ROLE IDENTIFIER ( $cdmc_owner_role );
GRANT CREATE SEQUENCE               ON SCHEMA IDENTIFIER
( $cdmc_owner_schema   ) TO ROLE IDENTIFIER ( $cdmc_owner_role );
GRANT CREATE FUNCTION               ON SCHEMA IDENTIFIER
( $cdmc_owner_schema   ) TO ROLE IDENTIFIER ( $cdmc_owner_role );
GRANT CREATE PROCEDURE              ON SCHEMA IDENTIFIER
( $cdmc_owner_schema   ) TO ROLE IDENTIFIER ( $cdmc_owner_role );
```

```
GRANT CREATE STREAM                ON SCHEMA IDENTIFIER
( $cdmc_owner_schema    ) TO ROLE IDENTIFIER ( $cdmc_owner_role );
GRANT CREATE MATERIALIZED VIEW   ON SCHEMA IDENTIFIER
( $cdmc_owner_schema    ) TO ROLE IDENTIFIER ( $cdmc_owner_role );
GRANT CREATE FILE FORMAT           ON SCHEMA IDENTIFIER
( $cdmc_owner_schema    ) TO ROLE IDENTIFIER ( $cdmc_owner_role );

GRANT ROLE IDENTIFIER ( $cdmc_owner_role ) TO USER <Your User Here>;
```

Change the role to cdmc_owner_role and define a control table:

```
USE ROLE        cdmc_owner_role;
USE DATABASE    CDMC;
USE WAREHOUSE   CDMC_WH;
USE SCHEMA      CDMC.cdmc_owner;

CREATE OR REPLACE TABLE cdmc_control
(
id                NUMBER          NOT NULL,
reference         STRING(255)     NOT NULL,
name              STRING(255)     NOT NULL,
description       STRING(2000)    NOT NULL,
last_updated      TIMESTAMP       NOT NULL
);

CREATE OR REPLACE SEQUENCE seq_cdmc_control_id START WITH 100000;
```

Add more control data within the associated script Chapter_10_CDMC.sql:

```
INSERT INTO cdmc_control
VALUES
( seq_cdmc_control_id.NEXTVAL, '1.1.1', 'Cloud Data Management business
cases are defined', 'As an organization moves its data and operations to
cloud environments, it is important to develop, communicate, cultivate, and
support business cases for cloud data management. An effective cloud data
management business case defines the objectives and expected outcomes of
the implementation. It is vital to develop an entire cloud business case
framework of metrics, measures and key performance indicators to articulate
the value of cloud data management.', current_timestamp() ),
```

312

(seq_cdmc_control_id.NEXTVAL, '1.1.2', 'Cloud Data Management business cases are syndicated and governed', 'Each cloud data management business case must be approved by an appropriate authority and sponsored by accountable stakeholders. Successfully managing data in cloud environments requires substantial support from both business and technology stakeholders within an organization. The interests of these groups must be aligned early and consistently represented through deployment.

Each cloud data management business case must be enforceable and periodically reviewed by sponsors throughout deployment and the cloud data management lifecycle. Reviews will ensure that the business cases meet requirements as the organization''s objectives evolve and the stakeholders change.', current_timestamp());

Use sample evidence data from tables to drive your controls:

```
CREATE OR REPLACE TABLE pmo_evidence
(
pmo_evidence_id            NUMBER         NOT NULL,
program_reference          STRING(4000)   NOT NULL,
executive_summary          STRING(4000),
planned_activities         STRING(4000),
expected_business_outcomes STRING(4000),
project_governance         STRING(4000),
resources                  STRING(4000),
timeline_text              STRING(4000),
timeline_gate              STRING(30),
project_status             STRING(30),
costs_text                 STRING(4000),
costs_currency             STRING(3),
costs_estimate             NUMBER,
costs_actual               NUMBER,
last_updated               TIMESTAMP      NOT NULL
);

CREATE OR REPLACE SEQUENCE seq_pmo_evidence_id START WITH 100000;
```

```
INSERT INTO pmo_evidence
VALUES
( seq_pmo_evidence_id.NEXTVAL, 'Archive System X.', 'This program seeks
to retire System X onto Snowflake and provide an archive repository for
ad-hoc query for the mandatory 10 year retention period.', 'The planned
activities include porting all System X schemas and objects into Snowflake,
one-off ingestion of all data, recreation of essential historical reports,
validation all data has migrated correctly, user acceptance signoff.',
'The expected business outcomes are to realize cost savings of GBP 100,000
per annum due to decommissioning on-prem hardware and an expected reduction
of 46% from reporting costs due to the on-demand capability Snowflake
brings.', 'Unavailability of key staff due to resource contention with
other high priority projects. Dependencies: Timely on-boarding of Product Y
to deliver reporting capability.', 'Key dependency upon Snowflake Subject
Matter Experts.', 'Timelines will be impacted by any unexpected divestments
where we are obliged to provision reporting services to acquiring
organizations.', 'Gate 2', 'Green', 'Total OpEx costs anticipated to be GBP
30,000, CapEx costs anticipated to be GPB 100,000.', 'GBP', 130000, 88000,
current_timestamp() ), ( seq_pmo_evidence_id.NEXTVAL, 'Ingest System Y.',
'This program seeks to ingest System Y into Snowflake for consolidated
reporting.', 'The planned activities include porting commonly used System Y
data into Snowflake, recreation of reports and user acceptance signoff.',
'The expected business outcomes are to ensure repeatable, consistent and
historically accurate reports with an expected reduction of 72% from
reporting costs due to the on-demand capability Snowflake brings.', 'No
risks identified, this is a low priority program. Dependencies: None
identified.', 'Key dependency upon reporting SME.', 'Delivery expected Q1
2022.', 'Gate 0', 'Amber', 'Total OpEx costs anticipated to be GBP 10,000,
CapEx costs anticipated to be GPB 10,000.', 'GBP', 20000, 12766, current_
timestamp() ), ( seq_pmo_evidence_id.NEXTVAL, 'Bad Test Case 1.', 'This is
a bad test case.', 'This test case is used to prove our controls deliver
correct results for incomplete pmo_evidence records.', 'The test case fails
due to missing attributes.', NULL, NULL, NULL, NULL, NULL, NULL, NULL,
NULL, NULL, current_timestamp() );
```

You now create three views that implement controls 1.1.1 and 1.1.2, noting they are incomplete and left for your enhancements. View v_pmo_evidence_summary generates attributes used to determine business case completeness and is the source for subsequent PMO-based views.

You use these views to store bespoke logic. Through experience gained while working through this chapter, we found the data extract using OBJECT_CONSTRUCT encountered later is programmatically far easier using a view as the source when extracting information using object tags.

```
CREATE OR REPLACE VIEW v_pmo_evidence_summary COPY GRANTS
AS
SELECT program_reference,
       10                     AS control_metrics_num,
       CASE LENGTH ( NVL ( executive_summary,          '' ))
          WHEN 0 THEN 0 ELSE 1
       END AS executive_summary_length,
       CASE LENGTH ( NVL ( planned_activities,         '' ))
          WHEN 0 THEN 0 ELSE 1
       END AS planned_activities_length,
       CASE LENGTH ( NVL ( expected_business_outcomes, '' ))
          WHEN 0 THEN 0 ELSE 1
       END AS expected_business_outcomes_length,
       CASE LENGTH ( NVL ( project_governance,         '' ))
          WHEN 0 THEN 0 ELSE 1
       END AS project_governance_length,
       CASE LENGTH ( NVL ( resources,                  '' ))
          WHEN 0 THEN 0 ELSE 1
       END AS resources_length,
       CASE LENGTH ( NVL ( timeline_text,              '' ))
          WHEN 0 THEN 0 ELSE 1
       END AS timeline_text_length,
       CASE LENGTH ( NVL ( costs_text,                 '' ))
          WHEN 0 THEN 0 ELSE 1
       END AS costs_text_length,
       CASE LENGTH ( NVL ( timeline_gate,             '' ))
          WHEN 0 THEN 0 ELSE 1
```

```
      END AS timeline_gate_length,
      CASE LENGTH ( NVL ( project_status,            '' ))
         WHEN 0 THEN 0 ELSE 1
      END AS project_status_length,
      CASE LENGTH ( NVL ( costs_currency,            '' ))
         WHEN 0 THEN 0 ELSE 1
      END AS costs_currency_length,
      control_metrics_num              -   // The number of expected
      populated attributes
      executive_summary_length         -
      planned_activities_length        -
      expected_business_outcomes_length -
      project_governance_length        -
      resources_length                 -
      timeline_text_length             -
      costs_text_length                -
      timeline_gate_length             -
      project_status_length            -
      costs_currency_length                      AS zero_count,
      TO_CHAR ( ROUND ((( control_metrics_num - zero_count )
      / control_metrics_num ) * 100, 2 )) AS control_metrics_pct,
      executive_summary,
      planned_activities,
      expected_business_outcomes,
      project_governance,
      resources,
      timeline_text,
      costs_text,
      NVL ( timeline_gate,   'Not Specified' )  AS timeline_gate,
      NVL ( project_status, 'Not Specified' )  AS project_status,
      NVL ( costs_currency, 'Not Specified' )  AS costs_currency,
      TO_CHAR ( NVL ( costs_estimate, 0    ))  AS costs_estimate,
      TO_CHAR ( NVL ( costs_actual,   0    ))  AS costs_actual,
      TO_CHAR ( costs_estimate - costs_actual ) AS costs_variance
FROM    pmo_evidence;
```

View v_pmo_evidence_summary_1_1_1 selects only those evidence attributes from view v_pmo_evidence_summary that support control 1.1.1. Note the join to the reference table cdmc_control:

```
CREATE OR REPLACE VIEW v_pmo_evidence_summary_1_1_1 COPY GRANTS
AS
SELECT 'v_pmo_evidence_summary_1_1_1' AS source_object,
       c.document                     AS cdmc_document,
       c.reference                    AS cdmc_control,
       v.program_reference,
       v.executive_summary,
       v.planned_activities,
       v.expected_business_outcomes,
       v.project_governance,
       v.resources,
       v.timeline_text,
       v.control_metrics_pct,
       v.costs_text,
       v.costs_currency,
       v.costs_estimate,
       v.costs_actual,
       v.costs_variance
FROM   v_pmo_evidence_summary v,
       cdmc_control           c
WHERE  c.reference           = 'CDMC_1_1_1';
```

View v_pmo_evidence_summary_1_1_2 selects only those evidence attributes from view v_pmo_evidence_summary that support control 1.1.2. Note the join to the reference table cdmc_control:

```
CREATE OR REPLACE VIEW v_pmo_evidence_summary_1_1_2 COPY GRANTS
AS
SELECT 'v_pmo_evidence_summary_1_1_2' AS source_object,
       c.document                     AS cdmc_document,
       c.reference                    AS cdmc_control,
       v.program_reference,
       v.timeline_gate,
       v.project_status
```

317

```
FROM    v_pmo_evidence_summary v,
        cdmc_control            c
WHERE   c.reference             = 'CDMC_1_1_2';
```

The skeleton repository is now ready for use and will be enhanced as you progress through the remainder of this chapter,

Object Tagging

A key deliverable of this chapter is to enable self-service of evidence for consumption by current and future mechanisms. With this in mind, object tagging offers capability to both identify stored attributes and facilitate self-service.

You first entitle your role cdmc_owner_role to access the Snowflake Account Usage Store:

```
USE ROLE securityadmin;
```

```
GRANT IMPORTED PRIVILEGES ON DATABASE snowflake TO ROLE cdmc_owner_role;
```

Then revert to role cdmc_owner_role:

```
USE ROLE cdmc_owner_role;
```

Now create tags for table and column tagging. The accompanying script Chapter_10_CDMC.sql has tag definitions for all chapters within this book, herewith a subset of tag definitions:

```
CREATE OR REPLACE TAG CDMC        COMMENT = 'Cloud Data Management
Capabilities';
CREATE OR REPLACE TAG CDMC_1      COMMENT = 'Governance & Accountability';
CREATE OR REPLACE TAG CDMC_1_1    COMMENT = 'Business cases are defined and
governed';
CREATE OR REPLACE TAG CDMC_1_1_1 COMMENT = 'Business cases are defined';
CREATE OR REPLACE TAG CDMC_1_1_2 COMMENT = 'Business cases are syndicated
and governed';
```

Apply tags to both table and columns. The accompanying script Chapter_10_CDMC.sql has tag assignments for all objects declared within this book, herewith a subset of tag assignments:

```
ALTER TABLE CDMC.cdmc_owner.v_pmo_evidence_summary_1_1_1
SET TAG CDMC        = 'Cloud Data Management Capabilities';
ALTER TABLE CDMC.cdmc_owner.v_pmo_evidence_summary_1_1_2
SET TAG CDMC        = 'Cloud Data Management Capabilities';
ALTER TABLE CDMC.cdmc_owner.v_pmo_evidence_summary_1_1_1
SET TAG CDMC_1      = 'Governance & Accountability';
ALTER TABLE CDMC.cdmc_owner.v_pmo_evidence_summary_1_1_2
SET TAG CDMC_1      = 'Governance & Accountability';
ALTER TABLE CDMC.cdmc_owner.v_pmo_evidence_summary_1_1_1
SET TAG CDMC_1_1    = 'Business cases are defined and governed';
ALTER TABLE CDMC.cdmc_owner.v_pmo_evidence_summary_1_1_2
SET TAG CDMC_1_1    = 'Business cases are defined and governed';
ALTER TABLE CDMC.cdmc_owner.v_pmo_evidence_summary_1_1_1
SET TAG CDMC_1_1_1 = 'Business cases are defined';
ALTER TABLE CDMC.cdmc_owner.v_pmo_evidence_summary_1_1_2
SET TAG CDMC_1_1_2 = 'Business cases are syndicated and governed';
ALTER TABLE CDMC.cdmc_owner.v_pmo_evidence_summary_1_1_1 MODIFY COLUMN
executive_summary         SET TAG CDMC_1_1_1 = 'Business cases are
defined';
ALTER TABLE CDMC.cdmc_owner.v_pmo_evidence_summary_1_1_1 MODIFY COLUMN
planned_activities        SET TAG CDMC_1_1_1 = 'Business cases are
defined';
ALTER TABLE CDMC.cdmc_owner.v_pmo_evidence_summary_1_1_1 MODIFY COLUMN
expected_business_outcomes SET TAG CDMC_1_1_1 = 'Business cases are
defined';
ALTER TABLE CDMC.cdmc_owner.v_pmo_evidence_summary_1_1_1 MODIFY COLUMN
project_governance        SET TAG CDMC_1_1_1 = 'Business cases are defined';
ALTER TABLE CDMC.cdmc_owner.v_pmo_evidence_summary_1_1_1 MODIFY COLUMN
resources                 SET TAG CDMC_1_1_1 = 'Business cases are defined';
ALTER TABLE CDMC.cdmc_owner.v_pmo_evidence_summary_1_1_1 MODIFY COLUMN
timeline_text             SET TAG CDMC_1_1_1 = 'Business cases are defined';
ALTER TABLE CDMC.cdmc_owner.v_pmo_evidence_summary_1_1_2 MODIFY COLUMN
timeline_gate             SET TAG CDMC_1_1_2 = 'Business cases are
                          syndicated and governed';
```

```
ALTER TABLE CDMC.cdmc_owner.v_pmo_evidence_summary_1_1_2 MODIFY COLUMN
project_status              SET TAG CDMC_1_1_2 = 'Business cases are
                            syndicated and governed';
ALTER TABLE CDMC.cdmc_owner.v_pmo_evidence_summary_1_1_1 MODIFY COLUMN
costs_text                  SET TAG CDMC_1_1_1 = 'Business cases are defined';
ALTER TABLE CDMC.cdmc_owner.v_pmo_evidence_summary_1_1_1 MODIFY COLUMN
costs_currency              SET TAG CDMC_1_1_1 = 'Business cases are defined';
ALTER TABLE CDMC.cdmc_owner.v_pmo_evidence_summary_1_1_1 MODIFY COLUMN
costs_estimate              SET TAG CDMC_1_1_1 = 'Business cases are defined';
ALTER TABLE CDMC.cdmc_owner.v_pmo_evidence_summary_1_1_1 MODIFY COLUMN
costs_actual                SET TAG CDMC_1_1_1 = 'Business cases are defined';
```

Note latency of up to 2 hours in Snowflake Account Usage Store before tags are visible. Latency may be avoided by using `information_schema` table functions

---

You can interrogate the Snowflake Account Usage Store to see which tags are declared:

```
SELECT *
FROM    snowflake.account_usage.tags
WHERE   deleted IS NULL
ORDER BY tag_name;
```

Document Ingestion

As an alternative to having all supporting data within Snowflake tables, with some additional work it is possible to ingest and tag data from unstructured documents.

Unstructured Ingestion Pattern

Figure 10-3 suggests a standard document ingestion pattern extracting data from a document and landing the extracted data into a Snowflake table.

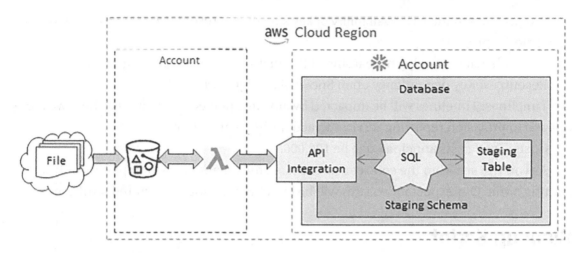

Figure 10-3. *CDMC capabilities*

We assume familiarity with the suggested components in Figure 10-3. To programmatically extract document content, we suggest using the Apache Tika content analysis toolkit deployed via AWS Lambda at `https://tika.apache.org/`. Amazon Textract offers capability to automatically extract printed text, handwriting, and data from any document. Further information can be found here: `https://aws.amazon.com/textract/`.

Template Document

The following document template and content is provided to illustrate how content for a business case should be structured to enable programmatic extract. Each section should be a Table of Content heading.

Program Reference: Archive System X.

Executive Summary: This program seeks to retire System X onto Snowflake and provide an archive repository for ad-hoc query for the mandatory 10-year retention period.

Planned Activities: The planned activities include porting all System X schemas and objects into Snowflake, one-off ingestion of all data, recreation of essential historical reports, validation all data has migrated correctly, and user acceptance signoff.

Expected Business Outcomes: The expected business outcomes are to realize cost savings of GBP 100,000 per annum due to decommissioning on-prem hardware and an expected reduction of 46% from reporting costs due to the on-demand capability Snowflake brings.

Project Governance: Unavailability of key staff due to resource contention with other high priority projects.

Dependencies: Timely on-boarding of Product Y to deliver reporting capability.

Resources: Key dependency upon Snowflake Subject Matter Experts.

Timelines: Timelines will be impacted by any unexpected divestments where we are obliged to provision reporting services to acquiring organizations.

Costs: Total costs anticipated to be 100,000.

As you will see from the code constructs later, the business case template is only a starting point. Our expectation is this will be built upon to add all required attributes.

Cleanup Script

All good test cases include a cleanup script, leaving your environment as you found it. These SQL commands remove your evidence repository:

```
USE ROLE sysadmin;

DROP DATABASE IDENTIFIER ( $cdmc_database );

DROP WAREHOUSE IDENTIFIER ( $cdmc_warehouse );

USE ROLE securityadmin;

DROP ROLE IDENTIFIER ( $cdmc_owner_role  );
```

Governance and Accountability

Reference: CDMC_Framework_V1.1.1.pdf, page 16.

Business Cases Are Defined and Governed

Reference: CDMC_Framework_V1.1.1.pdf, page 21

Business Cases Are Defined

Reference: CDMC_Framework_V1.1.1.pdf page 21.

Programmatic scoring of the business case definitions is not possible except to indicate the presence or absence of content.

Where content can be programmatically extracted, these metrics should be captured and scored:

- Business case lifecycle, control gates, and expected timeliness of deliverables

- Expected outcomes and savings realized compared to baseline metrics

- List of business problem types along with the executive responsible for resolution

- Progress against planned deliverables and control gates

Using the sample record within table pmo_evidence assumed to be sourced from either PMO tooling or the business case document, let's now examine how to provision summary reporting capability.

We assume each business case is uniquely identified by a "program reference," which will be either a PMO-defined label or the fully qualified business case file name.

From your business case you expect each heading to have content; therefore, your first metric is to ensure the corresponding Snowflake table attributes are populated and later checks address costs, progress, and status.

You first set your context:

```
USE ROLE      cdmc_owner_role;
USE DATABASE  CDMC;
USE WAREHOUSE CDMC_WH;
USE SCHEMA    CDMC.cdmc_owner;
```

With your context defined, you can begin to score your control.

The next SQL statement returns a small data set from your sample table-based test data. The SQL statement requires enhancing according to any changes made to table pmo_evidence and views v_pmo_evidence_summary and v_pmo_evidence_summary_1_1_1:

```
SELECT *
FROM   v_pmo_evidence_summary_1_1_1;
```

Figure 10-4 shows a partial summary of the expected output.

CDMC_CONTROL	REFERENCE	PCT	CURRENCY	ESTIMATE	ACTUAL	VARIANCE
1.1.1	Archive System X.	100.00	GBP	130000	88000	42000
1.1.1	Ingest System Y.	100.00	GBP	20000	12766	7234
1.1.1	Bad Test Case 1.	30.00	Not Specified	0	0	NULL

Figure 10-4. *1.1.1 Control proofs*

You can go further with extracting control proof data by converting the output to JSON. You will use JSON to store information into your archive table as the number of attributes is indeterminate for each control proof, making JSON the ideal format to store such information:

```
SELECT OBJECT_CONSTRUCT ( * )
FROM   v_pmo_evidence_summary_1_1_1;
```

Figure 10-5 shows a partial summary of the expected output.

OBJECT_CONSTRUCT (*)
{ "CDMC_CONTROL": "CDMC_1.1.1", "CONTROL_METRICS_PCT": "100.00", "COSTS_ACTUAL": "88000", "COSTS_CURRENCY": "GB
{ "CDMC_CONTROL": "CDMC_1.1.1", "CONTROL_METRICS_PCT": "100.00", "COSTS_ACTUAL": "12766", "COSTS_CURRENCY": "GBI
{ "CDMC_CONTROL": "CDMC_1.1.1", "CONTROL_METRICS_PCT": "30.00", "COSTS_ACTUAL": "0", "COSTS_CURRENCY": "Not Spec

Figure 10-5. *1.1.1 JSON control proofs*

Your ultimate aim is to extract control information using an object tag. You should not need to know where your control data is stored within Snowflake in order to extract the tagged data.

Furthermore, the ability to extract tagged data should not be restricted to just control data but instead be a generic utility for all object tag-based consumption.

The following stored procedure, sp_get_object_data_by_tags, takes a comma-separated list of tags and dynamically builds SQL statements to generate a JSON record containing the resultant dataset from all tagged tables. Note that the input string is tokenized using a comma as the field separator. You may wish to trim the resultant fields to remove spaces.

```
CREATE OR REPLACE PROCEDURE sp_get_object_data_by_tags ( P_CSV_TAGLIST
STRING )
RETURNS VARIANT
LANGUAGE javascript
EXECUTE AS CALLER
AS
$$
    var sql_stmt      = "";
    var sql_stmt_1    = "";
    var stmt          = "";
    var stmt_1        = "";
    var recset        = "";
    var recset_1      = "";
    var row_as_json   = {};
    var array_of_rows = [];
    var table_as_json = {};
    const taglist     = P_CSV_TAGLIST.split ( "," );

    for ( let num_tags = 0; num_tags < taglist.length; num_tags++ )
    {
       sql_stmt  = "SELECT DISTINCT LOWER ( object_name )\n"
       sql_stmt += "FROM    snowflake.account_usage.tag_references\n"
       sql_stmt += "WHERE   tag_name = '" + taglist[num_tags].toUpperCase()
       + "'\n";

       stmt = snowflake.createStatement ({ sqlText:sql_stmt });

       try
       {
          recset = stmt.execute();
          while(recset.next())
          {
             sql_stmt_1   = "SELECT OBJECT_CONSTRUCT ( * )\n"
             sql_stmt_1  += "FROM   " + recset.getColumnValue(1) + ";"
             stmt_1 = snowflake.createStatement ({ sqlText:sql_stmt_1 });

             try
```

```
                {
                    recset_1 = stmt_1.execute();
                    while(recset_1.next())
                    {
                        row_as_json = recset_1.getColumnValue(1)
                        array_of_rows.push(row_as_json);
                    }
                }
                catch ( err )
                {
                    result  = sql_stmt_1;
                    result += "\nCode: "          + err.code;
                    result += "\nState: "         + err.state;
                    result += "\nMessage: "       + err.message;
                    result += "\nStack Trace:\n" + err.stackTraceTxt;
                }
            }
        }
        catch ( err )
        {
            result  = sql_stmt;
            result += "\nCode: "          + err.code;
            result += "\nState: "         + err.state;
            result += "\nMessage: "       + err.message;
            result += "\nStack Trace:\n" + err.stackTraceTxt;
        }
    }
    table_as_json[P_CSV_TAGLIST] = array_of_rows;
    return table_as_json;
$$;
```

Account Usage Store latency may affect your results. Wait and then retry.

You can now invoke stored procedure sp_get_object_data_by_tags:

```
CALL sp_get_object_data_by_tags('CDMC' );
CALL sp_get_object_data_by_tags('CDMC_1' );
CALL sp_get_object_data_by_tags('CDMC_1_1' );
CALL sp_get_object_data_by_tags('CDMC_1_1_1' );
CALL sp_get_object_data_by_tags('CDMC_1_1_1,CDMC_1_1' );
CALL sp_get_object_data_by_tags('CDMC_1_1_2,CDMC_1_1_1,CDMC_1_1' );
```

Multiple tags may result in duplicate records within the output JSON. This is expected behavior.

Figure 10-6 shows a partial summary of the expected output.

SP_GET_OBJECT_DATA_BY_TAGS
{ "CDMC_1_1_2,CDMC_1_1_1,CDMC_1_1": [{ "CDMC_CONTROL": "CDMC_1_1_2", "CDMC_DOCUMENT": "CDMC_Fr;

Figure 10-6. *1.1.1 SP_GET_OBJECT_DATA_BY_TAGS output*

Click on the browser output to expand the JSON, from which the following section is shown. The source tag CDMC_1_1_2,CDMC_1_1_1,CDMC_1_1 has been added by sp_get_object_data_by_tags to reference the input parameter. Note that sections have been abbreviated or removed.

```
{
  "CDMC_1_1_2,CDMC_1_1_1,CDMC_1_1": [
    {
      "CDMC_CONTROL": "CDMC_1_1_2",
      "CDMC_DOCUMENT": "CDMC_Framework_V1.1.1.pdf",
      "PROGRAM_REFERENCE": "Archive System X.",
      "PROJECT_STATUS": "Green",
      "SOURCE_OBJECT": "v_pmo_evidence_summary_1_1_2",
      "TIMELINE_GATE": "Gate 2"
    },
```

```
{
    "CDMC_CONTROL": "CDMC_1_1_2",
    "CDMC_DOCUMENT": "CDMC_Framework_V1.1.1.pdf",
    "PROGRAM_REFERENCE": "Ingest System Y.",
    "PROJECT_STATUS": "Amber",
    "SOURCE_OBJECT": "v_pmo_evidence_summary_1_1_2",
    "TIMELINE_GATE": "Gate 0"
},
{
    "CDMC_CONTROL": "CDMC_1_1_2",
    "CDMC_DOCUMENT": "CDMC_Framework_V1.1.1.pdf",
    "PROGRAM_REFERENCE": "Bad Test Case 1.",
    "PROJECT_STATUS": "Not Specified",
    "SOURCE_OBJECT": "v_pmo_evidence_summary_1_1_2",
    "TIMELINE_GATE": "Not Specified"
},
{
    "CDMC_CONTROL": "CDMC_1_1_1",
    ...
    "TIMELINE_TEXT": "Timelines will be impacted by any unexpected
    divestments where we are obliged to provision reporting services to
    acquiring organizations."
},
{
... Sections removed ...
}
]
}
```

You can now score your results according to the following scale:

- **Not Initiated (1)**: No discussion or action related to relevant source data has occurred.

- **Conceptual (2)**: The need to capture and measure relevant source data is acknowledged.

- **Developmental (3)**: Capturing and measuring relevant source data requirements is underway.

- **Defined (4)**: Capturing and measuring relevant source data requirements has been defined.

- **Achieved (5)**: Source data has been captured, measured, and adopted.

- **Enhanced (6)**: Programmatic capture, measurement, and response to relevant source data is embedded within your organization's culture.

With reference to the expected output from the query above, you can draw your conclusion. The evidence suggests your control lies between Achieved (5) and Enhanced (6), where you could reasonably conclude Enhanced (6) as the outcome because you have implemented a complete control.

But to enable historical reporting, you must store the results for the control in your audit history table evidence_archive. To do this, you build another stored procedure named sp_store_object_data_by_tag, which wraps stored procedure sp_get_object_data_by_tag and stores the output into your audit table:

```
CREATE OR REPLACE TABLE evidence_archive
(
evidence_archive_id       NUMBER          NOT NULL,
tag_name                  STRING(4000)    NOT NULL,
payload                   VARIANT         NOT NULL,
last_updated              TIMESTAMP       NOT NULL
);
```

You also create a sequence for your evidence_archive table primary key:

```
CREATE OR REPLACE SEQUENCE seq_evidence_archive_id START WITH 100000;
```

In order to create an audit record for records returned from invoking stored procedure sp_get_object_data_by_tags, you now create another JavaScript stored procedure named sp_store_object_data_by_tags:

```
CREATE OR REPLACE PROCEDURE sp_store_object_data_by_tags ( P_CSV_TAGLIST
STRING )
RETURNS VARIANT
```

```
LANGUAGE javascript
EXECUTE AS CALLER
AS
$$
    var sql_stmt              = "";
    var stmt                  = "";
    var recset                = "";
    var row_as_json           = {};
    var array_of_rows         = [];
    var json_to_store         = {};
    var json_string           = "";

    sql_stmt = snowflake.createStatement
                (
                    {
                    sqlText: "call sp_get_object_data_by_tags ( :1 )", binds:[
                    P_CSV_TAGLIST ]
                    }
                );
    recset = sql_stmt.execute();
    recset.next();
    row_as_json = recset.getColumnValue(1)
    array_of_rows.push(row_as_json);
    json_to_store[P_CSV_TAGLIST] = array_of_rows;
    json_string = JSON.stringify(json_to_store);

    sql_stmt  = "INSERT INTO evidence_archive\n"
    sql_stmt += "(\n"
    sql_stmt += "evidence_archive_id,\n"
    sql_stmt += "tag_name,\n"
    sql_stmt += "payload,\n"
    sql_stmt += "last_updated\n"
    sql_stmt += ")\n"
    sql_stmt += "SELECT seq_evidence_archive_id.NEXTVAL,\n"
    sql_stmt += "        :1,\n"
```

```
sql_stmt += "          TO_VARIANT ( PARSE_JSON
                       ( '" + json_string + "' )),\n"
sql_stmt += "          current_timestamp();\n"

stmt = snowflake.createStatement ({ sqlText:sql_stmt, binds:[P_CSV_
TAGLIST] });

try
{
   stmt.execute();
   result = "Success";
}
catch ( err )
{
    result  = sql_stmt;
    result += "\nCode: "          + err.code;
    result += "\nState: "         + err.state;
    result += "\nMessage: "       + err.message;
    result += "\nStack Trace:\n" + err.stackTraceTxt;
}
return result;
$$;
```

Now invoke sp_store_object_data_by_tags:

```
CALL sp_store_object_data_by_tags('CDMC' );
CALL sp_store_object_data_by_tags('CDMC_1' );
CALL sp_store_object_data_by_tags('CDMC_1_1' );
CALL sp_store_object_data_by_tags('CDMC_1_1_1' );
CALL sp_store_object_data_by_tags('CDMC_1_1_1,CDMC_1_1' );
CALL sp_store_object_data_by_tags('CDMC_1_1_2,CDMC_1_1_1,CDMC_1_1' );
```

And prove your audit table evidence_audit has records:

```
SELECT *
FROM   evidence_archive;
```

You should see output similar to Figure 10-7.

EVIDENCE_ARCHIVE_ID	TAG_NAME	PAYLOAD	LAST_UPDATED
100003	CDMC	{ "CDMC": [{ "CDMC": [{ "CDMC_CONTROL": "CDMC_1_1_1", "CDMC_DOCUME...	2022-08-29 01:36:34.474
100004	CDMC_1	{ "CDMC_1": [{ "CDMC_1": [{ "CDMC_CONTROL": "CDMC_1_1_1", "CDMC_DOCU...	2022-08-29 01:36:38.853
100005	CDMC_1_1	{ "CDMC_1_1": [{ "CDMC_1_1": [{ "CDMC_CONTROL": "CDMC_1_1_1", "CDMC_DC...	2022-08-29 01:36:43.502
100006	CDMC_1_1_1	{ "CDMC_1_1_1": [{ "CDMC_1_1_1": [{ "CDMC_CONTROL": "CDMC_1_1_1", "CDMC)...	2022-08-29 01:36:48.306
100007	CDMC_1_1_1,CDMC_1_1	{ "CDMC_1_1_1,CDMC_1_1": [{ "CDMC_1_1_1,CDMC_1_1": [{ "CDMC_CONTROL": "...	2022-08-29 01:36:56.920
100008	CDMC_1_1_2,CDMC_1_1_1,CDMC_1_1	{ "CDMC_1_1_2,CDMC_1_1_1,CDMC_1_1": [{ "CDMC_1_1_2,CDMC_1_1_1,CDMC_1_1"...	2022-08-29 01:37:09.872

Figure 10-7. *Audit Table Content 1*

Business Cases Are Syndicated and Governed

Reference: CDMC_Framework_V1.1.1.pdf, page 26

Where PMO tooling is available and content can be programmatically extracted, these metrics should be captured and scored:

- Program control gates and corresponding artifact approval status

- Program review schedule and status updates

- Audit trail evidence

While creating the evidence repository, you also created view v_pmo_evidence_summary_1_1_2 to select only those evidence attributes from view v_pmo_evidence_summary, which support control 1.1.2.

You can now reuse the previously created infrastructure. Simply invoke sp_store_object_data_by_tags with your previously declared tag for the current control:

```
CALL sp_store_object_data_by_tags ( 'CDMC_1_1_2' );
```

And prove your audit table evidence_audit has records:

```
SELECT *
FROM   evidence_archive
WHERE  tag_name = 'CDMC_1_1_2';
```

You should see output similar to Figure 10-8.

EVIDENCE_ARCHIVE_ID	TAG_NAME	PAYLOAD	↓ LAST_UPDATED
100009	CDMC_1_1_2	{ "CDMC_1_1_2": [{ "CDMC_1_1_2": [{ "CDMC_CONTROL": "CDMC_1_1_2", "CDMC_DOCU...	2022-08-29 01:44:23.584

Figure 10-8. *Audit Table Content 2*

Data Sovereignty and Cross-Border Data Movement Are Managed

Reference: CDMC_Framework_V1.1.1.pdf, page 40

Data Sovereignty Is Tracked

Reference: CDMC_Framework_V1.1.1.pdf, page 40

Where content can be programmatically extracted, metrics for these objects should be identified:

- **All external stages**: External file systems where data can be either ingested from or unloaded to

- **All data transfers**: CSPs and regions where data is sent to

- **All data shares**: Data exchange name and consumer account information

- **All reader account usage**: Data extracts from entitled reader accounts

Out of scope for implementation here, you may implement functionality to resolve geolocation information for login IP addresses using a website such as www.iplocation.net/.

Account Usage Store views are subject to varying degrees of latency that may be up to 3 hours.

You first set your context:

```
USE ROLE      cdmc_owner_role;
USE DATABASE  CDMC;
USE WAREHOUSE CDMC_WH;
USE SCHEMA    CDMC.cdmc_owner;
```

With your context defined, you extend the evidence repository and associated control objects for points 1, 2, 3, and 4, as shown in Figure 10-9.

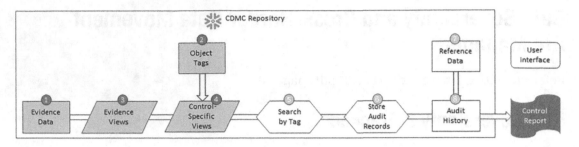

Figure 10-9. *Creating evidence repository objects*

Evidence Views

For this control you have several data sources, all internal to Snowflake and accessible via the Account Usage Store.

Your ultimate objective for this section is to produce a single composite control-specific reporting view named v_snowflake_evidence_summary_1_4_1 based upon several underlying Account Usage Store views and discussed shortly.

You now create several baseline views, one for each discrete dataset identified above. The intent is to facilitate later expansion and bespoke implementation for each view according to your need.

Remember these are template views to illustrate the "art of what is possible" and are not intended to be the end state.

For external stages, you create view v_account_usage_stages to resolve all required source information. To facilitate the creation of reporting view v_snowflake_evidence_summary_1_4_1, you add several nullable attributes that are populated by later views.

```
CREATE OR REPLACE VIEW v_account_usage_stages COPY GRANTS
AS
SELECT 'ACCOUNT_USAGE.STAGES'         AS data_source,
       stage_name                     AS object_name,
       NULL                           AS user_name,
       stage_catalog                  AS database,
       stage_schema                   AS schema,
       NULL                           AS client_ip,
       NULL                           AS event_timestamp,
```

```
        NULL                            AS query_id,
        NULL                            AS query_text,
        NULL                            AS execution_status,
        stage_url                       AS external_location,
        NULL                            AS outbound_cloud,
        stage_region                    AS external_region,
        NULL                            AS bytes_transferred,
        NULL                            AS transfer_type,
        created                         AS created
FROM    snowflake.account_usage.stages
WHERE   stage_type = 'External Named'
AND     deleted IS NULL;
```

Unless you have created a stage, the above query will not return data. However, the view declaration is correct and can be tested:

```
SELECT *
FROM    v_account_usage_stages;
```

For data transfers, create view v_account_usage_data_transfer, which resolves all required source information, noting the result set is filtered to return only the previous months information. You may wish to amend this.

```
CREATE OR REPLACE VIEW v_account_usage_data_transfer COPY GRANTS
AS
SELECT 'ACCOUNT_USAGE.DATA_TRANSFER_HISTORY' AS data_source,
        NULL                            AS object_name,
        NULL                            AS user_name,
        NULL                            AS database,
        NULL                            AS schema,
        NULL                            AS client_ip,
        NULL                            AS event_timestamp,
        NULL                            AS query_id,
        NULL                            AS query_text,
        NULL                            AS execution_status,
        NULL                            AS external_location,
        target_cloud                    AS outbound_cloud,
```

```
          target_region                 AS external_region,
          bytes_transferred             AS bytes_transferred,
          transfer_type                 AS transfer_type,
          start_time                    AS created
FROM      snowflake.account_usage.data_transfer_history
WHERE     start_time > ADD_MONTHS ( current_timestamp(), -1 )
ORDER BY start_time DESC;
```

Unless you have external data transfers, the above query will not return data. However, the view declaration is correct and can be tested:

```
SELECT *
FROM     v_account_usage_data_transfer;
```

For share usage, you create view v_data_sharing_usage_listing_access, which resolves all required source information:

```
CREATE OR REPLACE VIEW v_data_sharing_usage_listing_access COPY GRANTS
AS
SELECT 'DATA_SHARING_USAGE.LISTING_ACCESS_HISTORY' AS data_source,
          exchange_name                 AS object_name,
          NULL                          AS user_name,
          listing_global_name           AS database,
          NULL                          AS schema,
          NULL                          AS client_ip,
          NULL                          AS event_timestamp,
          NULL                          AS query_id,
          NULL                          AS query_text,
          NULL                          AS execution_status,
          consumer_account_locator      AS external_location,
          share_name                    AS outbound_cloud,
          cloud_region                  AS external_region,
          NULL                          AS bytes_transferred,
          NULL                          AS transfer_type,
          query_date                    AS created
FROM      snowflake.data_sharing_usage.listing_access_history;
```

Unless you have shares configured and in use, the above query will not return data. However, the view declaration is correct and can be tested:

```
SELECT *
FROM   v_data_sharing_usage_listing_access;
```

For reader account usage, you create view v_reader_account_usage, which resolves all required source information:

```
CREATE OR REPLACE VIEW v_reader_account_usage COPY GRANTS
AS
SELECT 'READER_ACCOUNT_USAGE'                 AS data_source,
       lh.reader_account_name                AS object_name,
       lh.user_name                          AS user_name,
       qh.database_name                      AS database,
       qh.schema_name                        AS schema,
       lh.client_ip                          AS client_ip,
       lh.event_timestamp                    AS event_timestamp,
       qh.query_id                           AS query_id,
       qh.query_text                         AS query_text,
       qh.execution_status                   AS execution_status,
       NULL                                  AS external_location,
       qh.outbound_data_transfer_cloud       AS outbound_cloud,
       qh.outbound_data_transfer_region      AS external_region,
       qh.outbound_data_transfer_bytes       AS bytes_transferred,
       NULL                                  AS transfer_type,
       qh.start_time                         AS created
FROM   snowflake.reader_account_usage.login_history lh,
       snowflake.reader_account_usage.query_history qh
WHERE  lh.reader_account_name       = qh.reader_account_name
AND    lh.user_name                 = qh.user_name
AND    lh.is_success                = 'TRUE'
AND    qh.reader_account_deleted_on IS NULL
ORDER BY lh.event_timestamp DESC;
```

Unless you have reader accounts configured and in use, the above query will not return data. However, the view declaration is correct and can be tested:

```
SELECT *
FROM   v_reader_account_usage;
```

Control Specific View

Recognizing that your underlying views defined above may be extended to include attributes to satisfy other CDMC controls, you create a control-specific view v_snowflake_evidence_summary_1_4_1 by selecting only those attributes required to satisfy CDMC 1.4.1 while joining to your reference table cdmc_control to derive common definitions for this control:

```
CREATE OR REPLACE VIEW v_snowflake_evidence_summary_1_4_1 COPY GRANTS
AS
SELECT 'v_snowflake_evidence_summary_1_4_1' AS source_object,
        c.document                          AS cdmc_document,
        c.reference                         AS cdmc_control,
        v.data_source,
        v.object_name,
        v.user_name,
        v.database,
        v.schema,
        v.client_ip,
        v.event_timestamp,
        v.query_id,
        v.query_text,
        v.execution_status,
        v.external_location,
        v.outbound_cloud,
        v.external_region,
        v.bytes_transferred,
        v.transfer_type,
        v.created
```

```
FROM    (
        /* External Stages */
        SELECT data_source,
                object_name,
                user_name,
                database,
                schema,
                client_ip,
                event_timestamp,
                query_id,
                query_text,
                execution_status,
                external_location,
                outbound_cloud,
                external_region,
                bytes_transferred,
                transfer_type,
                created
        FROM    v_account_usage_stages
        UNION
        /* External Data Transfers */
        SELECT data_source,
                object_name,
                user_name,
                database,
                schema,
                client_ip,
                event_timestamp,
                query_id,
                query_text,
                execution_status,
                external_location,
                outbound_cloud,
                external_region,
                bytes_transferred,
```

```
           transfer_type,
           created
    FROM   v_account_usage_data_transfer
    UNION
    /* Share Usage */
    SELECT data_source,
           object_name,
           user_name,
           database,
           schema,
           client_ip,
           event_timestamp,
           query_id,
           query_text,
           execution_status,
           external_location,
           outbound_cloud,
           external_region,
           bytes_transferred,
           transfer_type,
           created
    FROM   v_data_sharing_usage_listing_access
    UNION
    /* Reader Account Usage */
    SELECT data_source,
           object_name,
           user_name,
           database,
           schema,
           client_ip,
           event_timestamp,
           query_id,
           query_text,
           execution_status,
           external_location,
           outbound_cloud,
```

```
            external_region,
            bytes_transferred,
            transfer_type,
            created
      FROM    v_reader_account_usage
      ) v,
      cdmc_control              c
WHERE  c.reference              = 'CDMC_1_4_1'
ORDER BY v.data_source DESC;
```

With your control-specific view v_snowflake_evidence_summary_1_4_1 in place and assuming some system usage resulting in data movement, you can check if your new view returns appropriate output:

```
SELECT *
FROM    v_snowflake_evidence_summary_1_4_1;
```

You may also check if the control-specific view v_snowflake_evidence_summary_1_4_1 returns a JSON record:

```
SELECT OBJECT_CONSTRUCT ( * )
FROM    v_snowflake_evidence_summary_1_4_1;
```

Object Tags

In similar manner to the object tags created earlier within this chapter, you must tag your control-specific view v_snowflake_evidence_summary_1_4_1:

```
ALTER TABLE CDMC.cdmc_owner.v_snowflake_evidence_summary_1_4_1
SET TAG CDMC = 'Cloud Data Management Capabilities';
ALTER TABLE CDMC.cdmc_owner.v_snowflake_evidence_summary_1_4_1
SET TAG CDMC_1 = 'Governance & Accountability';
ALTER TABLE CDMC.cdmc_owner.v_snowflake_evidence_summary_1_4_1
SET TAG CDMC_1_4 = 'Data sovereignty and cross-border data movement are
managed';
ALTER TABLE CDMC.cdmc_owner.v_snowflake_evidence_summary_1_4_1
SET TAG CDMC_1_4_1 = 'Data sovereignty is tracked';
```

```
ALTER TABLE CDMC.cdmc_owner.v_snowflake_evidence_summary_1_4_1 MODIFY
COLUMN data_source                SET TAG CDMC_1_4_1 = 'Data sovereignty
                                  is tracked';
ALTER TABLE CDMC.cdmc_owner.v_snowflake_evidence_summary_1_4_1 MODIFY
COLUMN object_name                SET TAG CDMC_1_4_1 = 'Data sovereignty
                                  is tracked';
ALTER TABLE CDMC.cdmc_owner.v_snowflake_evidence_summary_1_4_1 MODIFY
COLUMN user_name                  SET TAG CDMC_1_4_1 = 'Data sovereignty
                                  is tracked';
ALTER TABLE CDMC.cdmc_owner.v_snowflake_evidence_summary_1_4_1 MODIFY
COLUMN database                   SET TAG CDMC_1_4_1 = 'Data sovereignty
                                  is tracked';
ALTER TABLE CDMC.cdmc_owner.v_snowflake_evidence_summary_1_4_1 MODIFY
COLUMN schema                     SET TAG CDMC_1_4_1 = 'Data sovereignty
                                  is tracked';
ALTER TABLE CDMC.cdmc_owner.v_snowflake_evidence_summary_1_4_1 MODIFY
COLUMN client_ip                  SET TAG CDMC_1_4_1 = 'Data sovereignty
                                  is tracked';
ALTER TABLE CDMC.cdmc_owner.v_snowflake_evidence_summary_1_4_1 MODIFY
COLUMN event_timestamp            SET TAG CDMC_1_4_1 = 'Data sovereignty
                                  is tracked';
ALTER TABLE CDMC.cdmc_owner.v_snowflake_evidence_summary_1_4_1 MODIFY
COLUMN query_id                   SET TAG CDMC_1_4_1 = 'Data sovereignty
                                  is tracked';
ALTER TABLE CDMC.cdmc_owner.v_snowflake_evidence_summary_1_4_1 MODIFY
COLUMN query_text                 SET TAG CDMC_1_4_1 = 'Data sovereignty
                                  is tracked';
ALTER TABLE CDMC.cdmc_owner.v_snowflake_evidence_summary_1_4_1 MODIFY
COLUMN execution_status           SET TAG CDMC_1_4_1 = 'Data sovereignty
                                  is tracked';
ALTER TABLE CDMC.cdmc_owner.v_snowflake_evidence_summary_1_4_1 MODIFY
COLUMN external_location          SET TAG CDMC_1_4_1 = 'Data sovereignty
                                  is tracked';
```

```
ALTER TABLE CDMC.cdmc_owner.v_snowflake_evidence_summary_1_4_1 MODIFY
COLUMN outbound_data_transfer_cloud SET TAG CDMC_1_4_1 = 'Data sovereignty
is tracked';
ALTER TABLE CDMC.cdmc_owner.v_snowflake_evidence_summary_1_4_1 MODIFY
COLUMN external_region               SET TAG CDMC_1_4_1 = 'Data sovereignty
                                     is tracked';
ALTER TABLE CDMC.cdmc_owner.v_snowflake_evidence_summary_1_4_1 MODIFY
COLUMN created                       SET TAG CDMC_1_4_1 = 'Data sovereignty
                                     is tracked';
```

Notwithstanding Account Usage Store latency, you now test if your data is accessible using previously defined JavaScript stored procedures:

```
CALL sp_get_object_data_by_tags ( 'CDMC_1_4_1' );

CALL sp_store_object_data_by_tags ( 'CDMC_1_4_1' );
```

And finally, confirm the generated JSON has been stored in your evidence_archive table:

```
SELECT *
FROM   evidence_archive
WHERE  tag_name = 'CDMC_1_4_1';
```

Control Reports

Mentioned for completeness, but out of scope for delivery within this book, is the requirement to substantiate compliance and overall score-card for each control over time. Our expectation is further development of the evidence_archive table will occur where reporting capability will deliver

- Evidence of current compliance per control

- Summary control score card information

- Historized control score card information and trends over time

With the basic information capture mechanisms demonstrated within this chapter, we leave it for you to determine how to consolidate the reporting with existing statistics.

Summary

You began this chapter by exploring Cloud Data Management Capabilities (CDMC) as a framework for delivering data standards and industry best practice.

Within your organization you should expect PMO tooling, data catalogues, and other repositories to provision useful information to support your CDMC control implementation, some of which may be held within unstructured documents. We propose an evidence repository to both work through the content within this book and for extension for your organization's specific needs.

You then built out a few controls demonstrating how to reference external data loaded into your evidence repository and the Snowflake Account Usage Store. We leave deriving evidence from external unstructured documentation for your further investigation and suggest *Building the Snowflake Data Cloud* offers a suitable starting point.

For the curious, we suggest the object tagging JavaScript stored procedures have wider utility within your organization to facilitate self-service by end users. We expect object tagging to assume increasing importance within our business community as they discover the power of classifying objects and attributes within Snowflake.

Naturally we are unable to provide complete coverage of every CDMC control and we hope the small sample supplied whets your appetite for further investigation.

In the absence of suitable system-derived control data, and to develop CDMC Framework controls, we proposed a suite of document templates containing content not readily available or programmatically accessible via Snowflake. They include

- Business Case

- Solution Design Document

- Feed Onboarding Contract

Each template is articulated in the "Appendix – Supporting Documentation" section.

Appendix – Supporting Documentation

In this section we offer document templates in support of the CDMC Framework.

Assuming information is recorded in document format, you must structure your documents with programmatic data extraction in mind as supported by using Snowflake unstructured features. Alternatively, where information is recorded using a cataloging tool, a programmatically accessible interface should be provisioned.

When developing documents, continual reference should be made to the CDMC Framework to ensure conformance to controls.

By making data and information programmatically accessible, you facilitate higher CDMC scoring for each control expressed as the proofs can be obtained using code-based automated processing.

Business Case

Building a business case provides a solid foundation for delivering successful business outcomes. For existing systems without a business case, you may not always have the original rationale for your systems. Retrospectively developing a business case is a good way for sponsors to revalidate their decision to continue sponsorship and provide a baseline document for future reference. A business case helps the organization avoid wasting resources on projects that do not yield sufficient successful business outcomes.

All new system proposals must have a solid business case, for which this section offers a guide, all in support of the CDMC Framework.

The business case content should be extended to meet your organizational needs. Bear in mind the business case structure must lend itself to programmatic content extraction.

Key points to remember:

- Keep the content concise. Convey the bare essentials only.

- Stick to plain English.

- Tell a story that is both interesting and clearly articulated.

- Describe the vision while bringing passion to your project.

- Demonstrate the project value to your business.

- Ensure the message is readable and consistent.

Some sections have a Snowflake-specific section offering guidance on the information required to satisfy CDMC controls. This can be ignored for non-CDMC scoring business cases developed using the template.

Executive Summary

An executive summary is typically a single page providing a concise overview of the problem statement and how the proposal will solve the problem.

This section is all about "why" change is needed and should be both short and to the point. Where appropriate, use a diagram, which conveys both meaning and context far more readily than the written word.

Every point within the executive summary should reference headings within the business case highlighting the major points. The final point should focus on the successful business outcome.

Keep it simple, stupid. The executive summary should make the "ask" self-evident to the audience and make the decision easy for the recipients.

Problem Statement

Having briefly outlined the problem statement within the executive summary, a more detailed deep dive into the problem statement provides both context and opportunity to introduce supporting information, analysis, and research.

You should also explain how the problem statement represents a disconnect from your organization's aims, goals, and objectives.

Draw attention to the key team members who will be involved with solving the problem statement and their expected contribution.

Everyone will want to know the cost, a timeline projection of anticipated spend will be required, and the anticipated return on the investment to achieve successful business outcomes.

List the available options for solving the problem statement along with pros and cons. Highlight the best option to resolve the problem statement along with a strong recommendation, not forgetting the "do nothing" option.

In this section, you have the opportunity to address "how" you will solve the problem statement and your proposal must resonate with your audience. For more senior audiences, summarize information, but be prepared to dive into the detail if asked required. Supporting detail should be available in appendices and not part of the core presentation.

The ideal outcomes for this section of your proposal are few questions from your audience and broad-based agreement.

Snowflake Specific

Every organization should have a standard process (or lifecycle) for the development and approval of business cases. You should also specify measures to determine how effective moving data to Snowflake has been,

Your business case should list the business problem types along with the responsible executive for overseeing the resolution delivery.

Planned Activities

For the planned activities, propose the implementation approach, highlighting the project delivery milestones with expected deliveries along with budget requirements and dependencies. At this stage, perfection is not the objective, nor is a highly detailed financial breakdown. Instead, focus on what tangible components will be delivered within the anticipated timeline and identify any incremental business benefits.

Your planned activities should include a DRAFT Plan on a Page (POAP) exposing benign benefits to other initiatives, and you should reference all external team support for your proposal. Likewise, you should expose all key dependencies.

Senior audiences are less interested in the detail. They want to know that behind the summary POAP there is a credible, deliverable, and fully costed plan, which should be available in the appendices.

While you should expect questions on what you propose to deliver, anticipated timelines involved, and costs, the key planned activities should be sensible, self-evident, and consistent. In project management, you can usually have two of cost, timeline, and functionality so be prepared to answer questions on how all three will be achieved.

Project planning activities:

- Determine project scope and activities.

- Identify data sources and support teams.

- Determine the cost estimates.

- Revise the risk assessment.

- Identify the critical success factors.

- Prepare the project charter.

- Create a high-level plan.

- Initiate the project.

347

Expected Business Outcomes

You must explain in non-technical terms what the successful business outcomes looks like. This may also include non-tangible benefits such as positive media exposure and brand recognition.

Include statements showing where the proposed solution aligns to the organizations aims, goals, and objectives.

The business outcomes should include financial benefits representing a return on your investment along with expected timeline for realization.

Key points to identify are the progressive successful business outcomes as the project delivers each milestone. If funding were to be removed part way through your program delivery, you must be in a position to demonstrate your achievements.

Project Governance

Effective project governance ensures your proposal will run smoothly with respect to structure, people, and information. You must also discuss the known issues, risks, and dependencies relating to your proposal.

In particular, be prepared to discuss known issues as these may represent current impediments to progress.

Risks should be mitigated wherever possible, which may include more frequent reporting cycles and closer management than otherwise may be the case.

Dependencies must be tracked to ensure delivery timelines are not impacted.

Resources

You should highlight those resources whose skill, knowledge, and expertise will be required to deliver the project according to cost, timeline, and functionality.

You may anticipate resource hiring timelines and initiate hiring in advance of project approval but we suggest such initiatives are limited in scope and non-committal.

Timelines

You should not expect to know detailed project delivery timelines at this point. Your POAP should be sufficient to give confidence your proposal is realistic, and ball-park estimates are fine with indicative delivery dates.

Timelines are often impacted by staff holidays, change freeze periods, and unexpected absences. As a rule of thumb, assume 70% loading on staff, so you should add appropriate contingency into your timeline.

Detailed project planning occurs when the project scope has been identified along with all deliverables identified and estimates provided.

Costs

With your POAP, you can estimate the cost of each deliverable, recognizing all estimates are subject to revision as your detailed knowledge increases. A useful metric is to apply t-shirt size costs according to the perceived complexity of each deliverable. You also use known employee costs, which must include a sum for the provision of infrastructure.

You should also know the signing off authority of those who can approve your program and any committees whose approval will be required for higher value signoff and adjust your project scope and deliverables accordingly.

Appendix

Throughout the development of your business case you will have developed a suite of supporting documentation.

Remember,

- The business case is a high-level summary of detailed analysis, research, and foundational supporting information.

- Keep the details in the appendix for reference when questioned.

- Stay focused on the high-level perspective of the audience.

Solution Design Document

A solution design document (SDD) provides the comprehensive contextual overview of the components required to deliver the business case. We use SDDs to explain both "how" and "what" is to be delivered from our business case (the "why") to both business and technical audiences, therefore SDDs are subject to scrutiny by our organization's approval bodies.

SDDs are living documents. They articulate the "as-is" technical estate and illustrate the "to-be" technical estate, so they **must be maintained and updated as the associated components change over time**.

For existing applications where no system documentation is present, we suggest SDDs be developed retrospectively to both articulate the "as-is" state of each component and provide a baseline for future enhancement.

You must draw a distinction between core Snowflake platform integration components and SDD scope. Your core Snowflake platform should conform to the LZ concept articulated earlier. An SDD relates to a specific business case to either create a new application or enhance an existing application on your Snowflake platform.

Business Case

The business case must be defined for each SDD as this is the basis for any change.

Governance approval provides valuable audit information for later replay and should likewise be recorded in the SDD.

Business Summary

You must list any new business services provisioned or existing business services enhanced by the SDD, and any business impact to your organization of the proposed changes should be fully explained.

You must respect data classifications and ensure all data transit meets or exceeds your organization minimum standards.

You may expect certain attributes are subject to privacy regulations and must identify both the data sources and regulatory frameworks in-scope for the SDD.

Your business summary must also describe how the new, or amended, business service impacts existing operations. For each identified component, you must identify the data custodians, data stewards, and executive sponsor.

Outline user stories should also be defined, providing context for the overall business solution.

System Context Diagram

You now define the system context diagram from which one or more Feed Onboarding Contracts (FOCs) will be derived. A system context diagram typically shows all interactions for an application and is the foundation for explaining data movement and usage. A sample system context diagram is shown in Figure 10-10.

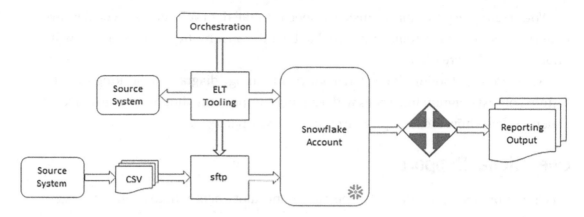

Figure 10-10. *Sample system context diagram*

From Figure 10-10, you can easily discern all data movement and system component interaction. Your business case should result in a much more complex system context diagram.

We prefer a left-to-right data flow for consistency and through experience have found conforming all system context diagrams to a common pattern helps all interested parties more easily understand proposals.

You must define the SDD scope to positively include in-scope items and exclude those items not addressed.

Technology Summary

All proposed technologies referenced within the system context diagram must be listed along with their internal organization references (if available) to facilitate later FOC creation.

Your Cyber Security colleagues will be most interested in any proposed change that affects a known and agreed security posture. Wherever possible, you should not disturb the security guard rails surrounding your Snowflake platform, and if a change is required, engage your Cyber Security colleagues at the earliest possible opportunity.

All technical risks must be identified with mitigating actions identified. Tooling decisions must fall within your organization's risk appetite along with details of their hosted environments, where possible internal system catalogue references should be supplied for each tooling decision.

Your technology summary must also specify initial data volumes, expected growth pattern, and the data in-transit network bandwidth consumption. These factors will inform later FOC creation.

All referenced tooling along with a network topology diagram illustrating how the system context diagram integrates with existing components should also be delivered, which explains how the enterprise architecture is impacted.

Operational Support

An Operational Support Manual is the most important document to facilitates system support.

Within operational support manual, the service availability hours and support team contact details must be recorded for each business service and dependent components. You also define incident severity classifications and their target resolution priorities; not all incidents are equal in impact and timeliness for resolution.

The incident escalation process will also be defined for incidents not resolved within the specified timeframe.

For all business services provided, all Key Performance Indicators (KPIs) will be defined providing measurement metrics over time.

Business continuity and disaster recovery (DR) procedures must also be defined with most recent test results available for inspection.

All periodic activities must be defined along with their frequency. Examples of which are

- Monthly user recertification

- Periodic DR test

- Year-end manual tasks

For all system activities, process logging must be available and your operations manual must define where logs are to be found for all components. System logging, monitoring, statistics, and alerting should also be defined, recorded, and the outcome distributed to authorized monitoring functions within your organization.

Data Lifecycle

Every data entity within your Snowflake account has a finite lifecycle that varies according to specified duration. You may set default data retention periods for data sets and also define policies to classify and protect data too.

For archived data, you must both define and enforce your specified retention period. You must not retain data beyond its lifetime and may also be required to delete data before its defined lifetime.

Different jurisdictions implement disparate rulings, so any and all such jurisdictional regulations that apply to data must be specified.

Feed Onboarding Contract

A Feed Onboarding Contract (FOC) is a technically focused document developed by the development team and forms the system of record for delivery of identified components. FOCs are not intended for approval by architecture boards or others; they exist for the development team's internal use only and **must be maintained and updated as the associated components change over time**.

We suggest an FOC be created for each new interaction between two endpoints. For preexisting endpoint connectivity, a FOC should be created retrospectively, and for new endpoints between a new source and your Snowflake application, you should mandate a FOC as a prerequisite before implementation.

For existing applications where no system documentation is present, we suggest FOCs be developed retrospectively to articulate the "as-is" state of each component.

FOCs must be maintained for consistency with the real-world implementation.

In simple terms, the FOC has two core components:

1. Articulates the pipework between two points in space

2. Describes the information flow through the pipework

Separating these two concepts is very important. Once the pipework is established, increasing the information flow is relatively easy.

The FOC content should be extended to meet your organizational needs. Bear in mind the FOC structure must lend itself to programmatic content extraction.

Business Case

The business case must be defined for each FOC as this is the basis for any change. The FOC is one of many, each referencing the corresponding SDD.

An approved change request reference for the deliverables identified within the FOC provides valuable audit information for later replay and should likewise be recorded in the FOC.

Feed Context Diagram

The SDD system context diagram can be broken down into one or more FOCs. A feed context diagram typically shows the interactions for single feed and is the foundation for explaining data movement and usage. Figure 10-11 represents a simple reporting solution.

Figure 10-11. *Feed context diagram*

In Figure 10-11 you see

- A single source system providing one or more data sets from source endpoint A to Snowflake target endpoint B via tooling C.

- A single consumption endpoint D into a Gateway E, which may host multiple connected reporting outputs at endpoint F.

Usage

The FOC is the artifact produced by low-level analysis of the interaction between two endpoints with the expectation a suitably skilled developer will readily understand the content and be able to deliver the corresponding implementation.

Various documents and processes are derived from the FOC, specifically the ones in the next sections.

Data Ownership

Every system from where data is sourced must be named, along with contextual ancillary information as viewed from your organization's perspective.

Each source system must have a specified business owner and technical owner. They should not be the same person, and their primary means of contact must also be available (i.e., email, phone number, etc.). You must also explain each person's role and responsibilities along with secondary contacts and escalation channels.

Component Delivery

Every work item identified within the FOC must be recorded into a software development lifecycle (SDLC) tool where the person responsible for delivering the component is identified along with an estimate of how many days of effort will be expended. Within the SDLC tooling you also map dependencies and, for reporting purposes, you group components into a hierarchical structure to provide context. When a component has been built, you then test and mark it ready for deployment.

Project Plan

Every work item recorded within your SDLC tooling must be abstracted into a project plan, which provides a more management-focused view of the work in hand. The project plan brings together all FOCs along with other dependencies such as resource management, governance, environment provisioning, architectural reviews, system change freezes, and such. You also use the project plan to highlight delivery bottlenecks, resource crunches, and project risks for downstream reporting into time sheeting systems and RAID reporting to steering committees.

System Catalog Maintenance

You rely upon the information contained with an FOC to maintain your organization's system catalog. Each FOC is an integral part of your overall enterprise architecture.

User Acceptance Testing

The FOC is the specification (golden source or system of record) for new feature build and therefore is the lowest level document from which your user community will sign off against as part of the SDLC and precursor to deployment into your production environment.

Operational Support

Just before the components specified within your FOC are delivered into your production environment, your operational support staff will need to derive their support procedures from the FOC. This is not an afterthought. Your operations staff must have the opportunity to add their requirements into every FOC as part of the SDLC.

Release into Production

The FOC is an essential document supporting the detail behind each change deployed into production and forms part of the audit trail for each delivery.

Endpoints

To design an effective system architecture, you need to know how data transits from point to point within your organization. You need the information relating to the endpoints for each data transit.

A source endpoint is an interface exposed by a communicating party or channel, a location from which a feed or data set is derived. Conversely, the target endpoint is the communicating channel or location into which a feed or data set is delivered to. Figure 10-12 illustrates where endpoints are located.

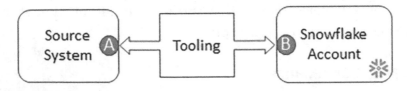

Figure 10-12. Endpoints

Endpoints provide a fixed point for the invoking process to make connection into or access a service. They may include but are not limited to

- Uniform Resource Locator (URL), otherwise known as a web address, which may also support application programming interfaces (APIs) or Reporting as a Service (RaaS)

- Fixed Internet Protocol (IP) address with a specific port

- Database object access

- File Transfer Protocol (FTP) / Secure File Transfer Protocol (SFTP) service

- Cloud service provider (CSP) storage

Key principles for endpoints in a service-orientated architecture (SOA) are to ensure consistent, repeatable behavior with known outcomes for specified inputs and encapsulate functionality. You should not need to know how the inner workings of a component implement capability.

Your FOC must include physical details of endpoints A and B along with contact details of endpoint A's provisioning team. The endpoint B Snowflake provisioning team is already known with the expectation that you, the reader, belong to this team.

Each source endpoint will also have a content delivery format such as CSV, Excel, JSON, PDF, JPEG, or PNG and may have an expected file name format often prefixed with a date stamp. We prefer date stamps to be in YYYYMMDD format to preserve file system ordering. Each file must have a suffix corresponding to its data content.

Some files have a header record, which must be removed before loading, and a footer record with record count or checksum. Your FOC must record this information too.

Authentication

Authentication is the process of verifying an identity (an end user, service user, or authorized token-based access) is valid and correctly entitled to access a system.

Figure 10-13 shows the endpoints of interest where, once authenticated, you must have identity provisioned to perform transactions. They are as follows:

- A must have entitlement provisioned to invoke the services provisioned from the source system.

- B must have entitlement provisioned to receive the inbound payload delivered from the source system.

- C must have capability to securely store separate identities for A and B.

- D must have entitlement provisioned to enable outbound services.

- E represents a gateway that must have embedded capability to securely store the identity for D.

- F represents reporting tooling dependent upon gateway E for serving data and must have embedded capability to securely store identity for E.

Figure 10-13. *Authentication points*

In Figure 10-13, for data ingestion into Snowflake you use tooling at point C to separately implement connections to point A for your data source and to point B for your data target. As a general principle, your tooling should not rely upon a username and password for either connection, but wherever possible should implement Single Sign On (SSO), Multi-Factor Authentication (MFA), and a variety of other secure token-based authentication mechanisms for the duration of the connection which may have a timeout applied.

For data consumption from Snowflake at point D, your authentication options may be restricted where a service user is provisioned with a fixed password, which should be randomly generated containing a minimum of 16 digits and held in gateway E.

The reporting output F authenticates against the gateway E where authentication options may also be restricted.

Your FOC must include every endpoint authentication details **except passwords** for all endpoints A through F.

Connectivity

With your endpoints defined and authentication method specified, you must now consider how data will transit between the endpoints from data sourcing at point A through to landing at point B. The same is true for data egress connectivity between point D and in this example the gateway at point E.

Whenever a connection is made into Snowflake, and regardless of the type of connection made, Snowflake converts the connection to HTTPS SSL TLS1.2, thus rendering the connection secure and encrypted. For most use cases, a guaranteed secure connection is sufficient. Please do confirm with your Cyber Security colleagues first.

You must state the connectivity patterns implemented for both data sourcing and data consumption within your FOC.

Tooling

You may choose to either develop your own tooling or implement third-party tools to both ingest and publish data sets.

In this example you have a single instance of tooling at point C for which you must record the type of tooling: ELT / ETL, SnowPipe, AWS Lambda, etc.

Where third-party tools are implemented, and for internal tooling supported by other teams, you must record the vendor, support website, support contact information, service-level agreements (SLAs), escalation procedure, internal procurement team email address, and any other pertinent information.

Operations Procedures

The most important documentation for any system is the Operations Manual. You must hold a record of where the Operations Manual is held, the internal escalation process, and actions in the event of a Snowflake system failure.

For each feed, you must know the frequency of delivery and the time window during which you should expect the feed to be available. You should also know the expected data volume along with any expected growth over time and tolerance which if exceeded should trigger an alert.

For business-critical processes, you should implement alerting either via inherent orchestration tooling capability or something home-grown. Regardless of choice, the alert mechanism should be registered within the FOC and periodically tested.

Your FOC must specify the actions to be carried out in the event of the feed failing. Example actions include

- Suspend all jobs, wait for the feed to be reinstated, and then rerun.

- Clear data loads from objects to the point immediately before failure.

- Reuse the previous day's data set to allow onward processing to complete.

Logging

Every system interaction must be logged and your FOC must specify the unique attribute(s) used for searching logs to quickly identify problems.

Your FOC must identify all locations within your feed handler where logging is performed along with any post-incident actions required to preserve logging for audit purposes. There are several places where logging is performed:

- Snowflake immutable 1 year query_history view in Snowflake Account Usage Store.

- Tooling logs such as Matillion and Coalesce, which may only be accessible via the tooling console (Matillion) or available externally (Coalesce) via flat file or written into a Snowflake logging table.

- Infrastructure logs for scheduling tools such as Control M and Autosys, which may be expressed as flat files onto servers.

- Snowflake stored procedures writing to a Snowflake logging table.

Data Sets

Each FOC may describe one or more data sets transiting between the endpoints. By way of example, imagine your source endpoint provisions service via APIs. Capability is provisioned to invoke multiple APIs from a single endpoint and your FOC should articulate the dataset delivered from each API according to the parameters passed for invocation.

Every discrete dataset described by an FOC must articulate each attribute name, datatype, size, and mandatory/optional status.

The Snowflake ingestion pattern is dependent upon the nature of the data delivery. The following list is not exhaustive and your ingestion patterns may differ:

- Daily, whole file submission

- Intra-day delta file ingestion

- Near real-time data stream ingestion

- One-off whole schema ingestion for archiving

Validations

Data validations may be specified for individual attributes and for groups of attributes. Some simple examples include

- Checking attributes declared as mandatory are populated

- Checking attribute values match their declared datatype or will successfully convert to declared datatype

- Ensuring attributes match values from local reference data

- For hierarchical data, ensure circular relationships do not exist.

A typical outcome for validation failure is to reject records and exclude from the dataset with the ultimate consequence of incomplete data

Transformations

You must specify any transformations applied to the whole or part of your dataset by tooling used to implement connectivity between endpoints.

Example transforms include

- Extracting part of an attribute to create a new attribute

- Manipulating two or more attributes to generate a new attribute

A less obvious requirement is to identify any temporary data caching implemented by tooling as this is of material interest to your Cyber Security colleagues.

Enrichments

Data enrichments may be specified to extend the range of attributes available for reporting against a data set ingested into Snowflake. Enrichments may be implemented within both external tooling and within Snowflake, but regardless of where enrichments are applied, your FOC must have all (or none) specified along with the supplemental data source and attribute join criteria.

Example enrichments include

- Adding ISO country code attributes for a supplied country code

- Embedding hard coded literal attributes

Data Management

Assuming you have correctly populated your FOC and then implemented your physical feed into Snowflake, you should also provision a metadata catalog. A metadata catalog is an organized inventory of data assets within your organization. Without a metadata catalog, you will be reliant upon your technical team to map data and identify relationships. Facilitating self-service will remain a dream.

Most, if not all, data management tools offer connectivity to Snowflake and, subject to appropriate entitlement, ingest metadata into their catalog. Some tools also provide data profiling, data lineage, and other services.

Your interest is in using data management tooling to link disparate datasets showing equivalence. One simple example is marking attributes containing email addresses as equivalent, meaning you can link data from disparate data domains. Your FOC should identify all known equivalents to existing data sets to ease the creation of your metadata catalog.

You may also specify data masking rules according to consumer entitlement to view fully obscured, partially obscured, or plain-text data.

Data access policies may require row-level security to be implemented to prefilter data sub-sets according to your entitlement.

And for ease of identifying objects and attributes, object tags should be declared where appropriate and recorded within the FOC.

Signoff

You must record the contact details of those providing technical signoff for the FOC.

CHAPTER 11

Behavior Change Management

Snowflake enables by default the easy management of changes to the data warehouse, including the ability to roll back changes if necessary and to track and monitor the impact of changes on performance and data integrity. Additionally, Snowflake offers built-in versioning and time travel capabilities, which allow users to easily view and revert to previous versions of data and schema.

Snowflake automatically deploys releases every week in a seamless, behind-the-scenes manner. In this way, Snowflake ensures that its software is maintained with the most up-to-date version with the latest features enabled. Further information can be found here: `https://docs.snowflake.com/en/user-guide/intro-releases.html`.

In addition to the automated weekly deployments, Snowflake periodically provides behavior change bundles, which are a collection of changes that are bundled together and deployed as a single unit. The Snowflake Knowledge Base states

> *The behavior change release process at Snowflake lets you test new behavior changes for two months before the changes are enabled across Snowflake accounts.*

The full article can be found here: `https://community.snowflake.com/s/article/Behavior-Change-Policy?r=398`.

© Andrew Carruthers and Sahir Ahmed 2023
A. Carruthers and S. Ahmed, *Maturing the Snowflake Data Cloud*, https://doi.org/10.1007/978-1-4842-9340-9_11

Each bundle may include changes to data schema, data replication settings, and data governance policies, as well as new features or bug fixes. The goal of a behavior change bundle is to

- Minimize the impact of changes on performance and data integrity by deploying all related changes at once, rather than deploying them in separate, smaller updates

- Provide the ability to test the behavior change bundle before deployment into production environments

- Selectively enable new features and/or product capabilities for participation in private previews and provision early access for development activities

The automated weekly deployments are a way of ensuring Snowflake is updated with the latest features, security patches, and bug fixes, while a behavior change bundle is a way of deploying a candidate set of related future changes for testing and rollback if needed. Once the bundle has been deemed stable and ready for deployment, it may be included in the next automated weekly deployment. With the exception of November and December every year, Snowflake typically deploy behavior changes two months after publication, usually within the third or fourth weekly release for the month. November and December are excluded due to most organizations having a change freeze period over the Christmas holiday period. No one wants to be fixing bugs while the turkey and trimmings are going cold!

We also use the term "patching" to describe the activities relating to applying an update (or a patch) to software.

Through experience, we found enabling one specific behavior change bundle in order to test a private preview feature led to incorrect results being returned from views in a very narrow and easily missed test scenario. Having captured the issue and raised a support ticket, the behavior change bundle was refactored where subsequent testing proved the issue was resolved.

Further details may be found within documentation found at `https://docs. snowflake.com/en/user-guide/intro-releases.html` and `https://docs.snowflake .com/en/user-guide/managing-behavior-change-releases.html`.

Explaining Behavior Change Bundles

Since you now understand how automated change deployment eventually subsumes behavior change bundles, let's discuss where to find current Snowflake feature information, how to identify upcoming changes, and how to deploy behavior change bundles.

An overview of the new features, enhancements, and important behavior changes introduced by Snowflake can be found here: `https://docs.snowflake.com/en/release-notes/new-features.html`. You should also check the Behavior Change Log found at `https://community.snowflake.com/s/article/Pending-Behavior-Change-Log` within which you can see four headings along with associated commentary:

- **Deprecated Features**: Pending or deprecated features

- **Upcoming Changes**: Features that may be enabled or disabled within a behavior change bundle

- **Recently Implemented Changes**: Previously pending changes that have subsequently been implemented

- **Other Implemented Changes**: Changes not associated with a behavior change bundle or release

Implementing Behavior Change Bundles

You should work with your Snowflake Support Engineering staff to determine applicability of behavior change bundles for participation in private preview activities. You may also prefer to adopt a defensive posture by proactively deploying behavior change bundles and regression testing your business-critical applications. Both approaches are appropriate and support different aims.

You must also recognize that all behavior change bundle content is eventually deployed via the automated weekly deployment process. There is a time imperative to proactively test well in advance of any anticipated automated deployment.

Naturally, each organization will operate according to its own needs, and within organizations, there may be very different local operational needs. We therefore offer general guidelines for local customization and amendment.

All or Nothing

Contents of a behavior change bundle cannot be applied individually. Application of the behavior change bundle occurs at the Snowflake account level and is all or nothing. With this in mind, you must consider the impact on all users of your test environment.

Multi-Tenant Considerations

Multi-tenant environments are those Snowflake accounts where several applications coexist. Naturally, any changes made at the account level may affect all tenant applications and behavior change bundles are no exception.

We do not advocate the immediate enabling of any behavior change bundle in production environments. Behavior change bundles should only be applied to lower environments and tested thoroughly.

Where common test environments exist, and prior to the application of behavior change bundles, you must confirm that each tenant accepts the change bundle.

Creating a Test Account

In order to test behavior change bundles, you should first create a Snowflake account dedicated to testing. Refer to Chapter 8 and follow the instructions to create a new account.

Once your Snowflake test account is available, clone your application into the new test account. There are different ways to clone an application depending upon your preferences, environments, and the degree of testing to be conducted. You may simply clone one or more databases into your new test account, or for more extensive testing, you may use Disaster Recovery tooling to deliver a fully functional environment.

Checking Behavior Change Bundle Status

All behavior change bundles are enabled or disabled using the accountadmin role:

```
USE ROLE accountadmin;
```

To check the deployment status of a specific behavior change bundle:

```
SELECT system$behavior_change_bundle_status('2022_06');
```

You should expect the returned status of RELEASED.

However, for an unreleased (at the time of writing) change bundle such as

```
SELECT system$behavior_change_bundle_status('2025_01');
```

you will receive an error message of `Invalid Change Bundle 2025_01`.

Further information on checking the behavior change bundle status can be found here: `https://docs.snowflake.com/en/sql-reference/functions/system_behavior_change_bundle_status.html`.

Enabling Behavior Change Bundles

In this example, we used 2022_01. You will need to change 2022_01 for the behavior change bundle under test.

Do not run this code directly into your production environment. This code is for education only.

After checking the behavior change bundle status, move on to enabling and disabling the behavior change bundle:

```
USE ROLE accountadmin;
```

To check the deployment status of a specific behavior change bundle:

```
SELECT system$behavior_change_bundle_status('2022_01');
```

Now enable this behavior change bundle:

```
SELECT system$enable_behavior_change_bundle('2022_01');
```

Confirm that the behavior change bundle is ENABLED:

```
SELECT system$behavior_change_bundle_status('2022_01'); --ENABLED
```

Capturing Test Case Output

After enabling a behavior change bundle, run your test cases and capture the output.

Disabling Behavior Change Bundles

Now disable the behavior change bundle:

```
USE ROLE accountadmin;

SELECT system$disable_behavior_change_bundle('2022_01');
```

Confirm that the behavior change bundle is DISABLED:

```
SELECT system$behavior_change_bundle_status('2022_01'); -- DISABLED
```

Baseline Test Case Output

If you have your baseline from a previous test, this step is optional. Otherwise, rerun your test cases, capture the output, and compare to the new test output from above.

Actions

Depending upon the outcome of your behavior change bundle testing, you should inform all Landing Zone-provisioned account owners and issue instructions on how to enable or disable each behavior change bundle.

All behavior change bundle testing should be recorded for audit purposes along with any subsequent actions.

In the event of an issue being identified, we suggest the follow-on actions below.

Behavior Change Summary

We do not advocate the immediate enabling of any behavior change bundle in production environments. We therefore recommend the following:

- An automated regression test suite be developed for each application deployed into Snowflake where "before" and "after" results can be compared.

- The application is cloned into an isolated test account.

- The "before" test is run to generate baseline results.

- The application clone is reset to the timestamp before the baseline run.

- The behavior change bundle is enabled on the isolated test account.

- The "after" test is run to generate comparison results.

- The "before" and "after" results are compared for differences.

- You inform all interested parties of the testing outcome.

- You issue instructions on how to enable and disable behavior change bundles.

Most of the recommendations can be automated but the crucial point is the test cases must be up to date and accurately reflect the application configuration.

Checklist

Behavior change bundle testing is a periodic activity which should be considered on a monthly basis. As the LZ team are not expected to know every nuance of each application, we expect a high degree of collaboration between the LZ team and each application team. Each application team must ensure their test cases are maintained along with a script to clone in preparation for the deployment and testing of patches.

We leave it for each organization to determine its approach with the expectation of clear boundaries for both LZ and application teams. Our preferred approach is for the teams to collaborate in delivering their test environments ready for patch testing.

Our checklist is shown in Table 11-1.

Table 11-1. *Behavior Change List Testing Checklist*

Snowflake Account	SNOW_SALES_DW_UAT
Behavior Change Bundle	2021_10: Change this to the behavior change bundle under test
Application Test Pack Version	Your local application test pack version
Pass/Fail	
Failure Scenario	Describe the scenario. Your test harness will expose the issue.
Test Harness Name	Create a script to expose the issue.
Snowflake Support Case	Register the issue with Snowflake, attaching the test harness and recording the ticket number here.

You may also wish to maintain a list of all Snowflake accounts and behavior change bundle statuses, recognizing the rolling nature of change inclusion into weekly patches. In our opinion, behavior change bundle testing is useful for

- Ensuring your applications are future proofed for upcoming weekly releases. That is, you can proactively test changes with a period of grace before actual deployment.

- Enabling private preview features for development and testing in advance of feature release by Snowflake.

Sample Test Harness

If an issue is found when testing a behavior change bundle, we have found Snowflake support is most responsive when a self-contained test case is delivered along with narrative explaining the scenario under which the issue arose.

Do not run this test harness directly into your production environment. The code is for education only.

The sample test harness creates a new database, warehouse, and role before checking the patch status. Naturally, your test case will focus on exposing the issue identified. The test harness may look like this:

```
SET test_database  = 'TEST';
SET test_schema    = 'TEST.test_owner';
SET test_warehouse = 'test_wh';
SET test_role      = 'test_owner_role';

CREATE OR REPLACE DATABASE IDENTIFIER ( $test_database )
DATA_RETENTION_TIME_IN_DAYS = 90;

CREATE OR REPLACE WAREHOUSE IDENTIFIER ( $test_warehouse ) WITH
WAREHOUSE_SIZE       = 'X-SMALL'
AUTO_SUSPEND         = 60
AUTO_RESUME          = TRUE
MIN_CLUSTER_COUNT    = 1
MAX_CLUSTER_COUNT    = 4
SCALING_POLICY       = 'STANDARD'
INITIALLY_SUSPENDED = TRUE;

CREATE OR REPLACE SCHEMA IDENTIFIER ( $test_schema );

USE DATABASE  IDENTIFIER ( $test_database );
USE WAREHOUSE IDENTIFIER ( $test_warehouse );

USE ROLE securityadmin;

CREATE ROLE IF NOT EXISTS IDENTIFIER ( $test_role )
COMMENT = 'TEST.test_owner Role';
 GRANT USAGE   ON DATABASE  IDENTIFIER ( $test_database  ) TO ROLE
IDENTIFIER ( $test_role );
GRANT USAGE   ON WAREHOUSE IDENTIFIER ( $test_warehouse ) TO ROLE
IDENTIFIER ( $test_role );
GRANT OPERATE ON WAREHOUSE IDENTIFIER ( $test_warehouse ) TO ROLE
IDENTIFIER ( $test_role );

GRANT MODIFY        ON DATABASE IDENTIFIER ( $test_database ) TO ROLE
IDENTIFIER ( $test_role );
```

```
GRANT MONITOR         ON DATABASE IDENTIFIER ( $test_database ) TO ROLE
IDENTIFIER ( $test_role );
GRANT USAGE           ON DATABASE IDENTIFIER ( $test_database ) TO ROLE
IDENTIFIER ( $test_role );
GRANT CREATE SCHEMA ON DATABASE IDENTIFIER ( $test_database ) TO ROLE
IDENTIFIER ( $test_role );
```

You may need to amend the entitlements granted to accommodate your test case:

```
GRANT USAGE                     ON SCHEMA IDENTIFIER ( $test_schema ) TO
ROLE IDENTIFIER ( $test_role );
GRANT MONITOR                   ON SCHEMA IDENTIFIER ( $test_schema ) TO
ROLE IDENTIFIER ( $test_role );
GRANT MODIFY                    ON SCHEMA IDENTIFIER ( $test_schema ) TO
ROLE IDENTIFIER ( $test_role );
GRANT CREATE TABLE              ON SCHEMA IDENTIFIER ( $test_schema ) TO
ROLE IDENTIFIER ( $test_role );
GRANT CREATE VIEW               ON SCHEMA IDENTIFIER ( $test_schema ) TO
ROLE IDENTIFIER ( $test_role );
GRANT CREATE SEQUENCE           ON SCHEMA IDENTIFIER ( $test_schema ) TO
ROLE IDENTIFIER ( $test_role );
GRANT CREATE STREAM             ON SCHEMA IDENTIFIER ( $test_schema ) TO
ROLE IDENTIFIER ( $test_role );
-- Assign to logged in user
GRANT ROLE IDENTIFIER ( $test_role ) TO USER <Your Name Here>;
```

Enable the behavior change bundle. In this example, we use 2021_10 which you will need to change for the behavior change bundle under test. Otherwise, this error occurs: Invalid Change Bundle '2021_10'.

```
USE ROLE accountadmin;
```

Check the BEFORE status:

```
SELECT system$behavior_change_bundle_status('2021_10');
```

Apply the change bundle:

```
SELECT system$enable_behavior_change_bundle('2021_10');
```

Check the AFTER status:

```
SELECT system$behavior_change_bundle_status('2021_10'); --ENABLED
```

Add your test case here:

```
USE ROLE       IDENTIFIER ( $test_role      );
USE DATABASE   IDENTIFIER ( $test_database  );
USE SCHEMA     IDENTIFIER ( $test_schema    );
USE WAREHOUSE  IDENTIFIER ( $test_warehouse );
```

Your test case goes here with the behavior change bundle enabled, which should create both the objects and the test case scenario to expose the issue.

Once your test is complete, you must disable the change bundle:

```
USE ROLE accountadmin;
```

Check the BEFORE status:

```
SELECT system$behavior_change_bundle_status('2021_10');
```

Disable the change bundle:

```
SELECT system$disable_behavior_change_bundle('2021_10');
```

Check the AFTER status:

```
SELECT system$behavior_change_bundle_status('2021_10'); -- DISABLED
```

You may need to repeat your test case to identify the difference in behavior and expose the issue seen:

```
USE ROLE       IDENTIFIER ( $test_role      );
USE DATABASE   IDENTIFIER ( $test_database  );
USE SCHEMA     IDENTIFIER ( $test_schema    );
USE WAREHOUSE  IDENTIFIER ( $test_warehouse );
```

Repeat your test case here with the behavior change bundle disabled, which provides the control state for comparison.

Finally, remove your test case:

```
USE ROLE sysadmin;

DROP DATABASE IDENTIFIER ( $test_database );

DROP WAREHOUSE IDENTIFIER ( $test_warehouse );

USE ROLE securityadmin;

DROP ROLE IDENTIFIER ( $test_role );
```

Summary

We began this chapter by explaining the relationship between Snowflake's weekly release deployment and behavior change bundles. We then discussed the implications of applying behavior change bundles into multi-tenant environments. After creating a test account dedicated to behavior change bundle testing, we investigated how to enable a behavior change bundle and later disable the same behavior change bundle. We offered a checklist of actions to record for audit purposes along with a sample test harness which may be used when raising a support ticket with Snowflake.

Our focus turns to Disaster Recovery (DR), the subject of our next chapter.

CHAPTER 12

Disaster Recovery

Disaster Recover (DR) is the one capability no one ever wants to invoke in a real-world scenario. Despite the highly unlikely event of ever needing to invoke DR, you must provision full capability along with periodic testing to demonstrate to others that your service and application capability is resilient. Many organizations conduct at least annual, and sometimes more frequent, DR testing. You must be prepared.

This chapter articulates Snowflake DR capability as available at the time of writing. The authors acknowledge that Snowflake is continually improving DR capability and we expect you to periodically revise your DR plans in accordance with the latest Snowflake capabilities. We also acknowledge Snowflake may not yet implement CSP-specific features from within the Snowflake environment. We discuss these features external to Snowflake later in this chapter. Our aim is to explain the current "art of the possible" and highlight areas for further consideration. Our investigation relies upon two Snowflake accounts implemented using Snowflake Organization features explained in Chapter 8 and also reuses the CATALOG database developed in Chapter 7.

DR relies upon features that are only available with the Snowflake Business Critical Edition, as illustrated in Figure 12-1. Failover and Failback are only available for the Business Critical Edition.

Standard
Fully functional SQL Database
Secure Data Shares
24 x 365 Support
1 day Time Travel
Encryption in transit and at rest
Virtual Warehouses
Federated Authentication
Replication Enabled

Enterprise
Standard +
Clustered Warehouses
90 Day Time Travel
Annual rekeying of data
Materialized Views

Business Critical
Enterprise +
Externally validated secure
Encrypted data everywhere
Tri-Secret Secure, BYOK
AWS PrivateLink support
Enhanced security posture
Failover and Failback support

Virtual Private Snowflake (VPS)
Business Critical +
Customer dedicated virtual servers
Customer dedicated metadata store
Some restrictions...

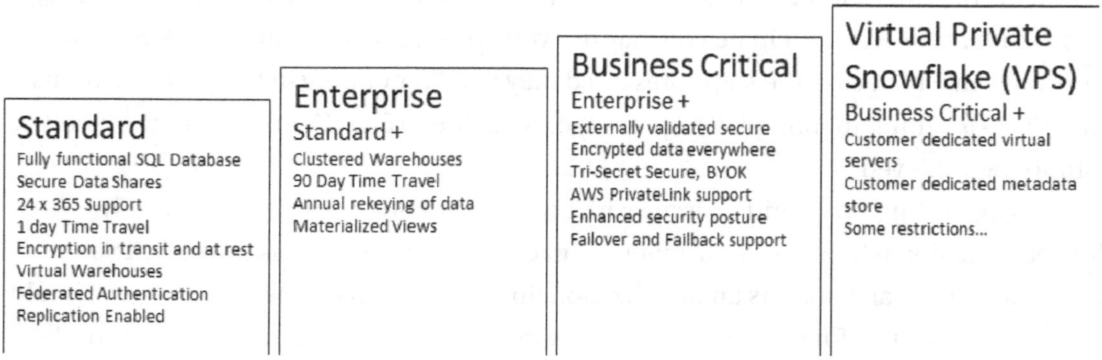

Figure 12-1. *Snowflake editions*

© Andrew Carruthers and Sahir Ahmed 2023
A. Carruthers and S. Ahmed, *Maturing the Snowflake Data Cloud*, https://doi.org/10.1007/978-1-4842-9340-9_12

To upgrade Standard or Enterprise edition to Business Critical Edition, please contact Snowflake Support. Further information can be found here: `https://community.snowflake.com/s/article/How-To-Submit-a-Support-Case-in-Snowflake-Lodge`.

In this chapter, we discuss these scenarios in turn:

- Single database replication

- Whole account replication including replication groups/failover groups

- Client redirect

It is imperative that you have a full understanding of each scenario, purpose, limitations, and options. To this end, we address each scenario in turn, providing a hands-on walk-through of each.

We start by discussing Snowflake availability and explain CSP high availability (HA) zones, along with providing a brief explanation of the components that constitute a CSP.

With your understanding of Snowflake availability, we discuss business continuity planning (BCP), which represents your organization's appetite for the timeliness of service recovery, potential for data loss, and a variety of other factors. Working through the features implemented by CSP HAs, we distill DR requirements down further before discussing failure scenarios and available mitigation options.

With the available information at hand, while taking into consideration organizational preferences and other limitations, we investigate the DR options available for implementing three scenarios: database replication, account replication, and replication groups/failover groups.

All testing was conducted using a free trial account available here: `https://signup.snowflake.com/`. We anticipate you may use your commercial organization's Snowflake accounts, in which case the steps presented may need amending in the event accounts already exist. Implementing DR requires the use of the `ORGADMIN` role, without which DR cannot be achieved.

During testing, we found an issue with account creation outside the currently logged-in region where the confirmation email with initial login was not sent. The confirmation email contains an initialization link required to set the default user. Should the same issue arise for your testing, raise a support ticket with Snowflake Support. We make reference to this scenario later in the section titled "Suspended Account."

No real-world DR scenario would be complete without considering the wider implications of single or multiple region failure. You must deliver a workable and pragmatic DR implementation catering for as many foreseeable failure scenarios as possible. We seed your thoughts within the "Holistic Approach" section by identifying a few scenarios and offering information for your further consideration.

Planning your DR test and then proving its effectiveness is essential in building confidence in both Snowflake and your capability to deal with any failure scenario.

To assist diagnosing your application failure scenario, we provide a basic troubleshooting guide to identify information useful to both yourself and to Snowflake when raising support tickets. We note the inability to connect to a Snowflake account will obviate some checks, but partial loss of service may necessitate DR invocation even if both the CSP region and your application are accessible.

Lastly, we deliver a template checklist as a starting point for your DR planning.

Working through this chapter will raise more questions for your organizations DR implementation than can be answered here. Nobody said designing and implementing DR was going to be easy...

Snowflake Availability

A key advantage of CSP hosting is their focus on high availability and resilience, where the CSP core implementation delivers capabilities to mitigate against invoking DR. In other words, a cloud-based implementation is inherently less fragile than its comparable on-prem equivalent but may still require DR capability. We begin our investigation with a recap of what each CSP provides out of the box, the core upon which Snowflake is built, the foundation upon which you deliver your applications.

CSPs claim around 99.999999999 percent durability (eleven 9s), equating to the loss of a single object out of a total 1 billion objects, once in around 100 years. CSPs also claim 99.99 percent availability (four 9s), equating to less than an hour of downtime per year, alternatively considered to be about 4 minutes per month. Differing views exist on whether 99.9 percent availability is acceptable or 99.999 percent availability is acceptable. The debate continues and is one for your further investigation according the business criticality of your applications and your organization's view of downtime.

Snowflake is implemented on a CSP-provisioned infrastructure, it therefore follows that Snowflake cannot claim higher durability or availability from a single CSP implementation within a given single region. Snowflake states that 99.9% availability and publishes annual reports that include incident information and outage minutes. Noting a Snowflake Community account is required, a sample report for AWS availability for 2022 may be found here: `https://community.snowflake.com/s/article/uptime-report-for-year-2022-snowflake-hosted-on-aws`.

Taking the CSP claims at face value is all well and good, but you should at least understand a little more of how these claims are substantiated, including the components and mitigating actions that deliver resilience.

Figure 12-2 illustrates services provisioned by a CSP, from which your Snowflake account is dependent.

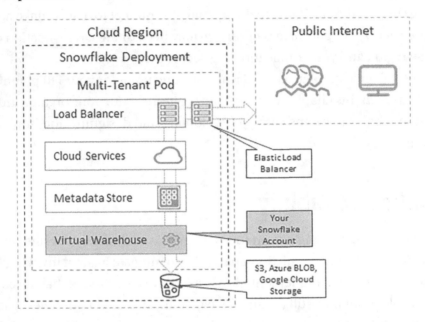

Figure 12-2. *Snowflake account provision*

Without going into too much detail, Figure 12-3 conceptually illustrates a single CSP implementation within a single region. In this example, you see three HA zones, each of which is a geographically isolated site within the CSP region. It also shows several service tiers which collectively provision the overall service with callouts showing the corresponding Snowflake services. We are unable to articulate all the interrelationships across each horizontal service suite, although we reasonably expect there to be several layers of resilience built in.

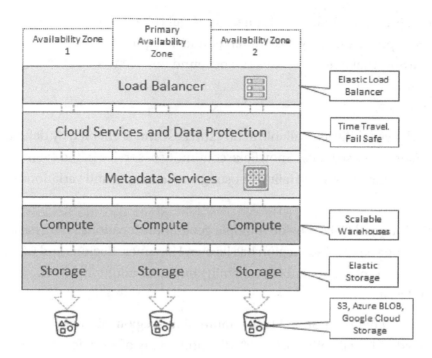

Figure 12-3. *Snowflake HA*

Clearly, an awful lot of thought has gone into the delivery of a CSP region, all of which underpins the DR implementation discussed within this chapter. With three HA zones and immutable micro-partition approach (discussed later), Snowflake provides protection against a number of scenarios including

- **Customer error**: Accidental deletion of both data and objects, loading one or more datafiles with bad data, or loading data out of sequence

- **Virtual machine failure**: Triple redundancy for critical cloud service, automatic retries for failed parts of a query

- **Storage/disk issues and failure**: Redundancy method not specified but guarantees eleven 9s durability

- **Single or dual availability zone failure**: (Availability zones on AWS, availability sets on Azure, regions and zones for Google Cloud Platform). Snowflake service may be degraded as load balancing will be across the remaining availability zone(s).

- **Single region failure**: Loss of service within a region, no HA zones are usable

- **Multi-region failure**: Loss of service across two or more regions within the same CSP

Snowflake is automatically provisioned across all HA zones within a region at the time of account provisioning. No user intervention is required.

Real-time Snowflake service status can be found here: `https://status.snowflake.com.`

In summary,

- High availability is available at no extra cost and is enabled by default. HA provides seamless Snowflake continuity of service in the event of component or service failure, hosting venue failure, and variations in load.

- Regions and CSPs provide multiple geographic locations for physical hosting venues. Each region has its own HA zones. Regions and CSPs provide replication and DR capability but incur additional work and cost.

We consider it highly improbable that multi-cloud region failure will occur. Based upon the stated CSP availability of 99.99%, the probability of a single region failing is therefore 0.01%. To determine the probability of multi-cloud region failure, we multiply each region probability together (i.e., 0.01% x 0.01%) resulting in a probability of 0.0001%. We also accept this is an overly simplistic calculation; a real-world calculation would have more factors and result in a higher calculated value. Caveat emptor.

The loss of service across two or more regions involving two different CSPs would indicate problems of a far more serious nature. We therefore suggest planning for every eventuality is difficult, if not impossible, to implement and test.

Lastly, backups to flat files within the context of CSP service-based provision are considered a position of last resort given the stated eleven 9s CSP durability. Should flat file backups be considered, a suitable target environment for both storage and later upload would need to be identified. Manual recovery from flat file backups is not something to be undertaken lightly and is beyond the scope of this book. We consider Snowflake in-built capabilities to be sufficient to implement a robust DR strategy, albeit with some manual work required.

Business Continuity Plan

Every system or application (henceforth collectively "application") within an organization will have a Business Continuity Plan (BCP) detailing strategies and timelines specific to the particular application.

We expect every BCP will contain support group contact information, escalation steps, and other organization-specific detail deliberately omitted from this section. Instead, we offer broad guidelines for your consideration and inclusion.

RPO and RTO

Two key deliverables of the BCP are to define the recovery point objective (RPO) and recovery time objective (RTO). We discuss each next.

The RTO defines the maximum time window your business colleagues will accept for application recovery for a single incident. As with RPO, this figure will differ from application to application, and within a multi-tenant account, consideration must be given to reconciling the differing perspectives.

The RPO describes the maximum amount of data loss your business colleagues will accept within an application for a single incident. This figure will differ from application to application, and within a multi-tenant account, consideration must be given to reconciling the differing perspectives.

As you will see when implementing replication groups, you have some flexibility in determining individual tenant RTOs and RPOs

Figure 12-4 may prove useful in visualizing RTO and RPO.

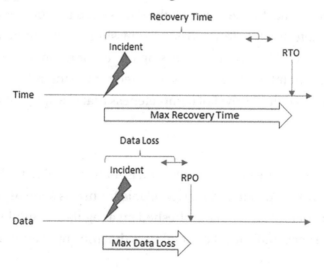

Figure 12-4. *RTO and RPO visualized*

Service-Level Agreement (SLA)

Your organization may offer differing levels of SLA according to the legal agreements in force with your customers, which may cover factors (amongst others) such as the ones in the following sections.

Accuracy

Accuracy refers to the degree of correctness of the data stored and processed by an application. It measures how closely the data conforms to the truth or reality. The SLA will typically specify the acceptable level of accuracy that the application must meet.

Integrity

Integrity refers to the completeness and consistency of the data stored and processed by an application. It measures the degree to which the data is free from errors, omissions, and unauthorized modifications. The SLA will typically specify the acceptable level of data integrity that the application must meet.

Completeness

Completeness refers to the degree to which the data stored and processed by the application is complete and free from errors or omissions. This can mean that data is complete in the sense that it is free from missing values, or it can mean that the data contains all the relevant information that is required for the intended use. The SLA will typically specify the acceptable level of completeness that the application must meet.

Timeliness

Timeliness refers to the degree to which the data is made available to the users in a timely manner. This can mean that data is available for use as soon as it is needed, or it can mean that the data is updated or refreshed on a regular schedule. The SLA will typically specify the acceptable level of timeliness that the application must meet.

Performance

Performance is a key metric that is used to measure the quality of the data services provided. The SLA will typically specify the acceptable levels of performance that the data service provider must meet, such as the maximum allowable response time for data queries or the minimum level of availability for the data services. The SLA may also specify the conditions under which performance guarantees are in effect, such as during normal business hours or during peak usage periods.

Summary

Your BCP must incorporate all contractually mandated factors, not just those highlighted above, which focus on the operational impacts of Snowflake service provision.

Organizational Considerations

Apart from RPO, RTO and SLA, your organization may have further requirements to comply with, such as

- External regulatory policies

- Data residency requirements

- Cyber security standards

- Conformance to approved design patterns and standards

- Key Performance Indicators (KPIs)

- Behavior Change Bundle status

- Defined exit strategy for the CSP and Snowflake

- Communications plan in the event of an outage

Your risk management team will direct your attention to your organization's requirements.

Resilience

We now briefly discuss resilience within the context of provisioning tooling in support of HA, DR, and Snowflake capabilities.

Data Recovery

Data recovery within Snowflake is reliant upon immutable micro-partitions. Snowflake does not add, change, or remove data from an existing micro-partition. Instead every data change is recorded by the creation of new micro-partitions.

Time Travel provides the capability to revert objects, conduct queries, and clone to a point in time in history. Note Time Travel is restricted to 1 day maximum for Standard Edition and up to 90 days for all higher editions and cannot be disabled at the account level, but can be disabled for database, schemas, and tables. Choose a setting for the account based on your business requirements.

We recommend setting `DATA_RETENTION_TIME_IN_DAYS` to 90 for your account using the `ACCOUNTADMIN` role

---

Fail Safe is an additional, non-configurable, 7-day retention period exclusively managed by Snowflake representing the final stage of storage lifecycle retention. When objects age out of Time Travel, the underlying micro-partitions are not immediately deallocated and returned to the cloud provider storage pool. Instead, Snowflake retains the micro-partitions, and by inference the associated metadata, and has the capability to restore objects.

We explain micro-partitions, Time Travel, and Fail Safe in great detail in our previous book, *Building the Snowflake Data Cloud*.

Fault Tolerance

Fault tolerance is the ability for your applications to seamlessly continue operation in the event of a component or service failure. As you will read below, CSPs provision redundant hardware and automation to mitigate against service outage. One example of fault tolerance is the Snowflake load balancer used in traffic management. It distributes incoming requests across multiple servers in order to prevent any one server from becoming overloaded and crashing. If one server fails, the load balancer redirects traffic to the remaining functional servers to ensure that the service remains available to users.

Availability

Availability relates to maintaining uninterrupted service provision. As stated by each CSP, four 9s is the expected availability (99,99%) and Snowflake reflects the underlying CSP availability figure. In the event a single HA zone fails, you may see a slow-down in service, which may manifest itself as increased response times for queries. If two HA zones fail, it is highly likely that your service will be degraded, although it is impossible to quantify beforehand.

It is possible that Snowflake core services could fail while the CSP infrastructure remains fully functional and usable. While this scenario is improbable, a possible valid scenario could occur while provisioning or commissioning a new Snowflake region.

Scalability

Scalability is the ability to increase or decrease resource allocation either through static or dynamic configuration capability. Within Snowflake, warehouses can be scaled up (increasing t-shirt size), out (increasing the number of clusters), or across (increasing the warehouse declarations for workload segregation). Naturally, there are cost implications to consider when scaling warehouses, and the over-provision of service can be an expensive mistake.

Self-Healing

Self-healing is provisioned by Snowflake management capability, an example of which is the seamless restart of warehouses in the event of failure. You may experience a slight delay while a warehouse spins up and requeries a result set; in practice, delays have proven to be minimal.

Automation

Another feature provisioned by Snowflake where components are automatically and seamlessly provisioned without human intervention. We may consider load balancing and data replication capabilities as automated capabilities.

Redundancy

As you would expect from the comprehensive service provision used by Snowflake, component redundancy is an inherent capability. An example of redundancy is data storage. One such implementation is RAID (`https://en.wikipedia.org/wiki/RAID`) where disk failure is tolerated and hot-swap replacements are made with subsequent automated array rebuild. Redundancy may also be implemented by replicating data sets across HA zones and CSP regions.

Monitoring and Alerting

The ability to detect and respond to a service outage is dependent upon having the means to identify non-self-healing fault conditions, preferably via events, alerts, and pro-active detection mechanisms. Your first line of defense is to identify the critical components to monitor, create alerting mechanisms, and then deliver information to the appropriate team for action.

Organizations often have multiple teams dealing with differing aspects of the technical estate. At the micro level, you consider monitoring and alerting for your applications, and for this you start with checking whether the Snowflake service is available: `https://status.snowflake.com`.

Occasionally, Snowflake will capture events from its own monitoring suite and deliver information via email from Snowflake Global Technical Support (`info@reply.snowflake.com`). Please check your spam folder to ensure receipt.

Resilience Summary

From a Snowflake service usage perspective, you are largely insulated from both data errors/object deletion and underlying HA zone fault conditions within a single component or service. Through the HA provision you can rely upon continuity of service if two of the three zones fail.

We discuss various DR scenarios and options next.

Scenarios and Options

Let's now look at the Snowflake scenarios and options available for both single region and multi-region failure.

The Snowflake Support Policy and SLA can be found here: `www.snowflake.com/wp-content/uploads/2017/07/Snowflake-Computing-Premier-Support-Policies-and-SLAs-2015-05-06.pdf`. Table 12-1 summarizes the Snowflake and CSP-provisioned options for business continuity.

Table 12-1. *Business Continuity Options*

Failure	Impact	Mitigation
Data Error	Data corruption, deletion, or object removal	Snowflake features: • Time Travel • Fail Safe
Virtual Machine Failure	Service layer failure	Snowflake built-in redundancy: • Triple redundancy for critical services • Automatic retries for query failure
Storage/Disk Failure	Storage layer failure	CSP HA features: • RAID • HA and region data replication
HA Zone Failure	Single or dual HA zone failure	Snowflake built-in redundancy • Load balancing • Multiple HA zone availability
Single Region Failure	Loss of CSP service within a single region	Snowflake features: • Cross-region replication • Cross-region failover
Multi-Region Failure	Loss of CSP service in multiple regions	Snowflake features: • Cross-cloud replication • Cross-cloud failover

From our discussions so far, both inbuilt Snowflake functionality and CSP HA capabilities provide acceptable cover and mitigation for most scenarios involving either single or dual HA zone failure.

Having considered how the Snowflake service is underpinned by HA zones, CSP services, durability, and availability, let's now consider how to implement DR for both single region (where all HA zones are unavailable), and multi-region failure scenarios.

Just as failover needs careful consideration and planning, so does failback.

As discussed, DR is not typically a single all-encompassing event affecting every aspect of service across multiple CSPs and regions. Furthermore, in a multi-tenant environment, there may be differing views of the severity of any incident where whole account failover is being proposed as a solution. One size does not always fit all, and perspective matters. We address multi-tenant accounts later in this chapter; see "Replication Groups/Failover Groups" for details.

Snowflake supports region group replication, but the terminology used is confusing. Snowflake use the terms "multi-tenant regions" and "Virtual Private Snowflake" to differentiate their offerings:

- Virtual Private Snowflake (VPS) offers a completely separate environment isolated from all other Snowflake accounts.

- All other Snowflake editions belong to multi-tenant regions.

Snowflake editions are described here: `https://docs.snowflake.com/en/user-guide/intro-editions.html#overview-of-editions`.

Data sharing and account migrations between VPS and multi-tenant regions are disabled by default. Snowflake Support can enable this feature; please refer to documentation found here for details: `https://docs.snowflake.com/en/user-guide/database-replication-config.html#region-support-for-database-replication-and-failover-failback`.

Supported Cloud Regions

We now discuss options for replication. We begin by examining the supported cloud regions as shown in Figure 12-5.

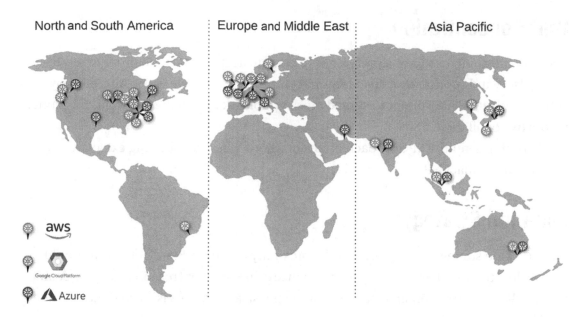

Figure 12-5. *Snowflake-supported cloud regions*

The latest information on supported cloud regions can be found here: https://docs.snowflake.com/en/user-guide/intro-regions.html#supported-cloud-regions.

There are several considerations to be borne in mind when developing your DR strategy. The following list is not exhaustive and you may find your organization has others, but for those identified, we will now briefly each discuss in turn.

Legal Agreements

Depending upon how your Snowflake Master Service Agreement (MSA) was created, there may be restrictions on the operation of accounts. For example, an MSA created under a CSP Marketplace agreement may prevent the operation of accounts on rival CSP platforms. Furthermore, prepurchased credits may not be transferrable, leading to the creation of another MSA to run in parallel with the original.

CSP Concentration

Your organization may have a preference to operate using one CSP in order to reduce the skillsets and support overhead required for running multiple CSPs. Alternatively, your organization may prefer to spread their vendor risk by placing Snowflake accounts across two or all three CSPs.

Adopting a multi-CSP approach is advantageous when considering exit planning from an individual CSP.

Location Strategy

Your location strategy must also consider proximity to other CSP regions. You should choose locations that are unlikely to experience civil unrest or (reasonably) foreseeable natural disaster. An optimal distance between physical locations is a few hundred miles.

No location strategy is complete without considering the external tooling that interacts with your applications. You may be fortunate in finding another region within the same CSP near your primary site geographical location or a region within a competitor CSP nearby, which resolves external tooling connectivity in terms of latency.

Another consideration for your customers is to reduce latency by implementing your Snowflake consumption-based accounts as close as possible to their points of consumption.

You may also encounter jurisdictional issues when transferring data across borders.

Costs

CSPs make money from consumption, that is, by selling both storage (disk) and compute. You may also incur charges when:

- Moving data into (ingress) Snowflake. Snowflake do not charge ingress fees, but the CSP might charge a data egress fee for transferring data from the provider to your Snowflake account.

- Moving data out (egress) of Snowflake.

Charges vary according to the source and target CSP and region.

Your implementation strategy should look to minimize not only ingress and egress costs, but execution costs too, noting there are differences in storage costs across regions within the same CSP.

Replication costs will vary according to the number of micro-partitions transferred between primary and secondary accounts. There are ways to minimize data transfer costs according to how the source objects are structured and data manipulation operations (DML) are performed. As an example, in one scenario we successfully reduced micro-partition churn for bulk operations by between 40% and 70% while reducing credit consumption by a similar amount.

For every replica, CSP storage costs will be incurred. Database replication incurs cost for the single database replicated, account replication incurs cost for the whole account, and replication groups incur costs for the objects replicated. For these scenarios, costs are easily calculated.

Further information on data transfer costs can be found here: `https://docs.snowflake.com/en/user-guide/cost-understanding-data-transfer.html#understanding-data-transfer-cost`.

Further information on understanding overall cost can be found here: `https://docs.snowflake.com/en/user-guide/cost-understanding-overall.html#understanding-overall-cost`.

Secure Direct Data Share

We discussed SDDS in Chapter 5 as a mechanism to enable consumers to access our data in real time and to capture consumption metrics. We mention the ability to securely share data where Snowflake accounts are colocated within both the same region and CSP to deliberately exclude as an option to provide resilience. Figure 12-6 conceptually illustrates the data interaction between two Snowflake accounts.

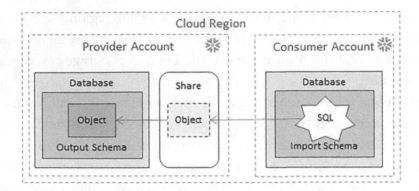

Figure 12-6. *SDSS conceptual model*

For those seeking a fuller explanation of how SDDS is achieved from within the same region and CSP, Figure 12-7 shows how micro-partitions are shared from the provider account to the consuming account, noting only the current micro-partitions are available to the consuming share. The net effect of data sharing is that all changes to data made by the provider are immediately available to the consumer, in real time.

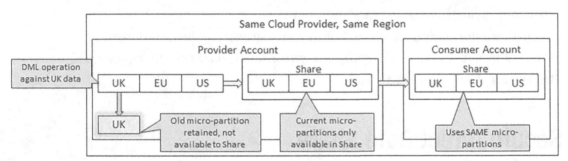

Figure 12-7. *SDDS physical implementation*

As both Snowflake accounts reside within the same region and CSP, it can readily be seen that the failure of all three HA zones will prevent both accounts from operating. We include SDDS to demonstrate the implementation pattern does not provide resilience and should therefore be excluded from further discussion on implementing DR.

Database Replication

Noting the information supplied is correct at the time of writing, Snowflake is constantly improving its DR offering and support. We conduct a hands-on walk through of database replication later within this section.

We strongly recommend you investigate the Snowflake documentation before defining your DR approach.

Figure 12-8 illustrates database replication alongside SDDS for comparison. The important point is to note is you can replicate databases both within a single region and across regions.

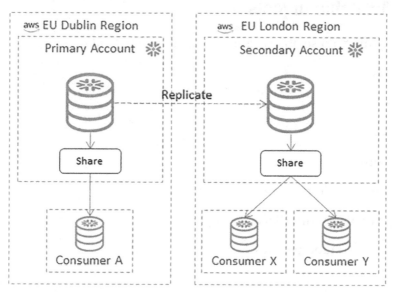

Figure 12-8. *Database replication conceptual representation*

The example shown uses the same AWS region for both accounts as we encountered an issue creating cross-region accounts when using a trial account. A ticket has been raised to Snowflake.

When discussing database replication, we refer to the account from which data is published as the primary and the account into which the data is consumed as the secondary. There may be more than one secondary account.

Database replication is used to replicate a single database only. Database refresh is asynchronous and implemented by each secondary pulling data from the primary, with the refresh cycle typically implemented by a task. If multiple databases are to be replicated, each one must be configured independently.

Database replication is serverless and does not rely upon a warehouse being declared or available.

Failover is manually invoked. There is no automated failover option. You remain responsible for external tooling connectivity, which is discussed later in this chapter.

Replicated Objects

The list of supported objects is evolving and the lists provided below should be checked before assuming correctness. Furthermore, each object state must also be verified. For example, stream contents should be checked to ensure consistency within a secondary database before failover in the event of primary database failure.

Currently Available Objects

At the time of writing, these objects are replicated within a database:

- Tables
- Sequences
- Views
- File formats
- Stored procedures
- SQL and JavaScript UDFs
- Row and column policies
- Tags
- Streams
- Tasks

Future Available Objects

These objects are not currently replicated within a database, so please check with your Snowflake Sales Engineer as replication for many of the below objects are in-flight:

- Temporary tables (may be erroneously indicated in documentation as unavailable, so please check with your Sales Engineer)
- External tables
- Stages
- Content replication for internal stages
- Pipes

- Copy history

- Storage integration in Snowflake

- Database roles

- Reader accounts

Unavailable Objects

While most object types are replicated within a database, there are some restrictions and notable exceptions:

- Objects cannot reference another object outside of the primary database.

- Stateful objects may have limitations.

- Entitlement is not automatically carried across.

- Objects referencing an explicitly excluded object (see below)

The above list is not exhaustive. Please refer to the latest list of replicated database objects which can be found here: `https://docs.snowflake.com/en/user-guide/account-replication-intro.html#replicated-database-objects.`

Excluded Objects

Some objects are explicitly identified as excluded from database replication within Snowflake documentation; there is no reference to future support. They include

- Temporary tables (may be erroneously indicated in documentation as unavailable, so please check with your Sales Engineer)

- Temporary stages

- Rose-based access control (entitlement does not transfer with the database)

- Shares (external to a database and therefore excluded; see SDDS above)

- Databases imported from shares

Several other objects are planned for a future version of database replication. Please refer to the latest list of replicated database objects at `https://docs.snowflake.com/en/user-guide/account-replication-intro.html#replicated-database-objects`.

Manual Configuration

Later in this chapter we discuss interaction with the CSP account. It is usual for a Snowflake account to have a corresponding CSP account to provision services and features not available within Snowflake. Some examples are

- Using storage integrations to map external storage (S3, Azure Blob Storage, GCP Cloud Storage) for Snowflake use via stages. A critical step for implementing any storage integration is the trust relationship with the CSP account.

- API integrations are used to implement external functions. For example, these may be AWS Lambdas which implement document content scraping capability interfacing to Textract.

- Notification integrations are a relatively new feature that implement email and queue functionality.

- SnowPipe AUTO_INGEST is reliant upon configuring CSP event notification. For AWS, you may use SNS and SQS.

- Account parameters are not replicated with database replication but schema- and object-level parameters are replicated.

- All secondary databases are read-only, and for those switched to become primary, there may be manual configuration work required to become fully functional and automated.

Your environment will differ, but it is important to capture all dependencies and address each in turn before attempting database replication. For further details, refer to documentation found here: `https://docs.snowflake.com/en/user-guide/database-replication-intro.html#overview-of-database-replication`.

Database Replication Limitations

The asynchronous/periodic refresh of a replicated database may result in data gaps appearing if data changes in the primary are not refreshed into the secondary before the primary fails. In other words, transactions may occur in the primary that are not synchronized with the secondary.

Source code should be scanned to identify object references external to the database being replicated before attempting to replicate a database. We suggest a code scanner be developed to interrogate the primary database `information_schema` to detect likely failure conditions before database replication is attempted. Naturally, this code scanner would need regular maintenance to match Snowflake monthly releases.

Objects in the primary referencing external objects may fail to replicate to the secondary. An example is where a view spans two databases in the primary account but only one database is replicated. The view will create but the dependency for the missing database will not be resolved. This is called a dangling reference. There are other types of dangling references, as documented here: `https://docs.snowflake.com/en/user-guide/database-replication-considerations.html#dangling-references`.

Entitlement is not automatically carried across from primary to secondary accounts for database replication. It is therefore possible for different RBAC entitlement policies to be implemented. While this approach may be desirable in some scenarios, for others the lack of automated application of primary RBAC may present challenges. Regardless of perspective, it is clear that careful consideration must be given to data access controls in every secondary account.

When a primary database is replicated to a secondary database, there is a default 10 TB size limit applied. Further information can be found here: `https://docs.snowflake.com/en/sql-reference/sql/alter-database.html#database-replication-and-failover-usage-notes`.

Database Replication Step by Step

For this walkthrough, we assume the code developed in Chapters 7 and 8 has been deployed and you have two or more Snowflake accounts available within your Snowflake organization.

In this section, the headings reflect where code should be executed, and do not refer to more than a single primary or secondary account. The numbering is used to indicate sequencing of usage, nothing more.

We only use two accounts, referenced as primary and secondary.

Suspended Account

If time has elapsed since creating an account, you may see this error message "Your account is suspended due to lack of payment method." You will need to add one more account (named for ease of reference) to your Snowflake organization to complete this section.

This is an optional section for the creation of a new secondary account.

Grant the role ORGADMIN to yourself:

```
USE ROLE accountadmin;

GRANT ROLE orgadmin TO USER <Your User Here>;
```

Switch the role to ORGADMIN:

```
USE ROLE orgadmin;
```

Then create a new account.

The region should differ from your currently logged-in account.

During testing, we found an issue with creating accounts outside the currently logged-in region where the confirmation email with initial login was not sent. The confirmation email contains an initialization link required to set the default user.

You can observe the new account exists when you execute SHOW ORGANIZATION ACCOUNTS; but the account is not accessible. Should the same issue arise for your testing, raise a support ticket with Snowflake Support; my ticket for reference is 00474730.

You may choose to create a new account within the current region as this approach allows database replication but not account replication or client redirect, which we address later:

```
CREATE ACCOUNT REPLICATION_TERTIARY
ADMIN_NAME      = ANDYC
ADMIN_PASSWORD = '<Your Password Here>'
FIRST_NAME      = Andrew
LAST_NAME       = Carruthers
EMAIL           = '<Your Email Here>'
EDITION         = BUSINESS_CRITICAL
REGION          = AWS_EU_WEST_1;
```

The CREATE ACCOUNT SQL statement returns "status" JSON, which when formatted using https://jsonformatter.org/ looks like this, noting your identifiers will differ:

```
{
  "accountLocator": "PC52900",
  "accountLocatorUrl": "https://pc52900.eu-west-1.snowflakecomputing.com",
  "accountName": "REPLICATION_TERTIARY",
  "url": "https://nuymclu-replication_tertiary.snowflakecomputing.com",
  "edition": "BUSINESS_CRITICAL",
  "regionGroup": "PUBLIC",
  "cloud": "AWS",
  "region": "AWS_EU_WEST_1"
}
```

Using a different browser, and referencing the email received from Snowflake, log into the new secondary account; you will use this later. You may wish to create a worksheet called SECONDARY to prevent confusion when switching between accounts.

A full explanation of the ORGADMIN role and capabilities is provided in Chapter 8.

Primary Account Setup 1

To prove you have several accounts within your Snowflake organization, run these commands:

```
USE ROLE orgadmin;

SHOW ORGANIZATION ACCOUNTS;
```

You should see two or more accounts listed in the result set. For the purposes of this walkthrough, we shall assume the primary account is the one created at initial sign-up, readily identified by the result set comment `Created by Signup Service`. My `account_name` is NP62160, as shown in Figure 12-9; yours will differ.

organization_l	account_name	snowflake_region	comment	account_loc	account_locator_url
NUYMCLU	REPLICATION_TERTIARY	AWS_EU_WEST_1	SNOWFLAKE	PC52900	https://pc52900.eu-w
NUYMCLU	NP62160	AWS_EU_WEST_2	Created by Signup Service	XL29287	https://xl29287.eu-we

Figure 12-9. *Account setup*

In the following example, you see two accounts created within the `organization_name` NUYMCLU including secondary `account_locator` PC52900.

Make a note of the `organization_name` and secondary `account_locator` values as you will need them shortly.

If you experience an error with the above SQL statements, we suggest you walk through Chapter 8, which explains the use of the `ORGADMIN` role and enables the creation of Snowflake accounts within your organization.

Assuming all is well, let's identify any existing replication databases:

```
SHOW REPLICATION DATABASES;
```

You should not see any results. In other words, no replication databases have been defined.

Your next objective is to enable replication for your primary account. For this you need the `organization_name` and `account_name` from Figure 12-9:

```
SELECT system$global_account_set_parameter (
        'NUYMCLU.NP62160',
        'ENABLE_ACCOUNT_DATABASE_REPLICATION',
        'true' );
```

You should see the response shown in Figure 12-10.

```
SYSTEM$GLOBAL_ACCOUNT_SET_PARAMETER ( 'NUYMCLU.NP62160', 'ENABLE_ACCOUNT_DATABASE_REPLICATION', 'TRUE' )     ...
[ "SUCCESS" ]
```

Figure 12-10. *Primary account enabled for replication*

Enable the secondary account for replication. This can only be performed using
ORGADMIN of the primary account. A user on the secondary account cannot be entitled to
use ORGADMIN as the role does not exist:

```
SELECT system$global_account_set_parameter (
        'NUYMCLU.REPLICATION_TERTIARY',
        'ENABLE_ACCOUNT_DATABASE_REPLICATION',
        'true' );
```

You should see SUCCESS for NUYMCLU.REPLICATION_TERTIARY, similar to Figure 12-10.

You must use the ACCOUNTADMIN role to enable and manage database replication and
failover:

```
USE ROLE accountadmin;
```

Your next step is to enable a database for replication to one of the secondary
accounts identified above. In this example, use CATALOG for the primary account
database. We use the account locator PC52900 for our secondary account, noting your
secondary account will differ.

```
SHOW DATABASES;
```

```
ALTER DATABASE catalog ENABLE REPLICATION TO ACCOUNTS NUYMCLU.REPLICATION_
TERTIARY;
```

Replication provides read-only access for the replicated database, not failover.
Confirm replication has been established:

```
SHOW REPLICATION DATABASES;
```

You should see a response similar to Figure 12-11, noting this image reflects a subset of returned information.

account_n	name	is_prima	primary	replication_allowed_to_accounts	...
NP62160	CATALOG	true	NUYMCLU.NP62160.CATALOG	NUYMCLU.NP62160, NUYMCLU.REPLICATION_TERTIARY	

Figure 12-11. *Confirm replication established*

To enable failover to the secondary account:

```
ALTER DATABASE catalog ENABLE FAILOVER TO ACCOUNTS NUYMCLU.REPLICATION_
TERTIARY;
```

Failover provides capability for the secondary database to become the primary and vice-versa.

Replication and failover provide point-in-time consistency for objects on the secondary account according to the status of the latest refresh.

Secondary Account Setup 1

With replication established, you must switch to the secondary account and then import the replicated database. For this, we prefer to use a different browser to keep the primary and secondary environments isolated. If you have not logged into your secondary account, you may need to search your email for the Snowflake confirmation and then log in using the sent link.

Once logged in, change the role to ACCOUNTADMIN and ensure you are using the secondary account:

```
USE ROLE accountadmin;

SELECT current_account();
```

The secondary account locator shown in the result set should match the account_ locator shown in Figure 12-12, which confirms you are logged into the secondary.

CURRENT_ACCOUNT()

PC52900

Figure 12-12. *Secondary account locator confirmation*

Create a database named IMPORT_CATALOG from the primary account database:

```
CREATE DATABASE import_catalog
AS REPLICA OF NUYMCLU.NP62160.catalog
DATA_RETENTION_TIME_IN_DAYS = 90;
```

You should see a response stating, "Database IMPORT_CATALOG successfully created."

Imported databases are read-only and point-in-time.

Check if database IMPORT_CATALOG exists in your replicated databases:

```
SHOW REPLICATION DATABASES;
```

You should now see two rows, as shown in Figure 12-13, reflecting your imported database and noting the status of each database.

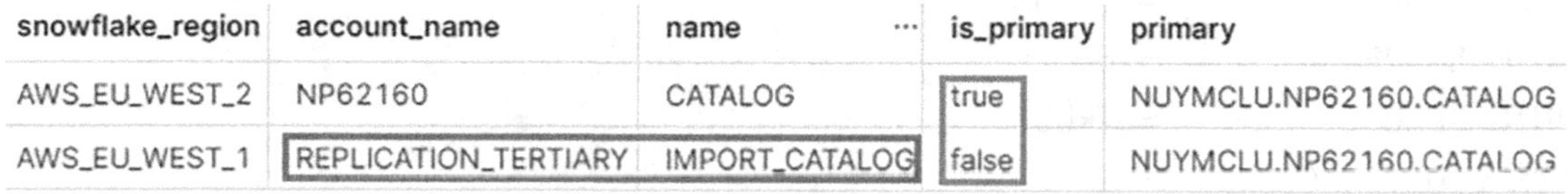

snowflake_region	account_name	name	...	is_primary	primary
AWS_EU_WEST_2	NP62160	CATALOG		true	NUYMCLU.NP62160.CATALOG
AWS_EU_WEST_1	REPLICATION_TERTIARY	IMPORT_CATALOG		false	NUYMCLU.NP62160.CATALOG

Figure 12-13. *Imported database status*

You should also check the presence of IMPORT_CATALOG within your available databases, as shown in Figure 12-14.

Figure 12-14. *Imported database status*

Refreshing the IMPORT_CATALOG database snapshot can be achieved by

```
ALTER DATABASE import_catalog REFRESH;
```

You can automate the refresh by creating a stored procedure but you must first create a database and schema.

Use ACCOUNTADMIN to create your new database as this is the only role that can enable and manage database replication and failover. For consistency, also create a schema and warehouse.

```
CREATE OR REPLACE DATABASE admin_tools DATA_RETENTION_TIME_IN_DAYS = 90;
```

Create a schema:

```
CREATE OR REPLACE SCHEMA    admin_tools.tools_owner;
```

Create a warehouse:

```
CREATE OR REPLACE WAREHOUSE admin_wh WITH
WAREHOUSE_SIZE        = 'X-SMALL'
AUTO_SUSPEND          = 60
AUTO_RESUME           = TRUE
MIN_CLUSTER_COUNT     = 1
MAX_CLUSTER_COUNT     = 4
SCALING_POLICY        = 'STANDARD'
INITIALLY_SUSPENDED   = TRUE;
```

Set the database, schema ,and warehouse context for your stored procedure:

```
USE DATABASE   admin_tools;
USE SCHEMA     tools_owner;
USE WAREHOUSE admin_wh;
```

Create a stored procedure:

```
CREATE OR REPLACE PROCEDURE admin_tools.tools_owner.sp_refresh_database (
P_DATABASE STRING ) RETURNS STRING
LANGUAGE javascript
EXECUTE AS CALLER
AS
```

```
$$
   var sql_stmt  = "ALTER DATABASE " + P_DATABASE + " REFRESH";
   var show_stmt = snowflake.createStatement ({ sqlText:sql_stmt });
   show_res = show_stmt.execute();

   return sql_stmt;
$$;
```

Confirm the stored procedure works:

```
CALL admin_tools.tools_owner.sp_refresh_database ( 'import_catalog' );
```

You should see this SQL statement: ALTER DATABASE import_catalog REFRESH returned.

Check the completion status for database IMPORT_CATALOG's last refresh:

```
SELECT phase_name,
       result,
       start_time,
       end_time,
       details
FROM TABLE ( information_schema.database_refresh_progress ( import_
catalog ));
```

Further information can be found here: https://docs.snowflake.com/en/ sql-reference/functions/database_refresh_progress.html#database-refresh- progress-database-refresh-progress-by-job.

Automate the database IMPORT_CATALOG refresh using a task:

```
CREATE TASK import_catalog_refresh_task
WAREHOUSE           = admin_wh
SCHEDULE            = '10 minute'
USER_TASK_TIMEOUT_MS = 14400000
AS
CALL sp_refresh_database ( 'import_catalog' );
```

By default, all tasks are created in SUSPENDED state, as the output of your next command illustrates:

```
SHOW tasks;
```

Enable the task:

```
ALTER TASK import_catalog_refresh_task RESUME;
```

You should consider adding logging to your stored procedure to provide evidence of the last refresh activity. Bear in mind that the logging will be inaccessible if you are logged to the local region in the event of failure. You must therefore consider replicating logging data into two or more accounts on either different regions within the same CSP or, preferably, across two CSPs for total resilience. We leave it for you to determine your strategy.

With replication established, confirm which is primary. To do this, refer to the is_primary value returned from this SQL command:

```
SHOW REPLICATION DATABASES;
```

You should see the same output as shown in Figure 12-13 (now repeated for your convenience as Figure 12-15) where you see is_primary is false for IMPORT_CATALOG.

snowflake_region	account_name	name	⋯	is_primary	primary
AWS_EU_WEST_2	NP62160	CATALOG		true	NUYMCLU.NP62160.CATALOG
AWS_EU_WEST_1	REPLICATION_TERTIARY	IMPORT_CATALOG		false	NUYMCLU.NP62160.CATALOG

Figure 12-15. *Imported database status*

You now set IMPORT_CATALOG to be primary:

```
ALTER DATABASE import_catalog PRIMARY;
```

And reexamine the IMPORT_CATALOG status, where you see the is_primary status has changed for the originating and replicated databases:

```
SHOW REPLICATION DATABASES;
```

As you can see in Figure 12-16, is_primary now indicates IMPORT_CATALOG is primary and is confirmed by the Primary listing.

snowflake_region	account_name	name	···	is_primary	primary
AWS_EU_WEST_2	NP62160	CATALOG		false	NUYMCLU.REPLICATION_TERTIARY.IMPORT_CATALOG
AWS_EU_WEST_1	REPLICATION_TERTIARY	IMPORT_CATALOG		true	NUYMCLU.REPLICATION_TERTIARY.IMPORT_CATALOG

Figure 12-16. *Primary database status*

You have not considered how IMPORT_CATALOG will be populated and content managed by end user tooling, feeds, and metadata managed by your cataloging tools.

To reset CATALOG to primary, you must switch back to your first account.

As with all sample code, once your testing is complete, you should suspend all tasks:

```
ALTER TASK import_catalog_refresh_task SUSPEND;
```

And clean up when finished:

```
USE ROLE accountadmin;
```

```
DROP TASK import_catalog_refresh_task;
DROP DATABASE import_catalog;
```

Primary Account Setup 2

Reverting back to the primary account, you can see the effect of switching your CATALOG database from your secondary account:

```
SHOW REPLICATION DATABASES;
```

Figure 12-17 repeats Figure 12-16, demonstrating the metadata for both primary and secondary is consistent.

snowflake_region	account_name	name	···	is_primary	primary
AWS_EU_WEST_2	NP62160	CATALOG		false	NUYMCLU.REPLICATION_TERTIARY.IMPORT_CATALOG
AWS_EU_WEST_1	REPLICATION_TERTIARY	IMPORT_CATALOG		true	NUYMCLU.REPLICATION_TERTIARY.IMPORT_CATALOG

Figure 12-17. *Primary database status*

Reverting to your original state, you now switch the primary database back:

```
ALTER DATABASE catalog PRIMARY;
```

Confirm you have successfully reverted back:

```
SHOW REPLICATION DATABASES;
```

Figure 12-18 shows the is_primary flag and primary information has reverted correctly.

snowflake_region	account_name	name	...	is_primary	primary
AWS_EU_WEST_2	NP62160	CATALOG		true	NUYMCLU.NP62160.CATALOG
AWS_EU_WEST_1	REPLICATION_TERTIARY	IMPORT_CATALOG		false	NUYMCLU.NP62160.CATALOG

Figure 12-18. *Primary database status*

Disable Database Replication

Disabling replication is achieved by issuing this SQL statement:

```
ALTER DATABASE catalog DISABLE REPLICATION;
```

Removing an account from allowable replicated databases requires the use of the ORGADMIN role:

```
USE ROLE orgadmin;

SELECT system$global_account_set_parameter (
        'NUYMCLU.NP62160',
        'ENABLE_ACCOUNT_DATABASE_REPLICATION',
        'false' );
```

Check the replicated databases:

```
SHOW REPLICATION DATABASES;
```

You should no longer see an entry for NUYMCLU.NP62160.

Database Replication Summary

From the limitations identified above, you can see there may be significant challenges to implementing database replication. Some of the limitations will prevent database replication from becoming a viable approach to implementing DR, while other limitations may either force rework in the primary database or additional engineering to create a clean database before replication becomes possible.

Recognizing Snowflake is continually improving database replicated features, we recommend all development teams be made aware of database replication limitations in order to develop their applications using techniques to mitigate the risk of replication failure.

All secondary databases are read-only, and for those switched to become primary, there may be manual configuration work required to become fully functional and automated. Each scenario will differ: we strongly recommend you prepare by practicing DR for each replicated database.

Database replication is only part of a DR plan. You must also consider external tooling connectivity, user interaction, and metadata management by your cataloging tooling. We recommend for each database where database replication and failover are considered, a single diagram articulating all the touchpoints is developed because every component needs to be properly considered and tested before a real-world scenario forces DR invocation.

You must also consider how to propagate database refresh logs to ensure resilience in the event of account or region failure as these logs will be important for informing how to recover a database and replay missing information. Furthermore, if a region fails, you must assume all external stages have become unavailable. You must therefore consider how to identify and replay missing files from external tooling into your new primary database.

Further information on database replication can be found at `https://docs.snowflake.com/en/user-guide/database-replication.html#database-replication` and `https://docs.snowflake.com/en/user-guide/database-replication-considerations.html#database-replication-considerations`.

Account Replication

At the time of writing, Account Replication and Failover is in public preview so the functionality outlined within this section must be considered subject to change by Snowflake.

Account replication and failover must be enabled for primary and all secondary accounts.

In contrast to database replication, where a single database is replicated, account replication occurs at the account level. That is, everything within the account is replicated, with some exceptions noted below. We discuss account replication by reusing the same patterns used for database replication.

We strongly recommend you investigate the Snowflake documentation before defining your DR approach.

Figure 12-19 illustrates account replication. The important point is to note is the cross-region nature of replication. In the example shown, we use AWS and Azure, showing that cross CSP replication is possible and may be preferred.

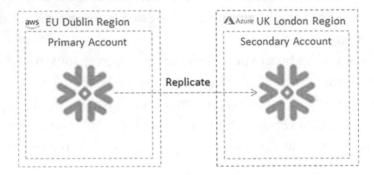

Figure 12-19. *Account replication conceptual representation*

You use account replication to mitigate against the failure of an entire CSP region, that is, the failure of all HA zones. Account replication is also used to mitigate against the failure of multiple regions within a single CSP. In this latter scenario, you would implement account replication across two or three CSPs.

When discussing account replication, we refer to the master account as the primary and all replicated accounts as the secondary. There may be more than one secondary account.

Account replication is serverless and does not rely upon a warehouse being declared or available.

Failover is manually invoked; there is no automated failover option. You remain responsible for external tooling connectivity, which is discussed later in this chapter.

Further information can be found here: `https://docs.snowflake.com/en/user-guide/account-replication-failover.html#account-replication-and-failover`.

Single Region Failure

We define single region failure as the loss of service within a region where all HA zones are unavailable. Your objective is to protect against the total loss of your application service to your customers, thus ensuring business continuity.

Mitigating against the loss of a single region within a single CSP, you have two options:

- Same-CSP replication to another region

- Cross-CSP replication to another region

If your DR choice is account replication within the same CSP, then the same code-base and integration/implementation patterns will be available to deploy across to the new region albeit with new trust relationships and CSP account tooling integrations (AWS Secrets Manager, AWS Key Management Service, etc.).

Alternatively, if your DR choice is account replication across different CSPs, then your code-base and integration patterns become more challenging when implementing equivalence to the primary. From a Snowflake perspective, new trust relationships and CSP account tooling integrations must be implemented along with new CSP account tooling integrations providing comparable capability to the primary (AWS Secrets Manager, AWS Key Management Service, etc.). You must also ensure your documentation is maintained to reflect the CSP-specific implementation details.

Multi-Region Failure

We define multi-region failure as the loss of service across two or more regions within the same CSP. Your objective is to protect against the total loss of your application service to your customers ensuring business continuity.

With this in mind, cross-CSP account replication is your only option to mitigate risk, noting your code-base and integration patterns become more challenging when implementing equivalence to the primary. We outlined a few considerations in the previous section.

You must also consider region groups, such as North America and Europe. These two region groups contain multiple Snowflake regions within their respective geographic areas. You may choose to store data in the North America region group to colocate with consumers located in North America, or within the Europe region group if most of your consumers are located in Europe.

Within each region group, you have the option to choose which specific Snowflake region to create your account in. Boundaries become a little blurred with regard to Europe since the United Kingdom left the European Union. Further information can be found here: `https://docs.snowflake.com/en/user-guide/admin-account-identifier.html#region-groups`.

Replicated Objects

The list of supported objects is evolving and the lists provided below should be checked before assuming correctness.

Currently Available Objects

At the time of writing, these objects are replicated within an account:

- Databases (see "Database Replication" above)

- Warehouses

- Shares

- Users

- Roles (RBAC; database roles to follow)

- Resource monitors

- Integrations

  - Security: SAML2, SCIM, OAuth (see link below)

  - API

  - Storage (to follow)

- Parameters

- Network policies

- Replication groups/failover groups (addressed separately below)

This list is not exhaustive. Please refer to the latest list of replicated account objects at `https://docs.snowflake.com/en/user-guide/account-replication-intro.html#replicated-objects`.

For security integration replication across accounts, please refer to the documentation found here: https://docs.snowflake.com/en/user-guide/account-replication-security-integrations.html#replication-of-security-integrations-network-policies-across-multiple-accounts.

Future Available Objects

These objects are not currently replicated within an account. Please check with your Snowflake Sales Engineer as replication for many of the below objects is in-flight.

- Roles (RBAC; database roles)
- Integrations (storage)

Unavailable Objects and Entitlements

While most object types are replicated within an account, there are some restrictions and notable exceptions:

- Reader accounts
- Network policy entitlements
- Integration entitlements

For each of the above, you may need to implement localized scripts to remediate.

This list is not exhaustive. Please refer to the latest list of replicated account objects at https://docs.snowflake.com/en/user-guide/account-replication-intro.html#introduction-to-account-replication-and-failover.

Excluded Objects

Some objects are explicitly identified as excluded from account replication as they exceed the scope of this chapter and are mentioned to prompt your further investigation:

- Databases imported from shares
- Tri-Secret Secure/Bring Your Own Key (BYOK)
- AWS Lambdas and Azure/GCP equivalents
- CSP account-specific tooling

With regard to Tri-Secret Secure, Snowflake documentation states replicated accounts do not need to share the same key. We therefore understand each secondary account may have its own local key to implement Tri-Secret Secure. Please contact your Sales Engineer for more details. Documentation can be found here: `https://docs.snowflake.com/en/user-guide/database-replication-intro.html#database-replication-and-encryption`.

Manual Configuration

We discussed the creation of a locally created monitoring capability earlier in this book. If implemented, the share usage must be updated to reflect that consumption is from a new primary account and any SDDS to a central monitoring store must be checked to ensure statistics are delivered correctly.

As indicated previously in this section, manual configuration steps must be performed to ensure both conformance to standards, and continuity of service. As an example and discussed later within the "Holistic Approach" section, your network policies may need amending for CSP-hosted tooling.

Account Replication Limitations

During replication, a check is made on the ownership of any new objects within the primary which are replicated to all secondary accounts. If the corresponding roles are not replicated to the secondary, then object ownership reverts to `ACCOUNTADMIN`.

However, for roles replicated from the primary account to the secondary accounts, future grants are also replicated. This includes future grants on currently unsupported replication objects.

Privilege grants are replicated for all account objects except for

- Network policies

- Integrations

When a primary account is replicated to a secondary, every object in the secondary is overwritten. Any local roles or users are deleted when the secondary account is refreshed. To avoid this scenario, ensure all objects are created in the primary for refresh to every secondary.

Account Replication Step by Step

For this walkthrough, we assume the code developed in Chapters 7 and 8 has been deployed and you have two or more Snowflake accounts available within our Snowflake organization.

Secondary accounts cannot be administered in the same way as primary accounts unless they become primary accounts.

Further information on developing overall approach to account replication can be found here: `https://docs.snowflake.com/en/user-guide/account-replication-config.html#replicating-account-objects`.

In this section, the headings reflect where code should be executed and do not refer to more than a single primary or secondary account. The numbering is used to indicate sequencing of usage, nothing more.

We only use two accounts, referenced as primary and secondary.

Replication Groups/Failover Groups

Replication groups support both full account failover and multi-tenant occupancy of a single account where you may prefer to replicate a specific subset of components in support of individual RTO and RPO objectives. Your DR strategy may not require an "all or nothing" approach. Replication groups and failover groups allow you to achieve both DR and BCP objectives in a selective manner.

Replication groups implement the capability to ring fence subsets of objects but are reliant on strict isolation of objects. If, for example, two replication groups rely upon common reference data held within a separate database, then each replication group DR plan must consider how they will address the dependency. Replication groups provide read-only access to the replicated objects and are the preferred solution to implement DR for multi-tenant accounts.

Figure 12-20 illustrates two replication groups, labelled as Tenant 1 and Tenant 2, each containing isolated groups of objects. Replication groups can span different cloud service provider (CSP) regions; in the example shown, we use AWS and Azure, showing cross CSP replication is possible and may be preferred for greater resilience.

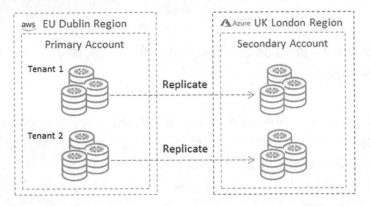

Figure 12-20. *Replication group/failover group conceptual representation*

Replication groups provide read-only access to replicated objects with the primary providing full transactional capability and the secondary providing read-only capability.

Failover groups provide the means for a secondary replication group to become the primary and to failback as required. While we illustrate a simple scenario in Figure 12-20, it should be noted multiple secondaries can be provisioned, and any one of the secondaries can become the primary.

Further information can be found here: `https://docs.snowflake.com/en/`
`user-guide/account-replication-intro.html#replication-groups-and-`
`failover-groups`.

Replication Group Limitations

Databases and shares can only belong to a single replication group. A database or share cannot belong to more than one replication group except where the replication group is replicated to different secondary accounts.

Account-level objects can only belong to a single replication group, which must contain every in-scope account level object. This represents an all-or-nothing proposition; when invoked, the account fails over.

Replicated objects within a secondary account cannot be added to a primary replication group.

Primary Account Setup 1

Prove you have two or more accounts within your Snowflake organization:

```
USE ROLE accountadmin;

SHOW REPLICATION ACCOUNTS;
```

Figure 12-21 shows sample replication accounts available within your Snowflake organization, noting the is_org_admin is set to true for the account used to administer your organization.

snowflake_region	account_name	account_locator	comment	organization_na	is_org_admin
AWS_EU_WEST_2	NP62160	XL29287	Created by Signup Service	NUYMCLU	true
AWS_EU_WEST_1	REPLICATION_TERTIARY	PC52900	SNOWFLAKE	NUYMCLU	false

Figure 12-21. *Available replication accounts*

In our example, NP62160 is the primary and REPLICATION_TERTIARY is the secondary.

To prove you are using the primary, fetch the Snowflake region and account locator:

```
SELECT current_region(), current_account();
```

Figure 12-22 shows the expected response for the account, confirming the use of the primary account, noting your response will differ.

CURRENT_REGION()	CURRENT_ACCOUNT()
AWS_EU_WEST_2	XL29287

Figure 12-22. *Current Snowflake region/account locator*

Having confirmed you are connected to the correct Snowflake account, you now set up replication groups and failover groups.

Before adding databases to a replication group or failover group, you must disable all database replication. In this example, you have a single database named catalog, and regardless of whether replication is enabled or not, you should adopt a "belt and braces" approach by explicitly disabling replication for every database in your account:

```
SELECT system$disable_database_replication('catalog');
```

You should see a response similar to Figure 12-23.

SYSTEM$DISABLE_DATABASE_REPLICATION('CATALOG')

Successfully disabled replication on database 'CATALOG' and secondary databases

Figure 12-23. *Disabled database replication confirmation*

Then prove all replication databases have been disabled:

```
SHOW REPLICATION DATABASES;
```

You should not see any results from the above query. If results are found, disable each database and recheck before proceeding.

Now create a failover group named `tenant_1` containing a single named database catalog.

Amend the replication frequency according to your requirements:

```
CREATE FAILOVER GROUP tenant_1
OBJECT_TYPES         = account parameters, databases,
                       network policies,
                       resource monitors,
                       roles, users, warehouses
ALLOWED_DATABASES    = catalog
ALLOWED_ACCOUNTS     = NUYMCLU.REPLICATION_TERTIARY
REPLICATION_SCHEDULE = '10 MINUTE';
```

Adding `OBJECT_TYPES` of shares and integration also requires the corresponding object names to be declared:

- To failover shares, add `ALLOWED_SHARES = <your share name here>`.

- To failover integrations, add `ALLOWED_INTEGRATION_TYPES = <your integration type here>`.

You should see a response indicating failover group TENANT_1 has been successfully created.

During the preparation of this chapter, Snowflake released support for new OBJECT_TYPES. In the event of a CREATE FAILOVER GROUP error, check Snowflake documentation.

Supported OBJECT_TYPES can be found here: https://docs.snowflake.com/en/sql-reference/sql/create-failover-group.

Check that your failover group exists:

```
SHOW FAILOVER GROUPS;
```

You should see a response similar to Figure 12-24.

account_n	name	type	is_prir	primary	object_types	allowed_accounts
NP62160	TENANT_1	FAILOV	true	NUYMCLU.NP62160.TEN	DATABASES, ROLES	NUYMCLU.NP62160, NUYMCLU.REPLICATION_TERTI

Figure 12-24. *Failover group creation confirmation*

Further detail on failover groups can be found here: https://docs.snowflake.com/en/sql-reference/sql/create-replication-group.html#create-replication-group.

Secondary Account Setup 1

Switching your focus to the secondary account, you now create a replica of failover group named TENANT_1:

```
USE ROLE accountadmin;

CREATE FAILOVER GROUP replica_tenant_1
AS REPLICA OF NUYMCLU.NP62160.tenant_1;
```

Check your failover group exists:

```
SHOW FAILOVER GROUPS;
```

You should see a response similar to Figure 12-25 where both primary and secondary accounts are shown.

snowflake_region	account_name	name	type	comment	is_primary	primary	object_types
AWS_EU_WEST_2	NP62160	TENANT_1	FAILOVER	NULL	true	NUYMCLU.NP62...	DATABASES, ROL...
AWS_EU_WEST_1	REPLICATION_TERTIARY	REPLICA_TENANT_1	FAILOVER	NULL	false	NUYMCLU.NP62...	

Figure 12-25. *Replication failover group creation confirmation*

Global Identifiers

Global identifiers are unique object identifiers used within a Snowflake organization to match primary and secondary account objects with the same name. An account object must be included in a replication or failover group and is limited to

- Resource monitors

- Roles

- Users

- Warehouses

To implement global identifiers for replica failover group REPLICA_TENANT_1:

```
SELECT system$link_account_objects_by_name('replica_tenant_1');
```

You should see a JSON object in response that looks like this:

```
{
  "numRolesLinked" : 7,
  "numUsersLinked" : 2,
  "numWarehousesLinked" : 2,
  "numResourceMonitorsLinked" : 0
}
```

You use global identifiers to prevent objects from the above list, which preexist in the secondary, from being dropped during refresh. Objects created in the secondary by any means other than replication have no global identifier assigned by default as there is no equivalent object in the primary. When the secondary is refreshed from the primary, all objects without a global identifier of the type listed above will be dropped.

For further information on applying global identifiers please refer to documentation found at https://docs.snowflake.com/en/sql-reference/functions/system_link_ account_objects_by_name.html and https://docs.snowflake.com/en/user-guide/ account-replication-config.html#step-5-apply-global-ids-to-objects-created- by-scripts-in-target-accounts-optional.

Primary Account Setup 2

Reverting to your primary account, if you have not already done so, you must enable replication for all accounts in your Snowflake organization. To do so, switch roles to ORGADMIN and check whether replication is active:

```
USE ROLE orgadmin;
```

Enable both primary and secondary accounts for replication:

```
SELECT system$global_account_set_parameter (
        'NUYMCLU.NP62160',
        'ENABLE_ACCOUNT_DATABASE_REPLICATION',
        'true' );
```

```
SELECT system$global_account_set_parameter (
        'NUYMCLU.REPLICATION_TERTIARY',
        'ENABLE_ACCOUNT_DATABASE_REPLICATION',
        'true' );
```

Enable stream and task replication for the primary account:

```
USE ROLE accountadmin;
```

```
ALTER ACCOUNT NUYMCLU.NP62160
SET enable_stream_task_replication = true;
```

Please also refer to the documentation found here: https://docs.snowflake.com/ en/sql-reference/parameters.html#enable-stream-task-replication.

Secondary Account Setup 2

Refresh Secondary

To refresh the secondary account, you must refresh the failover group:

```
USE ROLE accountadmin;

ALTER FAILOVER GROUP replica_tenant_1 REFRESH;
```

You must set a warehouse before running a query:

```
USE WAREHOUSE compute_wh;
```

You can examine the last refresh details using this query:

```
SELECT phase_name,
       start_time,
       end_time,
       progress,
       details
FROM TABLE (snowflake.information_schema.replication_group_refresh_
progress('replica_tenant_1'));
```

You should see results similar to those shown in Figure 12-26.

PHASE_NAME	START_TIME	END_TIME	PROGRESS	DETAILS
SECONDARY_SYNCHRONIZING_MEMB...	2023-02-07 08:23:01.66...	2023-02-07 08:23:11.1...	NULL	NULL
SECONDARY_UPLOADING_INVENTORY	2023-02-07 08:23:11.17...	2023-02-07 08:23:12....	NULL	NULL
PRIMARY_UPLOADING_METADATA	2023-02-07 08:23:12.22...	2023-02-07 08:23:14....	NULL	{ "primarySnapshotTimestamp": 1675
PRIMARY_UPLOADING_DATA	2023-02-07 08:23:14.38...	2023-02-07 08:23:15....	100%	{ "bytesUploaded": 0, "databases": [
SECONDARY_DOWNLOADING_METADA...	2023-02-07 08:23:15.13...	2023-02-07 08:23:15....	100%	{ "completedObjects": 119, "objectTyp
SECONDARY_DOWNLOADING_DATA	2023-02-07 08:23:15.75...	2023-02-07 08:23:16....	100%	{ "bytesDownloaded": 0, "databases"
COMPLETED	2023-02-07 08:23:16.30...	NULL	NULL	NULL

Figure 12-26. *Replication refresh progress*

This query shows the refresh history for the past 7 days:

```
SELECT phase_name,
       start_time,
       end_time,
       total_bytes,
       object_count
FROM TABLE (snowflake.information_schema.replication_group_refresh_
history('replica_tenant_1'))
WHERE  start_time >= current_date - interval '7 days';
```

Figure 12-27 shows refreshes occurring every 10 minutes as specified during replication group configuration.

PHASE_NAME	START_TIME	END_TIME	TOTAL_BYTES	OBJECT_COUNT
COMPLETED	2023-02-07 10:04:02.17...	2023-02-07 10:04:11.311 -0800	{ "bytesDownloaded": 0, "b:	{ "completedObjects": 119, "obje
COMPLETED	2023-02-07 09:53:55.00...	2023-02-07 09:54:03.441 -0800	{ "bytesDownloaded": 0, "b:	{ "completedObjects": 119, "obje
COMPLETED	2023-02-07 09:43:45.00...	2023-02-07 09:43:53.216 -0800	{ "bytesDownloaded": 0, "b:	{ "completedObjects": 119, "obje
COMPLETED	2023-02-07 09:33:41.47...	2023-02-07 09:33:49.495 -0800	{ "bytesDownloaded": 0, "b:	{ "completedObjects": 119, "obje
COMPLETED	2023-02-07 09:23:33.33...	2023-02-07 09:23:40.230 -0800	{ "bytesDownloaded": 0, "b:	{ "completedObjects": 119, "obje
COMPLETED	2023-02-07 09:13:25.00...	2023-02-07 09:13:33.013 -0800	{ "bytesDownloaded": 0, "b:	{ "completedObjects": 119, "obje
COMPLETED	2023-02-07 09:03:17.64...	2023-02-07 09:03:25.645 -0800	{ "bytesDownloaded": 0, "b:	{ "completedObjects": 119, "obje

Figure 12-27. *Replication refresh history*

Refresh Costs

This query shows the refresh costs for the past 7 days:

```
SELECT start_time,
       end_time,
       replication_group_name,
       credits_used,
       bytes_transferred
FROM TABLE (snowflake.information_schema.replication_group_usage_history())
WHERE  start_time >= current_date - interval '7 days';
```

Figure 12-28 shows replication costs per hour for the past 7 days.

START_TIME	END_TIME	REPLICATION_GROUP_NAME	CREDITS_USED	BYTES_TRANSFERRED
2023-02-07 04:20:12.000 -0800	2023-02-07 05:20:12.000 -08...	REPLICA_TENANT_1	0.000436875	0
2023-02-07 05:20:12.000 -0800	2023-02-07 06:20:12.000 -08...	REPLICA_TENANT_1	0.001639169	0
2023-02-07 06:20:12.000 -0800	2023-02-07 07:20:12.000 -08...	REPLICA_TENANT_1	0.002010106	0
2023-02-07 07:20:12.000 -0800	2023-02-07 08:20:12.000 -08...	REPLICA_TENANT_1	0.001889481	0
2023-02-07 08:20:12.000 -0800	2023-02-07 09:20:12.000 -08...	REPLICA_TENANT_1	0.001562138	0
2023-02-07 09:20:12.000 -0800	2023-02-07 10:20:12.000 -0800	REPLICA_TENANT_1	0.001677137	0

Figure 12-28. *Replication costs*

All of the above metrics should be captured centrally for each of the secondary accounts refreshed.

We have exposed several metrics, although more exist. We recommend thorough reading of the documentation found here: `https://docs.snowflake.com/en/user-guide/account-replication-config.html#replicating-account-objects`.

Promote Secondary to Primary

Having proven that your secondary is periodically refreshed, you now promote your secondary to primary. You recheck your failover groups to confirm `TENANT_1` is primary as indicated by `is_primary` being set to `true`:

```
SHOW FAILOVER GROUPS;
```

You should see a response similar to Figure 12-29, which confirms `TENANT_1` is the primary.

snowflake_region	account_name	name	type	comment	is_primary	primary	object_types
AWS_EU_WEST_2	NP62160	TENANT_1	FAILOVER	NULL	true	NUYMCLU.NP62...	DATABASES, ROL...
AWS_EU_WEST_1	REPLICATION_TERTIARY	REPLICA_TENANT_1	FAILOVER	NULL	false	NUYMCLU.NP62...	

Figure 12-29. *Primary replication failover group confirmation*

```
ALTER FAILOVER GROUP replica_tenant_1 PRIMARY;

SHOW FAILOVER GROUPS;
```

Figure 12-30 shows the expected response where you see REPLICA_TENANT_1 is now the primary. Note the new primary is in a suspended state (not shown).

snowflake_region	account_name	name	type	comment	is_primary	primary	object_types
AWS_EU_WEST_2	NP62160	TENANT_1	FAILOVER	NULL	false	NUYMCLU.NP62...	
AWS_EU_WEST_1	REPLICATION_TERTIARY	REPLICA_TENANT_1	FAILOVER	NULL	true	NUYMCLU.NP62...	DATABASES, ROL...

Figure 12-30. *Secondary to primary replication failover group confirmation*

With your new primary active, you can conduct general administration tasks previously unavailable to a secondary account. One such administrative task is to change passwords of users or create new users in accordance with your cyber security approved profile.

Many other tasks require your attention, which we discuss within the DR plan below.

Primary Account Setup 3

With your accounts switched, you revert back to the old primary account where you check failure group status:

```
SHOW FAILOVER GROUPS;
```

You should see the same response as in Figure 12-30, confirming your account is no longer the primary but noting the TENANT_1 secondary_state (not shown) is SUSPENDED.

To reinstate account replication, you must resume the service:

```
ALTER FAILOVER GROUP TENANT_1 RESUME;
```

Then confirm the TENANT_1 secondary_state (not shown) is STARTED.

```
SHOW FAILOVER GROUPS;
```

You now revert the new primary to become the secondary, thus restoring the original state:

```
ALTER FAILOVER GROUP tenant_1 PRIMARY;
```

Check that your replica group status is as expected:

```
SHOW FAILOVER GROUPS;
```

You should see TENANT_1 is_primary is true and REPLICA_TENANT_1 secondary_state (not shown) is SUSPENDED.

Secondary Account Setup 3

Switching to your secondary, you must reenable replication:

```
ALTER FAILOVER GROUP replica_tenant_1 RESUME;
```

And confirm REPLICA_TENANT_1 secondary_state (not shown) is STARTED:

```
SHOW FAILOVER GROUPS;
```

Account Replication Summary

Using the refresh history provides information on when the most recent refresh completed for each refresh group. From these date stamps, you can determine how stale your secondary is, and when combined with other external tooling logs, should be able to determine compensating recovery actions.

When your original primary site is restored, you must prove both primary and secondary account data sets are consistent. We recommend the same checks are carried out across all secondaries, particularly for those accounts used for client-facing read-only service.

To ensure transactional consistency, follow these steps:

1. On the secondary account, fetch the timestamp of the latest complete refresh.

2. For a single table, fetch the HASH_AGG value for all rows.

3. On the primary account, using the same table, fetch the HASH_AGG value for all rows for the timestamp returned in step 1.

4. Compare the HASH_AGG values.

If the HASH_AGG values are the same, then the table contents are equivalent. If the HASH_AGG values differ, the table contents differ.

Please also refer to the documentation found here: `https://docs.snowflake.`
`com/en/user-guide/account-replication-config.html#comparing-data-sets-in-`
`primary-and-secondary-databases.`

Client Redirect

Historically, every time an application failed, all inbound and outbound connections
were (usually manually) reconfigured to point to the new application location. Only
when all service affecting connections were reinstated could the application resume
service. Historically, and for highly complex environments, both RTO and RPO have
been impacted by an inability to centrally redirect application connections.

Snowflake provides a centralized mechanism to redirect client applications to a new
failed over primary within seconds. In this section we walk through how to implement
Client Redirect.

Documentation on Client Redirect may be found here: `https://docs.snowflake.`
`com/en/user-guide/client-redirect.html#label-intro-to-client-redirect.`

Redirect Scenarios

We show two examples where a third party service connects to a primary account.
Figure 12-31 illustrates a service running on an AWS account within the same region
as our primary account. In this example, the region is entirely dependent upon the HA
zones for continuity of service, and Client Redirect will not protect our application in the
event of region failure.

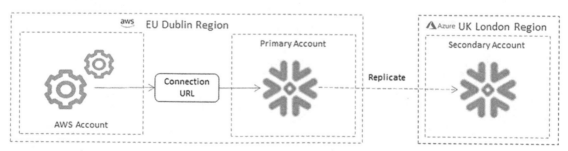

Figure 12-31. *Same region connection*

In the scenario depicted within Figure 12-31, we would be faced with losing two
components, which will complicate our efforts to recover.

In contrast, Figure 12-32 shows the same AWS account but this time provisioned on a different CSP region

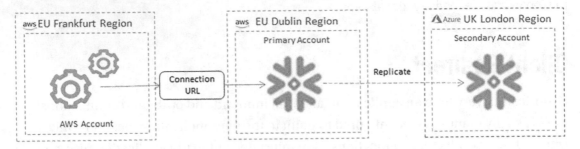

Figure 12-32. *Multi-region connection*

You can readily see our services are isolated and the loss of a single component is much more easily resolved, with two possible outcomes:

- The loss of the AWS account hosting the external service in the Frankfurt region will be handled with its own DR plan.

- The loss of AWS Dublin region will be handled by failover to Azure London region by Client Redirect.

We might reasonably expect multiple services, hosted on disparate environments, making connections into our Snowflake account. A slightly more realistic scenario is shown in Figure 12-33 where we have inbound data from our AWS account service and outbound PowerBI reporting, hosted on a multi-user Azure O365 account in the North Europe region.

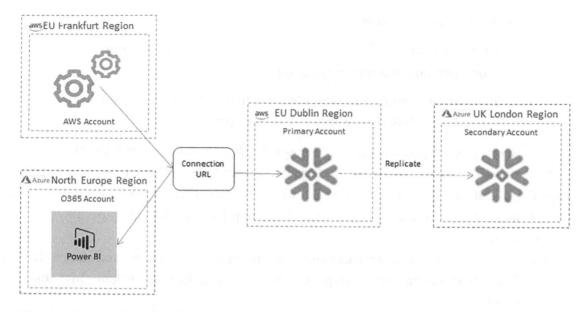

Figure 12-33. *Multi-CSP connections before failover*

Should an incident occur necessitating account failover, our secondary account is switched to become our primary and the old primary become unavailable, as shown in Figure 12-34.

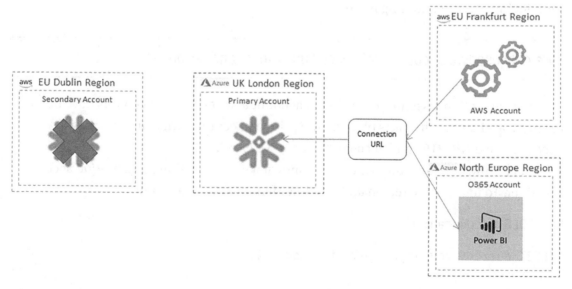

Figure 12-34. *Multi-CSP connections after failover*

Failover is a two-stage process:

- We must determine which of our secondary accounts is to become the new primary account and then invoke failover.

- When our new primary account is established, we must invoke Client Redirect to establish connectivity to the new primary account.

As you can see, the primary account is now on Azure UK London, with the old primary now becoming the failed secondary. We can only have a single primary account, regardless of how many secondaries exist. Note replication ceases between the old primary account and new primary account. This is implicit with switching to a new primary account.

Despite the presence of several inbound and outbound connections, you can easily see how Client Redirect provides a simple mechanism for delivering resilience into your Snowflake estate.

Client Redirect Step By Step

In this section, the headings reflect where code should be executed and do not refer to more than a single primary or secondary account. The numbering is used to indicate sequencing of usage, nothing more.

We only use two accounts, referenced as primary and secondary.

For this section, assume your primary account and secondary accounts have been restored, and all activity takes place on the primary account. which in our case is `account_name` NP62160 and `account_locator` XL29287.

Every Snowflake account within your Snowflake organization has a unique account identifier. To identify the region and account identifier for an account:

```
USE ROLE accountadmin;
```

```
SELECT current_region(), current_account();
```

You should see a response containing two values:

- The CSP region where the account is located (in our example AWS_ EU_WEST_2, noting yours may differ)

- The unique account identifier (mine is XL29287)

These two pieces of information are usually sufficient to identify an account and allow the URL to be constructed. Using the above information, the URL would look like this:

xl29287.eu-west-2.aws.snowflakecomputing.com

As an aside, the URL converts to

https://xl29287.eu-west-2.aws.snowflakecomputing.com/console/login#/

For legacy reasons, these two pieces of information may not be sufficient and extended information may be needed. Please refer to documentation found here: https://docs.snowflake.com/en/user-guide/admin-account-identifier.html#non-vps-account-locator-formats-by-cloud-platform-and-region

Why is this important?

Configuring Client Redirect requires the creation of a CONNECTION object.

You can check whether any CONNECTION objects have been declared by issuing this SQL command:

SHOW CONNECTIONS;

You should not see any CONNECTION objects returned.

Next, you require the account_name of all accounts enabled for failover:

SHOW REPLICATION ACCOUNTS;

Both the account_name and account_locator are highlighted within the returned result set shown in Figure 12-35.

snowflake_region	account_name	account_locator	comment	organization_name	is_org_admin
AWS_EU_WEST_2	NP62160	XL29287	Created by Signup Service	NUYMCLU	true
AWS_EU_WEST_1	REPLICATION_TERTIARY	PC52900	SNOWFLAKE	NUYMCLU	false

Figure 12-35. *Account name and locator*

Please also refer to the documentation at `https://docs.snowflake.com/en/user-guide/admin-account-identifier.html#account-identifiers` and `https://docs.snowflake.com/en/user-guide/admin-account-identifier.html#using-an-account-locator-as-an-identifier`.

Primary Account Setup 1

Ensuring you are connected to the primary account, you now create a `CONNECTION` object. In this example, use the `organization_name` and `account_locator` to identify the `CONNECTION` object. Regardless of the naming convention used, the `CONNECTION` object name must be unique within your Snowflake organization.

```
CREATE CONNECTION IF NOT EXISTS nuymclu_xl29287;
```

You should receive a response confirming the CONNECTION has been created.

Attempting to set up failover to your `REPLICATION_TERTIARY` will result in an error:

```
ALTER CONNECTION nuymclu_xl29287
ENABLE FAILOVER TO ACCOUNTS nuymclu.REPLICATION_TERTIARY;
```

You cannot enable failover to an account in the same region, attempting to do so results in this error: `The connection cannot be failed over to an account in the same region`. This is logical and consistent with our earlier explanation of regional outage within the "Redirection Scenarios" section.

To fix this issue, switch the role to `ORGADMIN` and then create a new account in a different region to the primary; please revert to Chapter 8 for details.

You should now see your `CONNECTION` has been created:

```
SHOW CONNECTIONS;
```

Figure 12-36 shows the important connection information from the result set.

name	is_prim	primary	failover_allowed_to_accounts	connection_url
NUYMCLU_XL29:	true	NUYMCLU.NP62160.NUYMCLU_XL:	NUYMCLU.NP62160, NUYMCLU.REPLICATION_TER	nuymclu-nuymclu_xl29287.snow

Figure 12-36. *Connection created*

Please also refer to the documentation found here: `https://docs.snowflake.com/en/user-guide/client-redirect.html#create-a-primary-connection`.

Secondary Account Setup 1

Within your secondary account, confirm the same information can be seen, as shown in Figure 12-36 above:

```
USE ROLE accountadmin;

SHOW CONNECTIONS;
```

Within your secondary account you must create a secondary connection linked to the primary connection. The connection names must match, so use the same name as the primary connection:

```
CREATE CONNECTION nuymclu_xl29287
AS REPLICA OF nuymclu.np62160.nuymclu_xl29287;
```

Then confirm you have two connections of the same name:

```
SHOW CONNECTIONS;
```

Figure 12-37 shows the new connection information from the result set, noting the is_primary setting for each connection.

account_name	is_primary	primary	connection_url
NP62160	true	NUYMCLU.NP62160.NUYMCLU_XL29287	nuymclu-nuymclu_xl29287.snowflakecomputing.com
REPLICATION_TERTIARY	false	NUYMCLU.NP62160.NUYMCLU_XL29287	nuymclu-nuymclu_xl29287.snowflakecomputing.com

Figure 12-37. *Connection created*

Please also refer to the documentation found here: https://docs.snowflake.com/en/user-guide/client-redirect.html#create-a-secondary-connection.

Private Connectivity

Private Connectivity relates to the use of AWS PrivateLink, Azure Private Link, and GCP Cloud Private Service Connect.

If your organization is not using Private Connectivity, Snowflake manages the CNAME record.

If Private Connectivity is in use, a DNS CNAME record must also be created and managed for your connection URL. For further information, refer to the documentation found here: `https://docs.snowflake.com/en/user-guide/client-redirect.html#label-client-redirect-configuring-the-dns-cname-record`.

Redirecting Client Connections

Figure 12-38 illustrates the example scenario by repeating Figure 12-33 where you have inbound data from your AWS account service and outbound PowerBI reporting hosted on a multi-user Azure O365 account in the North Europe region.

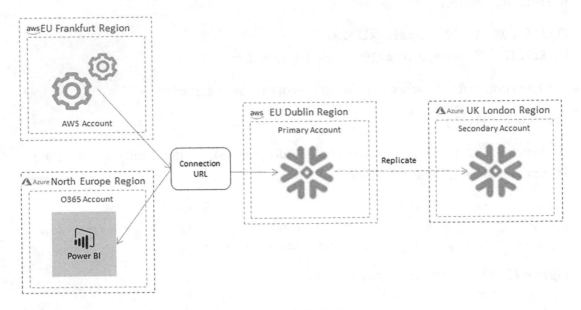

Figure 12-38. *Multi-CSP connections before failover*

The test scenario represents total service outage in the EU Dublin Region where all HA zones have been lost, causing service loss from the primary account.

Figure 12-39 shows the desired target state where Client Redirect has been invoked and service restored.

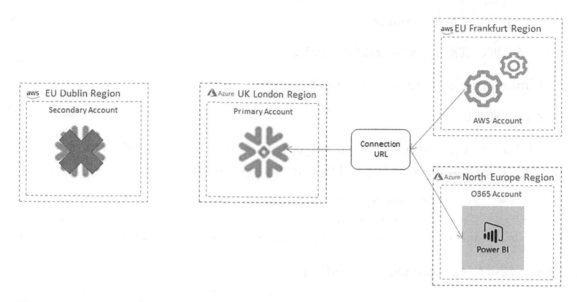

Figure 12-39. *Multi-CSP connections after failover*

As the primary account is unavailable, we now walk through the steps to invoke Client Redirect by connecting to the secondary account.

Secondary Account Setup 2

Having configured connections for both primary and secondary, prepare to switch connections:

```
USE ROLE accountadmin;
```

```
SHOW CONNECTIONS;
```

Confirm your primary `account_name` (NP62160) and `is_primary` (true) attributes are correctly set, as shown in Figure 12-40.

account_name	is_primary	primary	connection_url
NP62160	true	NUYMCLU.NP62160.NUYMCLU_XL29287	nuymclu-nuymclu_xl29287.snowflakecomputing.com
REPLICATION_TERTIARY	false	NUYMCLU.NP62160.NUYMCLU_XL29287	nuymclu-nuymclu_xl29287.snowflakecomputing.com

Figure 12-40. *Connections before failover*

Now invoke Client Redirect:

```
ALTER CONNECTION nuymclu_xl29287 PRIMARY;
```

Confirm your connection has switched over:

```
SHOW CONNECTIONS;
```

And confirm your secondary `account_name` (`REPLICATION_TERTIARY`) and `is_primary` (true) attributes are correctly set, as shown in Figure 12-41.

account_name	is_primary	primary	connection_url
NP62160	false	NUYMCLU.NP62160.NUYMCLU_XL29287	nuymclu-nuymclu_xl29287.snowflakecomputing.com
REPLICATION_TERTIARY	true	NUYMCLU.NP62160.NUYMCLU_XL29287	nuymclu-nuymclu_xl29287.snowflakecomputing.com

Figure 12-41. Connections after failover

Primary Account Setup 2

When your primary account service is restored, you must reverse the connection. To do this, from your original primary account, follow these steps. We assume familiarity with the expected outcome as these steps should be familiar.

```
USE ROLE accountadmin;

SHOW CONNECTIONS;

ALTER CONNECTION nuymclu_xl29287 PRIMARY;

SHOW CONNECTIONS;
```

Please also refer to the documentation found here: `https://docs.snowflake.com/en/user-guide/client-redirect.html#label-redirecting-client-connections`.

Client Redirect Considerations

Use `ORGADMIN` to provision Snowflake accounts from your LZ in order to ensure consistency of delivery for your Snowflake platform. You adopt the LZ approach to reduce the administrative overhead of implementing Snowflake and enable development teams to focus on delivering their applications, rather than focus on the one-off exercises associated with platform maintenance.

The use of `ORGADMIN` results in all provisioned Snowflake accounts belonging to a single Snowflake organization. Your organization may have a federated Snowflake landscape insofar as each operating division may run their own LZ delivered Snowflake environments, with little interaction between the other Snowflake teams. We suggest adoption of the account naming standards proposed in Chapter 8 at the earliest opportunity to readily identify accounts.

All `ORGADMIN`-provisioned accounts must have replication enabled at the point of delivery in order to later enable Client Redirect by the acquiring team. This is particularly important when centralizing monitoring solutions into a single account.

There are two approaches to identifying an account when using SQL commands. The preferred approach utilizes the organization name and account name as the account identifier. An alternative approach using the legacy account locator may require additional segments adding. Please refer to the documentation found here for details: `https://docs.snowflake.com/en/user-guide/admin-account-identifier. html#account-identifiers-for-replication-and-failover`.

Client Redirect Summary

Within this section, Client Redirect has been addressed in isolation from account failure, but in a real-world situation both Client Redirect and account failure must be considered together. The tools demonstrated within this chapter provide the means to develop your DR plan, which we address later in this chapter.

OAuth integration for client connection URLs force reauthentication when the connection URL is redirected using Client Redirect. Please refer to the documentation found here: `https://docs.snowflake.com/en/user-guide/client-redirect. html#label-oauth-redirect-behavior`.

As discussed in "Account Replication Summary," before reversing Client Redirect, you must ensure your table data matches in primary and secondary accounts.

Holistic Approach

In this chapter we have discussed replication options from a Snowflake-centric perspective and considered a few limited DR scenarios: database failure, account failure, and Client Redirect. We now consider the wider implications of losing a CSP region and in general terms describe the real-world actions and support required to restore service.

For each component that interfaces to Snowflake within your organization, there must also be a DR plan. Your Snowflake account, whether single-tenant or multi-tenant, will either ingest data from sources or produce data for consumption. You must align every ingestion/consumption DR plan with your account DR plan and then define an approach to quickly identify dependencies and actions to restore service.

External Data Source Failure

You start with depicting the loss of an inbound service, as shown in Figure 12-42, where a single component fails within a remote CSP (in this example, an AWS account in Frankfurt). Let's imagine the lost service is an application that delivers real-time data. While your Snowflake account remains available, the loss of data will impact your SLA for delivery to your consumers.

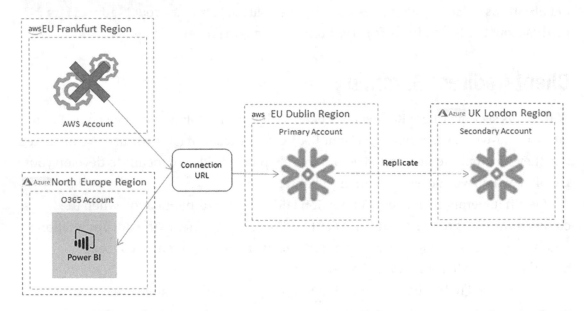

Figure 12-42. Loss of inbound data

DR failover for your account is not required, but you have a vested interest in the upstream data provider recovering as soon as possible, which you might reasonably assume will be from the same region and IP range.

What happens if the upstream application cannot be restored on the same infrastructure, but instead is failed over to an alternative CSP and region? Will your network policy allow connections from the failed-over IP range?

CSP Tooling

As part of your DR planning, you must consider the CSP-specific tooling and equivalents required to implement both same-CSP and cross-CSP DR scenarios, particularly for those tools where your Snowflake account is directly dependent for provision of service.

As an example, your Cyber Security colleagues may insist upon Tri-Secret Secure to hold your most sensitive data, with approved integration patterns for each CSP account where your customer key will be held. You must therefore ensure your Snowflake Secondary accounts are correctly secured before you consider failing over and perform sufficient testing to meet all cyber security requirements.

All external stages must likewise be secured with appropriate policies to prevent unauthorized access.

Feed Onboarding Contracts

As discussed in Chapter 10, we prefer every data interchange between two endpoints to be documented via a Feed Onboarding Contract (FOC). Adopting FOCs describes every component and interface along with actions in the event of a failure.

We strongly recommend every FOC is maintained and incorporated into your DR plan to facilitate both rapid identification of dependencies and enable rapid resolution of issues.

Network Topology

You must be mindful of your network topology and vulnerability to single points of failure when defining your DR approach. For each component, or application, we strongly advise developing one or more diagrams to illustrate the potential failure scenarios and impact. A picture paints a thousand words.

It should be readily apparent that FOCs form the basis for describing every internal and external interaction with your Snowflake environments.

SLA Impact

You must ensure your Snowflake-based application RTO and RPO are sufficiently broad to ensure any loss of upstream service will not impact your SLA. In other words, all data ingestion RTO and RPO must be less than or equal to your Snowflake application RTO and RPO.

Additionally, you must have both transparency and early visibility of relevant upstream and downstream issues as soon as possible.

Account Migration

You may decide to migrate an account for a few reasons:

- As part of a planned exit from one CSP to another

- Relocation of service from one region to another in support of business objectives such as proximity to customers or regulatory considerations

- Merger and Acquisition activity where colocation of accounts is desired

Account migration is distinctly different to a DR scenario, where the available options are dictated by the nature of the outage. As account migration is a preplanned, one-way activity, there is no failback unless the migration fails, in which case the primary is expected to be restored to service.

The steps outlined above in this chapter are expected to form the basis of any account migration, and the checklist may prove invaluable in planning an account migration.

As discussed in "Account Replication Summary," you must ensure your table data matches in both old and new accounts before finalizing your migration.

Account migration typically requires the Business Critical Edition (or higher). Please refer to the documentation found here for further details: `https://docs.snowflake.com/en/user-guide/replication-intro.html#account-migration`.

DR Plan

Every DR plan follows a common theme, which we outline within this section. Invoking DR failover is a highly disruptive and time-consuming activity so you must be clear on the need to failover before you invoke your DR plan.

Assess The Impact

When a service affecting incident occurs, you must assess the impact. You only consider invoking DR where the loss of service affects either the whole CSP region or a single tenant, which you can assume will be encapsulated within a single replication group.

Gathering information is time consuming, particularly in complex environments with multiple secondary accounts. We recommend all refresh statistics are made available in two or more accounts using SDDS as a minimum, or preferably Private Listings, to aid data discovery. You are looking for the most recent refreshed secondary as the most likely candidate to become the primary, notwithstanding other organization-specific considerations.

The troubleshooting guide towards the end of this chapter may assist your investigations.

Your Network Topology diagram, FOCs, and SLAs should also be consulted and the communications plan invoked to inform all interested parties of the situation. You must also contact your counterparts either at the CSP or upstream systems to ascertain their situation and estimated time to restore their service.

You may establish a bridge call or other communications channel to facilitate information exchange.

Invoke DR Action

Having conducted your impact assessment and established communications channels, you are in a position to determine the most appropriate DR action to restore service as quickly as possible. Your impact assessment will include all affected components along with their business criticality, their respective RTO and RPO, to prioritize service recovery in the order of importance.

You must first determine the data gap between the most recent refresh of your chosen secondary site or replication group to instantiate as primary. For each upstream data provider, you must consider

- How to repoint each data provider to your new primary

- Determine the order to apply missing data to the new primary

- Swapping a secondary to become the new primary

- Refreshing all other secondaries to refresh from the new primary

- Conduct system checks to confirm service is restored

- Communicate status to all stakeholders

Knowing the data gap between the last refresh time and the time DR is invoked provides an estimate for both RTO and RPO.

You make an assumption that all source data is available for replay, and like all assumptions, must be tested for validity. Equally, you assume staff are available to assist with DR invocation.

Update the DR plan to reflect any changes or discrepancies found since the last invocation.

Conduct a postmortem to identify what worked well, and more importantly, what did not work well.

Service Restoration

At some point, your original primary (now switched to become a secondary) will be restored to service.

A valid scenario is to retain the failed-over site as the primary, in which case no further action is necessary. However, for a variety of reasons your organization may prefer to failback to the original primary.

Failback is a preplanned activity, typically conducted during times of least system usage and in coordination with all upstream data providers and downstream data consumers.

As with the original failover, you must ensure there is no data gap between the current primary and target secondary. We suggest the following activities as a starting point for your failback:

- All data ingestion into the current primary is suspended.

- All data consumption from the current primary be disabled.

- Confirm both accounts or replication groups contain identical data sets.

- Reinstate the original primary from the chosen secondary.

- Enable data ingestion into the primary.

- Refresh all other secondaries to refresh from the primary.

- Reinstate data consumption from the primary.

- Conduct system checks to confirm service is restored.

- Communicate status to all stakeholders.

Service restoration should meet the RTO and RPO objectives as the inertia encountered when addressing unplanned outages should not occur.

DR Test

If you have been diligent in your preparation and maintained your DR plan according to changes made to your account and database configurations, your DR test should be straightforward.

You should runt DR tests at a minimum on an annual basis; some organizations prefer quarterly DR tests.

When planning a DR test, it is important to consider different scenarios that reflect more usual "real-world" occurrence. One outcome of DR testing is to educate your operations staff to expect the unexpected; to adapt, improvise, and overcome situations as they occur; and not to condition your staff to always expect the same DR scenario.

DR Checklist

Table 12-2 shows a template DR checklist for your consideration and further extension. Our aim is not to provide a fully comprehensive checklist as we recognize your organization will have its own requirements and each application will have its own tooling.

Table 12-2. *Disaster Recovery Checklist*

Item	Value
Snowflake Edition	Must be Business Critical Edition
Recovery Point Objective (RPO)	
Recovery Time Objective (RTO)	
Service Level Agreement (SLA)	Link to document
Exit Strategy	Link to document
Primary Account Locator	Unique account identifier
Secondary Account Locator 1	Unique account identifier
Secondary Account Locator n	Unique account identifier
Inbound Data Source 1	Link to FOC
Inbound Data Source n	Link to FOC
Outbound Consumer 1	Link to FOC
Application Documentation	Link to document
Network Topology	Link to document
DR Plan	Link to document
Last DR Test Date	
Last DR Test Outcome Report	Link to document

Troubleshooting

In this section, we identify a few items to check as part of your troubleshooting to be conducted before considering invoking DR, noting the full Snowflake service may not be available.

Check Snowflake Availability

To check whether the Snowflake service is available: `https://status.snowflake.com`.

Check For Snowflake Email

Occasionally, Snowflake will capture events from its own monitoring suite and deliver information via email from Snowflake Global Technical Support (`info@reply.snowflake.com`). Please check your email, including the spam folder, to ensure receipt.

Has Snowflake Released a New Version?

Occasionally Snowflake releases may cause problems. A useful step is to identify the Snowflake version:

```
SELECT current_version();
```

Your result should look something like Figure 12-43.

CURRENT_VERSION()

7.3.0

Figure 12-43. *Snowflake version*

Has a Change Bundle Been Enabled?

Another possibility could be a Behavior Change Bundle has been applied and has caused an issue. The list of new features can be found here: `https://docs.snowflake.com/en/release-notes/new-features.html`. The list of Behavior Change Bundles can be found here: `https://community.snowflake.com/s/article/Pending-Behavior-Change-Log`.

```
USE ROLE accountadmin;
```

Check status, noting your Change Bundle will differ from 2021_10:

```
SELECT system$behavior_change_bundle_status('2021_10');
```

We discuss Behavior Change Bundles in more depth in Chapter 11.

Failure Scope

When failure occurs, it is essential to quickly identify the scope of the problem. These questions will assist in determining the appropriate actions:

- What services, SLAs, and clients are affected?

- Does the failure impact the CSP region?

- Is the failure within the Snowflake account?

- Is the failure external to Snowflake?

- Do we have logs showing the last replication refresh to Secondaries?

- What options do we have to restore service?

- Is our documentation up to date?

Capturing information is essential to diagnosing the problem. We advise a quarterly review of every Snowflake account is undertaken to ensure your DR plan is up to date.

You must also engage your Fix on Fail Team, advise Service Management colleagues, raise a problem ticket, contact your client managers, and escalate to management where appropriate.

Summary

You began your investigation into DR by reading an overview of how CSPs provision service and the protection HA zones offer, before identifying CSP failure scenarios and subsequent options available. You learned the importance of BCP and SLAs by exploring a wide variety of subjects to consider when preparing BCP.

All available DR options were then discussed, including SDDS, database replication, account replication, and Client Failover, noting each has limitations and pending delivery by Snowflake to increase component coverage. With the exclusion of SDDS, you worked through a hands-on implementation of each DR option.

You explored the wider implication of both individual application failure and account single tenant failure in an attempt to more accurately reflect probable real-word scenarios.

Our skeleton DR plan, testing approach, and checklist offer a starting position when developing your DR approach and are left for your further consideration.

Given variations within each Snowflake account configuration, it is not possible to articulate all manual configuration steps required for all DR scenarios, and we hope this chapter has provided sufficient information and insight for you to develop your own DR plans.

Index

A

Printed in the United States
by Baker & Taylor Publisher Services

Printed in the United States
by Baker & Taylor Publisher Services